THEORY AND DESIGN OF PLATE AND SHELL STRUCTURES

THEORY AND DESIGN OF PLATE AND SHELL STRUCTURES

Maan H. Jawad
Nooter Corporation

Chapman & Hall
New York • London

First published in 1994 by
Chapman & Hall
One Penn Plaza
New York, NY 10119

Published in Great Britain by
Chapman & Hall
2-6 Boundary Row
London SE1 8HN

© 1994 Chapman & Hall, Inc.

Printed in the United States of America

All rights reserved. No part of this book may be reprinted or reproduced or utilized in any form or by any electronic, mechanical or other means, now known or hereafter invented, including photocopying and recording, or by an information storage or retrieval system, without permission in writing from the publishers.

Library of Congress Cataloging in Publication Data

Jawad, Maan H.
Theory and design of plate shell structures / Maan H. Jawad.
p. cm.
Includes index.
ISBN 0-412-98181-5
1. Plates (Engineering) 2. Shells (Engineering) I. Title.
TA660.P6J39 1994
624.1'776—dc20 93-48727
CIP

British Library Cataloguing in Publication Data

Please send your order for this or any other Chapman & Hall book to **Chapman & Hall, 29 West 35th Street, New York, NY 10001, Attn: Customer Service Department.** You may also call our Order Department at 1-212-244-3336 or fax your purchase order to 1-800-248-4724.

For a complete listing of Chapman & Hall titles, send your request to **Chapman & Hall, Dept. BC, One Penn Plaza, New York, NY 10119.**

To
Jennifer and Mark
Who Taught Me A Lot

Contents

Preface

The design of many structures such as pressure vessels, aircrafts, bridge decks, dome roofs, and missiles is based on the theories of plates and shells. The degree of simplification needed to adopt the theories to the design of various structures depends on the type of structure and the required accuracy of the results. Hence, a water storage tank can be satisfactorily designed using the membrane shell theory, which disregards all bending moments, whereas the design of a missile casing requires a more precise analysis in order to minimize weight and materials. Similarly, the design of a nozzle-to-cylinder junction in a nuclear reactor may require a sophisticated finite element analysis to prevent fatigue failure while the same junction in an air accumulator in a gas station is designed by simple equations that satisfy equilibrium conditions.

Accordingly, this book is written for engineers interested in the theories of plates and shells and their proper application to various structures. The examples given throughout the book subsequent to derivation of various theories are intended to show the engineer the level of analysis required to achieve a safe design with a given degree of accuracy.

The book covers three general areas. These are: bending of plates; membrane and bending theories of shells; and buckling of plates and shells. Bending of plates is discussed in five chapters. Chapters 1 and 2 cover rectangular plates with various boundary and loading conditions. Chapter 3 develops the theory of circular plates of uniform and variable thickness as well as plates on an elastic foundation. Chapter 4 presents approximate analyses such as the energy and yield-line methods for evaluating plates of different shapes. Chapter 5 discusses the bending of plates with various shapes and the bending of orthotropic plates.

Shell theory is presented in four chapters. Chapters 6 and 7 cover the membrane theory and its application to spherical and conical shells as well as other configurations. Bending of cylindrical shells is discussed in Chapter

8. Both long and short cylinders are evaluated due to mechanical as well as thermal loads. Examples combining circular plates and cylindrical shell components are given to illustrate the design of some actual structures. Bending of shells with double curvature is discussed in Chapter 9 and numerical examples are given.

Buckling of plates and shells is discussed in Chapters 10, 11, and 12. General Buckling theory of plates is given in Chapter 10 with approximations used in various design codes. Chapter 11 covers buckling of cylindrical shells with design applications. Chapter 12 discusses buckling of spherical, conical, and other miscellaneous shapes.

The discussion of plate and shell theories is incomplete without a brief mention of two topics. The first is shell roof structures, and the second is finite element formulations. A complete coverage of these two topics is beyond the scope of this book. However, a brief summary of the analysis of various roof structures is given in Chapter 13. Chapter 14 presents a summary of the finite element formulation as used in solving complicated plate and shell configurations.

Most of the chapters in this book can be covered in a two-semester course in "plate and shell theory." Also, a special effort was made to make the chapters as independent from each other as possible so that a course in "plate theory" or "shell theory" can be taught in one semester by selecting appropriate chapters.

In order to study and use the theory of plates and shells, the engineer is assumed to have a good working knowledge of differential equations and matrix analysis. In addition two appendices are given at the end of the book to make the book as "self-contained" as possible. The first appendix is for Fourier Series and the second one is for Bessel Functions.

Maan Jawad
St. Louis, MO
1994

Acknowledgments

The author is indebted to many people and organizations for their help in writing this book. A special thanks is given to the Nooter Corporation for its generous support with various stages of the book. A special thanks is also extended to Dr. James Hahn of the St. Louis Graduate Engineering Center of the University of Missouri-Rolla for his continual help; to Mr. Russ Doty for spending long hours reviewing and critiquing the manuscript; and to Mr. Paul Rose for sketching numerous figures.

The author would also like to thank all the engineering graduate students who took the time to comment on the manuscript. The author's appreciation is also given to Messrs. Basil T. Kattula, William J. O'Donnell, and all others who helped with various aspects of the book. Last but not least, a special thanks is given to my wife Dixie for her patience.

Acknowledgements

Abbreviations for Organizations

AASHTO, American Association of State Highway and Transportation Officials
ACI, American Concrete Institute
AISC, American Institute of Steel Construction
AISI, American Iron and Steel Institute
API, American Petroleum Institute
ASME, American Society of Mechanical Engineers
AWWA, American Water Works Association
NASA, National Aeronautics and Space Administration
TEMA, Tubular Exchangers Manufacturers Association

1

Bending of Simply Supported Rectangular Plates

1-1 Introduction

Many structures such as powerplant duct assemblies (Fig. 1-1), submarine bulkheads, ship and barge hulls, building slabs, (Fig. 1-2), and machine parts are designed in accordance with the bending theory of plates. The analysis of most plate configurations consists of solving a differential equation that is a function of deflection, applied loads, and stiffness of the plate. The solution of this differential equation results in an expression for the deflection of the plate. Other quantities such as forces and moments must then be determined from the calculated deflection. In this chapter, equations that express moments and forces in terms of deflection are developed first. Next, the basic differential equation for the bending of rectangular plates is established together with corresponding boundary conditions. These expressions are then used to solve various plate configurations and loading conditions.

The basic assumptions made in the derivation of the equations for the bending of thin plates are:

1. The thickness of the plate is substantially less than the lateral dimensions.
2. The plate is homogeneous and isotropic.
3. Loads are applied perpendicular to the middle surface of the plate.
4. The deflection of the plate due to applied loads is small.
5. Lines perpendicular to the middle surface before deformation remain perpendicular to the deformed middle surface.

With these assumptions, the basic relationships can now be derived.

Figure 1-1. Duct assembly. (Courtesy of the Nooter Corporation, St. Louis, MO.)

1-2 Strain–Deflection Equations

The relationship between strain and deflection of a thin plate is available from geometric considerations. We begin the derivation by letting an infinitesimal section (Fig. 1-3) undergo some bending deformation. The change in length at a distance z from the middle surface is expressed as

$$\frac{dx}{r_x} = \frac{dx + \varepsilon_x\,dx}{r_x + z} \tag{1-1}$$

or

$$\varepsilon_x = \frac{z}{r_x} \tag{1-2}$$

$$\varepsilon_x = \chi_x z \tag{1-3}$$

Figure 1-2. Reinforced concrete buliding. (Courtesy of the Portland Cement Association, Chicago, IL.)

where

r_x = radius of curvature in the x-direction;
ε_x = strain in the x-direction;
χ_x = curvature in the x-direction.

Similarly, in the y-direction,

$$\varepsilon_y = \frac{z}{r_y} \tag{1-4}$$

$$\varepsilon_y = \chi_y z \tag{1-5}$$

where

r_y = radius of curvature in the y-direction;
ε_y = strain in the y-direction;
χ_y = curvature in the y-direction.

The quantity χ_x is related to the deflection, w, and slope, dw/dx, by the

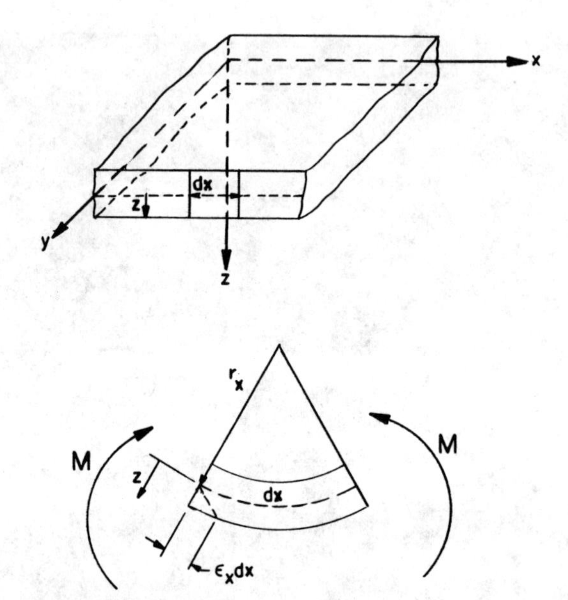

Figure 1-3.

expression (Shenk 1979)

$$\chi_x = \frac{-\dfrac{d^2w}{dx^2}}{\left[1 + \left(\dfrac{dw}{dx}\right)^2\right]^{3/2}}$$

or, for small deflections,

$$\chi_x = -\frac{d^2w}{dx^2}. \tag{1-6}$$

Similarly, in the y-direction,

$$\chi_y = -\frac{d^2w}{dy^2}. \tag{1-7}$$

Substituting Eqs. (1-6) and (1-7) into Eqs. (1-3) and (1-5) gives

$$\varepsilon_x = -z\frac{d^2w}{dx^2} \tag{1-8}$$

$$\varepsilon_y = -z\frac{d^2w}{dy^2}. \tag{1-9}$$

The shearing strain–deformation relationship can be obtained from Fig. 1-4. If an infinitesimal element of length dx and width dy undergoes shearing deformations due to in-plane shearing forces and twisting moments then from Fig. 1-4a

$$\sin \alpha \approx \alpha \approx \frac{\dfrac{\partial u}{\partial y}\, dy}{\left(1 + \dfrac{\partial v}{\partial y}\right) dy}$$

or, for small shearing angles,

$$\alpha = \frac{\partial u}{\partial y}.$$

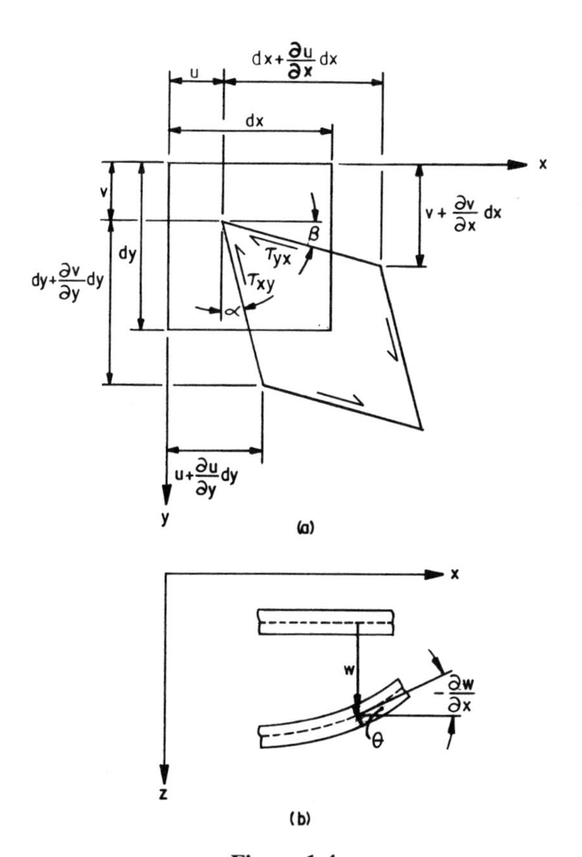

Figure 1-4.

Similarly,

$$\sin \beta \approx \beta \approx \frac{\frac{\partial v}{\partial x}\, dx}{\left(1 + \frac{\partial u}{\partial x}\right) dx}$$

$$\beta = \frac{\partial v}{\partial x}.$$

Hence,

$$\gamma_{xy} = \alpha + \beta = \frac{\partial u}{\partial y} + \frac{\partial v}{\partial x} \tag{1-10}$$

where

u = deflection in the x-direction;
v = deflection in the y-direction;
γ_{xy} = shearing strain;
$\frac{\partial u}{\partial y}, \frac{\partial v}{\partial x}$ = shearing strains due to twisting.

The rotation of the middle surface is shown in Fig. 1-4b and is given by $\partial w/\partial x$. Due to this rotation, any point at distance z from the middle surface will deflect by the amount

$$u = z \tan \theta \approx z\theta$$

or

$$u = -z\frac{\partial w}{\partial x} \qquad v = -z\frac{\partial w}{\partial y}.$$

Hence, Eq. (1-10) becomes

$$\gamma_{xy} = -2z\frac{\partial^2 w}{\partial x\, \partial y}. \tag{1-11}$$

Equations (1-8), (1-9), and (1-11) can be written as

$$\begin{bmatrix} \varepsilon_x \\ \varepsilon_y \\ \gamma_{xy} \end{bmatrix} = -z \begin{bmatrix} 1 & 0 & 0 \\ 0 & 1 & 0 \\ 0 & 0 & 2 \end{bmatrix} \begin{bmatrix} \dfrac{\partial^2 w}{\partial x^2} \\ \dfrac{\partial^2 w}{\partial y^2} \\ \dfrac{\partial^2 w}{\partial x\, \partial y} \end{bmatrix} \tag{1-12}$$

and are sufficiently accurate for developing the bending theory of thin plates. More precise strain expressions that are a function of the three displacement functions u, v, and w will be derived later when the buckling theory of thin plates is discussed.

1-3 Stress–Deflection Expressions

Our next step is to express Eq. (1-12) in terms of stress rather than strain because it is easier to work with stress. The relationship between stress and strain, excluding thermal loads, in a three-dimensional homogeneous and isotropic element (Fig. 1-5) is obtained from the theory of elasticity (Sokolnikoff 1956) as

$$\begin{bmatrix} \varepsilon_x \\ \varepsilon_y \\ \varepsilon_z \\ \gamma_{xy} \\ \gamma_{yz} \\ \gamma_{zx} \end{bmatrix} = \frac{1}{E} \begin{bmatrix} 1 & -\mu & -\mu & 0 & 0 & 0 \\ -\mu & 1 & -\mu & 0 & 0 & 0 \\ -\mu & -\mu & 1 & 0 & 0 & 0 \\ 0 & 0 & 0 & 2(1+\mu) & 0 & 0 \\ 0 & 0 & 0 & 0 & 2(1+\mu) & 0 \\ 0 & 0 & 0 & 0 & 0 & 2(1+\mu) \end{bmatrix} \begin{bmatrix} \sigma_x \\ \sigma_y \\ \sigma_z \\ \tau_{xy} \\ \tau_{yz} \\ \tau_{zx} \end{bmatrix} \tag{1-13}$$

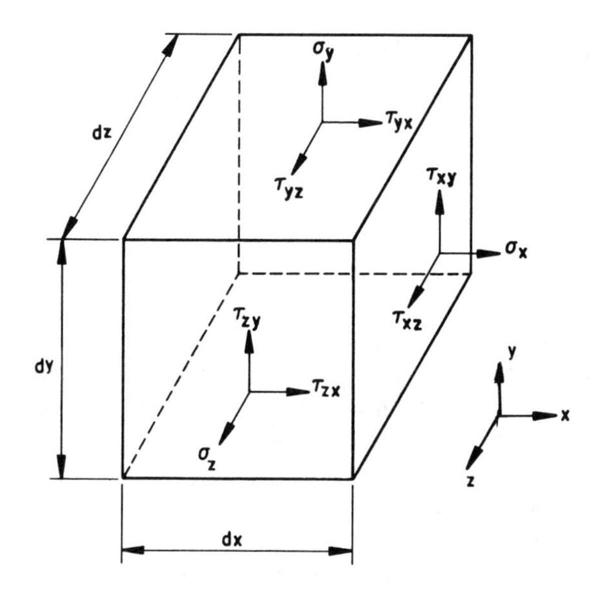

Figure 1-5.

where

ε = axial strain;
σ = axial stress;
γ = shearing strain;
τ = shearing stress;
E = modulus of elasticity;
μ = Poisson's ratio.

The quantities x, y, and z refer to the directions shown in Fig. 1-5. The quantity $2(1 + \mu)/E$ is usually written as $1/G$ where G is called the shearing modulus.

The stress perpendicular to the surface, i.e., in the z-direction, has a maximum value equal to the applied pressure. For the majority of plate applications in bending, the stress σ_z in the z-direction is small compared to the stress in the other two directions and thus can be neglected. In addition, the shearing stresses τ_{yz} and τ_{zx} are not needed in the formulation of a two-dimensional state of stress. Hence, for this condition, Eq. (1-13) can be written as

$$\begin{bmatrix} \sigma_x \\ \sigma_y \\ \tau_{xy} \end{bmatrix} = \frac{E}{1 - \mu^2} \begin{bmatrix} 1 & \mu & 0 \\ \mu & 1 & 0 \\ 0 & 0 & \dfrac{1 - \mu}{2} \end{bmatrix} \begin{bmatrix} \varepsilon_x \\ \varepsilon_y \\ \gamma_{xy} \end{bmatrix}. \tag{1-14}$$

Substituting Eq. (1-12) into Eq. (1-14) gives

$$\begin{bmatrix} \sigma_x \\ \sigma_y \\ \tau_{xy} \end{bmatrix} = \frac{-Ez}{1 - \mu^2} \begin{bmatrix} 1 & \mu & 0 \\ \mu & 1 & 0 \\ 0 & 0 & (1 - \mu) \end{bmatrix} \begin{bmatrix} \dfrac{\partial^2 w}{\partial x^2} \\ \dfrac{\partial^2 w}{\partial y^2} \\ \dfrac{\partial^2 w}{\partial x\, \partial y} \end{bmatrix}. \tag{1-15}$$

The elastic moduli of elasticity and Poisson's ratio for some commonly used materials are given in Table 1-1. The value of Poisson's ratio is relatively constant at various temperatures for a given material and is thus listed only for room temperature in Table 1-1.

1-4 Force–Stress Expressions

Equation (1-15) can be utilized better when the stress values are replaced by moments. This is because the moments at the edges of the plate are

Table 1-1. Moduli of elasticity and Poisson's ratio

Material	Poisson's Ratio	Room Temperature	Modulus of Elasticity[a] Temperature, °F 200	400	600	800	1000
Aluminum (6061)	0.33	10.0	9.6	8.7			
Brass (C71000)	0.33	20.0	19.5	18.8	17.8		
Bronze (C61400)	0.33	17.0	16.6	16.0	15.1		
Carbon Steel (C < 0.3)	0.29	29.5	28.8	27.7	26.7	24.2	20.1
Copper (C12300)	0.33	17.0	16.6	16.0	15.1		
Cu–Ni (70–30) (C71500)	0.33	22.0	21.5	20.7	19.6		
Nickel alloy C276	0.29	29.8	29.1	28.3	27.6	26.5	25.3
Nickel alloy 600	0.29	31.0	30.2	29.5	28.7	27.6	26.4
Stainless steel (304)	0.31	28.3	27.6	26.5	25.3	24.1	22.8
Titanium (Gr.1,2)	0.32	15.5	15.0	14.0	12.6	11.2	
Zirconium alloys	0.35	14.4	13.4	11.5	9.9		
Concrete	0.15	3.1[b]					
Wood, hard		2.1					
Wood, soft		1.3					

[a]In million psi.
[b]For 3000 psi concrete.

needed to satisfy some of the boundary conditions in solving the differential equation. The relationship between moment and stress is obtained from Fig. 1-6a. The moments shown in Fig. 1-6b are positive and are per unit length. By definition, the sum of the moments about the neutral axis due to the internal forces is equal to the sum of the moments of the external forces. Hence,

$$\begin{bmatrix} M_x \\ M_y \\ -M_{xy} \end{bmatrix} = \int_{-t/2}^{t/2} \begin{bmatrix} \sigma_x \\ \sigma_y \\ \tau_{xy} \end{bmatrix} z\, dz. \tag{1-16}$$

The negative sign of M_{xy} in Eq. (1-16) is needed since the direction of M_{xy} in Fig. 1-6b results in a shearing stress τ_{xy} that has a direction opposite to that defined in Fig. 1-4a in the positive z-axis. Substituting Eq. (1-15) into Eq. (1-16) results in

$$\begin{bmatrix} M_x \\ M_y \\ M_{xy} \end{bmatrix} = -D \begin{bmatrix} 1 & \mu & 0 \\ \mu & 1 & 0 \\ 0 & 0 & -(1-\mu) \end{bmatrix} \begin{bmatrix} \dfrac{\partial^2 w}{\partial x^2} \\ \dfrac{\partial^2 w}{\partial y^2} \\ \dfrac{\partial^2 w}{\partial x\, \partial y} \end{bmatrix} \tag{1-17}$$

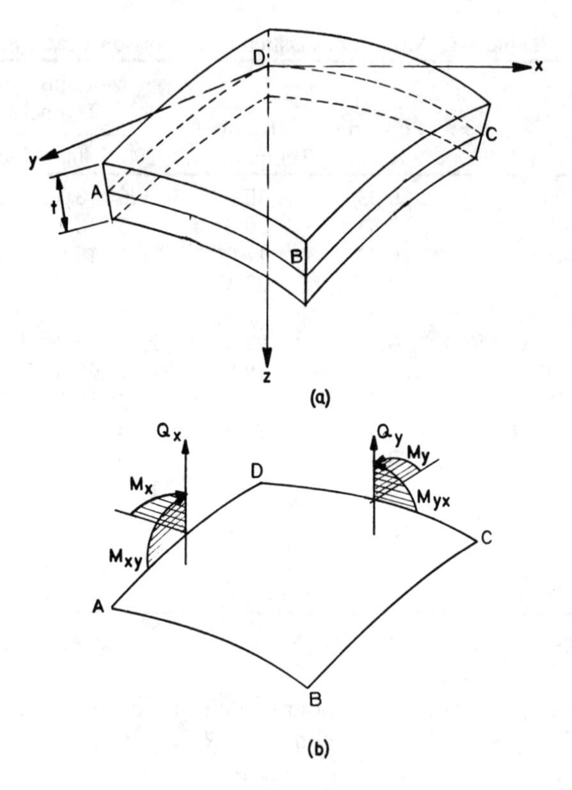

Figure 1-6.

where

$$D = \frac{Et^3}{12(1 - \mu^2)}. \tag{1-18}$$

The quantity D is the bending stiffness of a plate. It reduces to the quantity EI, which is the bending stiffness of a beam of unit width, when we let $\mu = 0$.

Problems

1-1 The finite element formulation for the stiffness of a solid three-dimensional element is based on the strain–stresse matrix Eq. (1-13). Rewrite this equation as a stress–strain matrix.

1-2 A strain gage rosette is mounted on the flat inside surface of a valve casting as shown in Fig. P1-2. The valve is then pressurized and the fol-

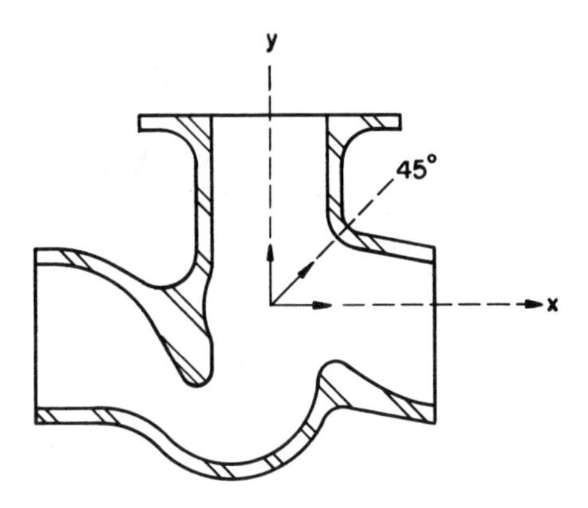

Figure P1-2.

lowing strain values were measured

$$\varepsilon_x = 300 \times 10^{-6} \text{ inches/inch}$$
$$\varepsilon_y = 150 \times 10^{-6} \text{ inches/inch}$$
$$\varepsilon_{45} = 600 \times 10^{-6} \text{ inches/inch.}$$

Calculate the maximum stresses if $E = 20{,}000$ ksi and $\mu = 0.15$. *Hint*: first, calculate the shearing strain γ_{xy} at the location of the strain gage from Mohr's circle which is expressed as

$$\varepsilon_\theta = \frac{\varepsilon_x + \varepsilon_y}{2} + \frac{\varepsilon_x - \varepsilon_y}{2} \cos 2\theta + \frac{\gamma_{xy}}{2} \sin 2\theta$$

where $\theta = 45°$ in this case. Then, calculate the principal strains from Beer and Johnson (1981)

$$\varepsilon_{\substack{\max \\ \min}} = \frac{\varepsilon_x + \varepsilon_y}{2} + \sqrt{\left(\frac{\varepsilon_x - \varepsilon_y}{2}\right)^2 + \left(\frac{\gamma_{xy}}{2}\right)^2}$$

$$\frac{1}{2}\gamma_{\max} = \pm \sqrt{\left(\frac{\varepsilon_x - \varepsilon_y}{2}\right)^2 + \left(\frac{\gamma_{xy}}{2}\right)^2}$$

$$\theta = \frac{1}{2} \tan^{-1} \frac{\gamma_{xy}}{\varepsilon_x - \varepsilon_y}$$

where 2θ is the orientation of the plane of maximum strain with respect to the plane of given strains. The last step is to use Eq. (1-14) to obtain maximum stress.

1-3 The stress in the x-direction of a point on the surface of a plate is equal to 35 MPa. The stress in the y-direction is equal to 70 MPa and that in the z-direction is equal to 35 MPa. Determine the maximum shearing stress by Mohr's circle and show the plane on which it acts.
1-4 Determine the maximum bending stress values σ_x and σ_y in a plate with length $a = 100$ cm and width $b = 75$ cm. The deflection is approximated by

$$w = k \sin \frac{\pi x}{a} \sin \frac{\pi y}{b}$$

where k is a constant equal to 0.462 cm. Let $t = 1.2$ cm, $\mu = 0.3$, and $E = 200{,}000$ MPa.
1-5 A simply supported rectangular plate with dimensions $a = 30$ inches and $b = 20$ inches (Fig. P1-5) is subjected to a uniform pressure of 15 psi. Determine the maximum bending moment in the middle of the plate by taking unit strips in the middle of the plate in the x- and y-directions. Assume the strips to be connected at point A. Compare the results with the more accurate solution obtained from the plate theory in Example 1-2.

1-5 Governing Differential Equations

The differential equation for the bending of a beam

$$\frac{d^2w}{dx^2} = -\frac{M(x)}{EI} \tag{1-19}$$

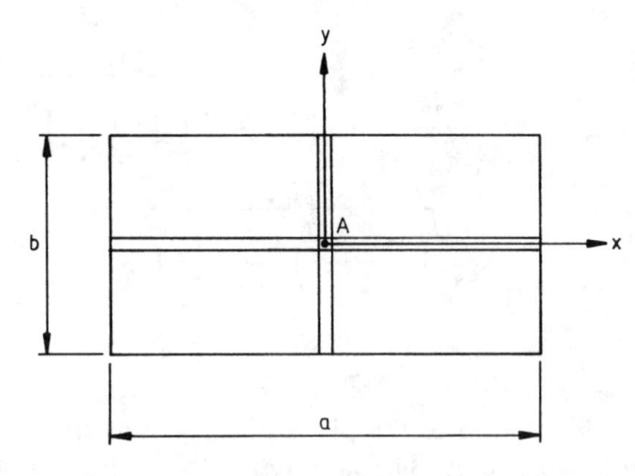

Figure P1-5.

can be expressed in terms of applied loads by

$$\frac{d^4w}{dx^4} = \frac{p(x)}{EI}. \tag{1-20}$$

A similar equation can be written for the bending of a plate. The corresponding differential equation for the bending of a plate is more complicated because it must include terms for the bending in the x- and y-directions as well as torsional moments that are present in the plate. Lagrange (Timoshenko 1983) was the first to develop the differential equation for the bending of a rectangular plate in 1811. We begin the derivation of the governing equations by considering an infinitesimal element dx, dy in Fig. 1-7 subjected to lateral loads p. The forces and moments, per unit length, needed for equilibrium are shown in Fig. 1-8 and are positive as shown. Also, downward deflection is taken as positive. It is of interest to note that two shearing forces, Q_x and Q_y, and two torsional moments, M_{xy} and M_{yx}, are needed to properly define the equilibrium of a rectangular plate.

Summation of forces in the z-direction gives the first equation of equilibrium:

$$p(x, y)\,dx\,dy - Q_x\,dy + \left(Q_x + \frac{\partial Q_x}{\partial x}dx\right)dy - Q_y\,dx + \left(Q_y + \frac{\partial Q_y}{\partial y}dy\right)dx = 0.$$

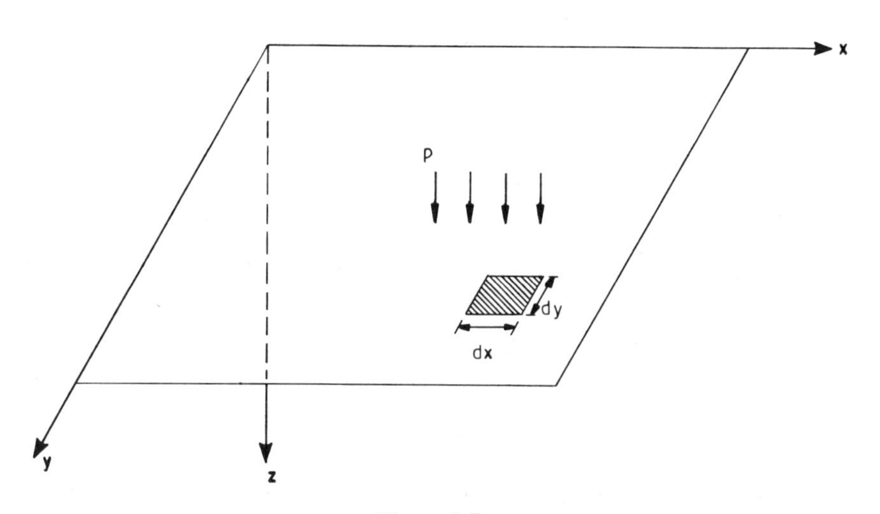

Figure 1-7.

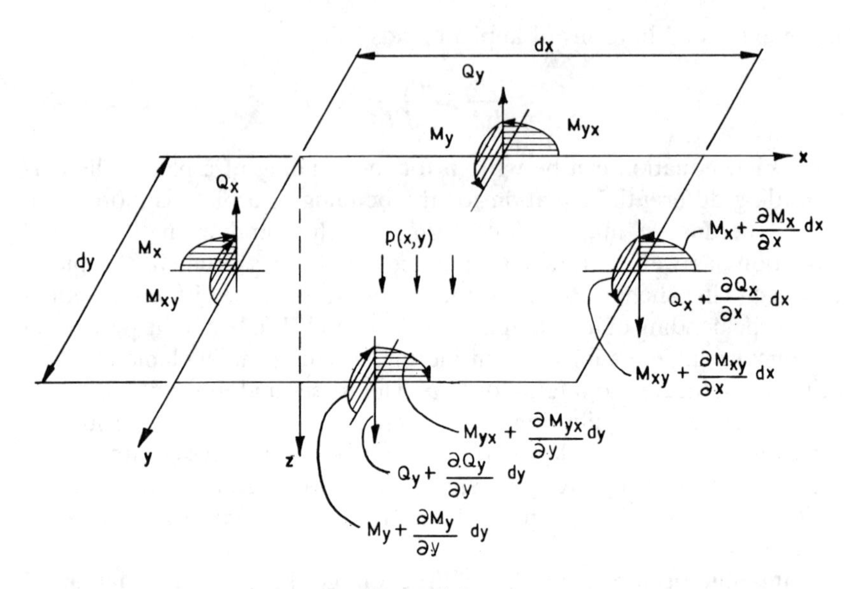

Figure 1-8.

This equation reduces to

$$p(x, y) + \frac{\partial Q_x}{\partial x} + \frac{\partial Q_y}{\partial y} = 0. \tag{1-21}$$

Summation of moments around the x-axis gives the second equation of equilibrium

$$\begin{aligned} M_y\,dx &- \left(M_y + \frac{\partial M_y}{\partial y}\,dy\right) dx - M_{xy}\,dy + \left(M_{xy} + \frac{\partial M_{xy}}{\partial x}\,dx\right) dy \\ &+ \left(Q_y + \frac{\partial Q_y}{\partial y}\,dy\right) dx\,dy - Q_x\,dy\,dy/2 \\ &+ \left(Q_x + \frac{\partial Q_x}{\partial x}\,dx\right) dy\,dy/2 + p\,dx\,dy\,dy/2 = 0. \end{aligned}$$

Simplifying this equation gives

$$Q_y + \frac{\partial M_{xy}}{\partial x} - \frac{\partial M_y}{\partial y} + \left(\frac{\partial Q_y}{\partial y} + \frac{1}{2}\frac{\partial Q_x}{\partial x} + \frac{1}{2}p\right) dy = 0.$$

The bracketed term in this equation is multiplied by an infinitesimal quantity dy. It can thus be deleted because its magnitude is substantially less than that of the other three terms. The equation becomes

$$\frac{\partial Q_y}{\partial y} = \frac{\partial^2 M_y}{\partial y^2} - \frac{\partial^2 M_{xy}}{\partial x\, \partial y}. \tag{1-22}$$

Summation of moments around the y-axis gives the third equation of equilibrium

$$\frac{\partial Q_x}{\partial x} = \frac{\partial^2 M_x}{\partial x^2} - \frac{\partial^2 M_{yx}}{\partial x\, \partial y}. \tag{1-23}$$

Substituting Eqs. (1-22) and (1-23) into Eq. (1-21) gives

$$p(x, y) + \frac{\partial^2 M_x}{\partial x^2} - 2\frac{\partial^2 M_{xy}}{\partial x\, \partial y} + \frac{\partial^2 M_y}{\partial y^2} = 0. \tag{1-24}$$

In this equation it was assumed that $M_{xy} = M_{yx}$ because at any point on the plate the shearing stress $\tau_{xy} = -\tau_{yx}$.

Substituting Eq. (1-17) into this equation gives

$$\frac{\partial^4 w}{\partial x^4} + 2\frac{\partial^4 w}{\partial x^2\, \partial y^2} + \frac{\partial^4 w}{\partial y^4} = p(x, y)/D. \tag{1-25}$$

A comparison of this equation with Eq. (1-20) for the bending of beams indicates that Eq. (1-25) is considerably more complicated because it considers the deflection in the x- and y-directions as well as the shearing effects in the xy plane.

Equation (1-25) can also be written as

$$\nabla^2\nabla^2 w = \nabla^4 w = p(x, y)/D \tag{1-26}$$

where

$$\nabla^2 w = \frac{\partial^2 w}{\partial x^2} + \frac{\partial^2 w}{\partial y^2}$$

and

$$\nabla^4 w = \frac{\partial^4 w}{\partial x^4} + \frac{\partial^4 w}{\partial x^2\, \partial y^2} + \frac{\partial^4 w}{\partial y^4}$$

Equation (1-26) is the basic differential equation for rectangular plates in bending. A solution of this equation yields an expression for the deflection, w, of the plate. The moment expressions are obtained by substituting the deflection expressions into Eq. (1-17). The shear forces are obtained from Eqs. (1-22), (1-23), and (1-17), and are given by

$$Q_x = -D\left(\frac{\partial^3 w}{\partial x^3} + \frac{\partial^3 w}{\partial x\, \partial y^2}\right) \tag{1-27}$$

$$Q_y = -D\left(\frac{\partial^3 w}{\partial x^2\, \partial y} + \frac{\partial^3 w}{\partial y^3}\right) \tag{1-28}$$

For sign convention we will assume a downward deflection as positive in Eq. (1-26). All other quantities are assumed positive as shown in Fig. 1-8.

Problems

1-6 Find M_x, M_y, M_{xy}, Q_x, and Q_y of a rectangular plate whose deflection is given by

$$w = k \sin \frac{m\pi x}{a} \sin \frac{n\pi y}{b}$$

where k, a, b, n, and m are constants.
1-7 Derive Eqs. (1-25), (1-27), and (1-28).

1-6 Boundary Conditions

The most frequently encountered boundary conditions for rectangular plates are essentially the same as those for beams. They are either fixed, simply supported, free, or partially fixed as shown in Fig. 1-9.

(a) Fixed Edges: For a fixed edge (Fig. 1-9), the deflection and slope are zero. Thus,

$$w|_{y=b} = 0 \tag{1-29}$$

$$\left.\frac{\partial w}{\partial y}\right|_{y=b} = 0. \tag{1-30}$$

(b) Simply Supported Edge: For a simply supported edge (Fig. 1-9), the deflection and moment are zero. Hence,

$$w|_{y=0} = 0 \tag{1-31}$$

and, from Eq. (1-17),

$$M_y|_{y=0} = -D\left(\frac{\partial^2 w}{\partial y^2} + \mu \frac{\partial^2 w}{\partial x^2}\right)\bigg|_{y=0} = 0. \tag{1-32}$$

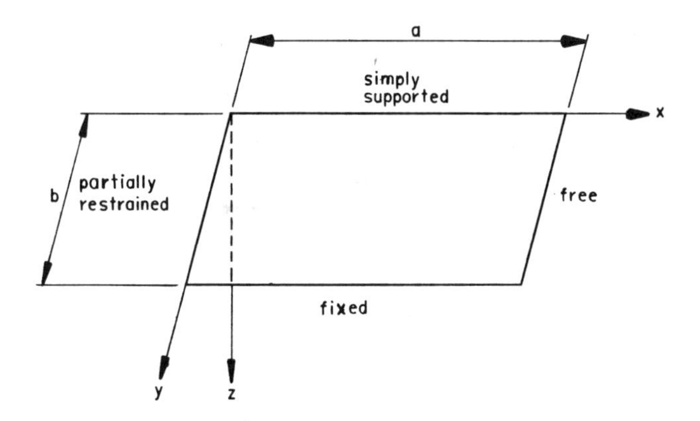

Figure 1-9.

The expression $\mu \dfrac{\partial^2 w}{\partial x^2}$ in Eq. (1-32) can be written as $\mu \dfrac{\partial}{\partial x}\left(\dfrac{\partial w}{\partial x}\right)$ which is the rate of change of the slope at the boundary. But the change in slope along the simply supported edge $y = 0$ is always zero. Hence the quantity $\mu \dfrac{\partial^2 w}{\partial x^2}$ vanishes and the moment boundary condition becomes

$$M_y\big|_{y=0} = \left.\frac{\partial^2 w}{\partial y^2}\right|_{y=0} = 0. \tag{1-33}$$

(c) Free Edge: At a free edge, the moment and shear are zero. Hence,

$$M_x\big|_{x=a} = M_{xy}\big|_{x=a} = Q_x\big|_{x=a} = 0.$$

From the first of these boundary conditions and Eq. (1-17) we get

$$\left.\left(\frac{\partial^2 w}{\partial x^2} + \mu \frac{\partial^2 w}{\partial y^2}\right)\right|_{x=a} = 0. \tag{1-34}$$

The other two boundary conditions can be combined into a single expression. Referring to Fig. 1-10, it was shown by Kirchhoff (Timoshenko and Woinowsky-Krieger 1959) that the moment M_{xy} can be thought of as a series of couples acting on an infinitesimal section. Hence, at any point along the edge

$$Q' = -\left.\left(\frac{\partial M_{xy}}{\partial y}\right)\right|_{x=a}$$

This equivalent shearing force, Q', must be added to the shearing force

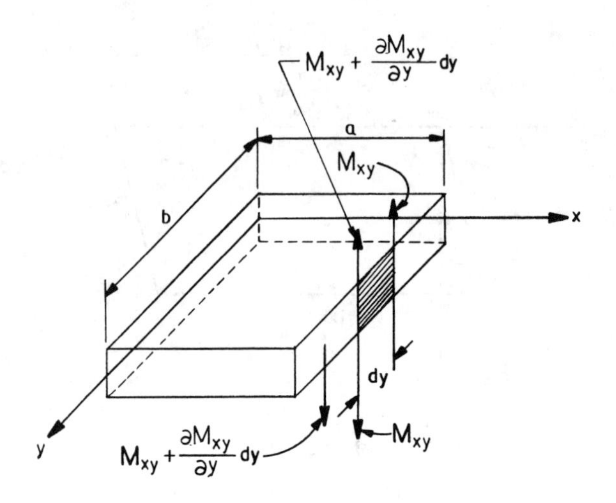

Figure 1-10.

Q_x acting at the edge. Hence the total shearing force at the free edge is given by Q' and Eq. (1-27) as

$$V_x = \left(Q_x - \frac{\partial M_{xy}}{\partial y}\right)\bigg|_{x=a} = 0$$

Substituting the values of Q_x and M_{xy} from Eqs. (1-27) and (1-17) into this equation gives

$$\left(\frac{\partial^3 w}{\partial x^3} + (2 - \mu)\frac{\partial^3 w}{\partial x\, \partial y^2}\right)\bigg|_{x=a} = 0. \tag{1-35}$$

Equations (1-34) and (1-35) are the two necessary boundary conditions at a free edge of a rectangular plate.

(d) Partially Fixed Edge: A partially fixed edge occurs in continuous plates or plates connected to beams. For this latter condition, Fig. 1-11 shows that the two boundary conditions are given by

$$V|_{\text{plate}} = V|_{\text{beam}}$$

$$D\left[\frac{\partial^3 w}{\partial x^3} + (2 - \mu)\frac{\partial^3 w}{\partial x\, \partial y^2}\right]\bigg|_{x=0} = EI\left(\frac{\partial^4 w}{\partial y^4}\right)\bigg|_{x=0} \tag{1-36}$$

and

$$M|_{\text{plate}} = M|_{\text{beam}}$$

$$D\left(\frac{\partial^2 w}{\partial x^2} + \mu\frac{\partial^2 w}{\partial y^2}\right)\bigg|_{x=0} = GJ\left(\frac{\partial^3 w}{\partial x\, \partial y^2}\right)\bigg|_{x=0}. \tag{1-37}$$

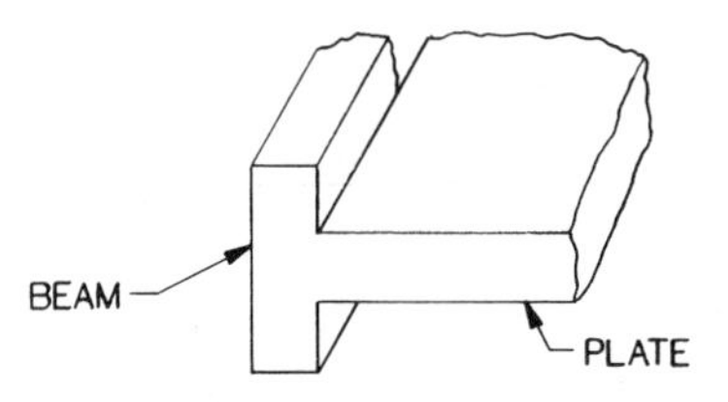

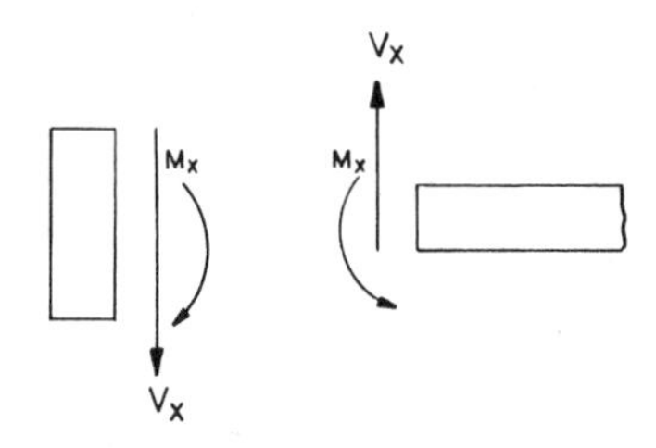

Figure 1-11.

(e) Corner Reactions: It was shown in the derivation of Eq. (1-35) that the torsion moment M_{xy} shown in Fig. 1-10 can be resolved into a series of couples. At any corner, say $x = a$ and $y = b$ in Fig. 1-12, the moment M_{xy} results in a downward force and so does M_{yx} as shown in the figure. Hence the total reaction at $x = a$ and $y = b$ is given by

$$R = 2(M_{xy})\Big|_{\substack{x=a\\y=b}} = 2D(1 - \mu)\left(\frac{\partial^2 w}{\partial x\, \partial y}\right)\Big|_{\substack{x=a\\y=b}} \tag{1-38}$$

Equation (1-38) is normally used to determine the force in corner bolts of rectangular cover plates of gear transmission casings, flanges, etc.

To summarize, Eqs. (1-29) and (1-30) are used for fixed edges whereas Eqs. (1-31) and (1-33) are utilized for simply supported edges. Free edges are expressed by Eqs. (1-34) and (1-35) and boundaries of plates with beam edges are given by Eqs. (1-36) and (1-37). Corner loads are expressed by Eq. (1-38).

Example 1-1

Find the moment and reaction expressions for a simply supported rectangular plate (Fig. 1-13a) of length a, width b, and subjected to a sinusoidal

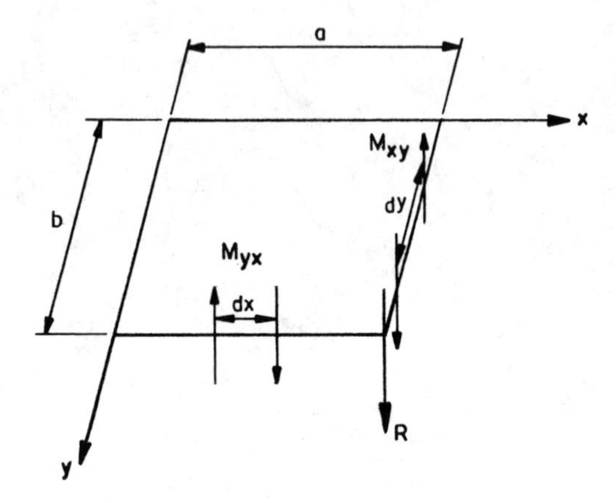

Figure 1-12.

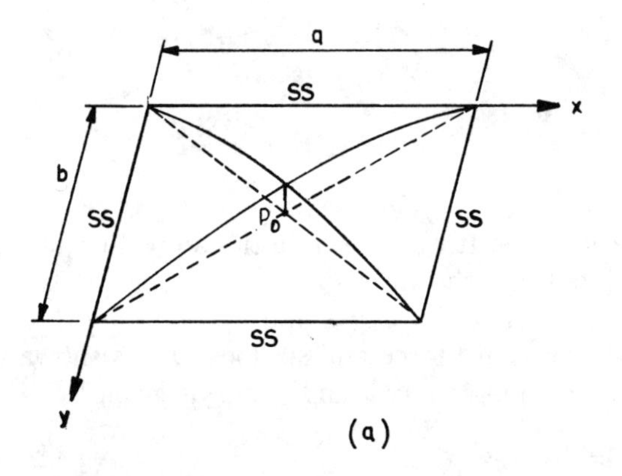

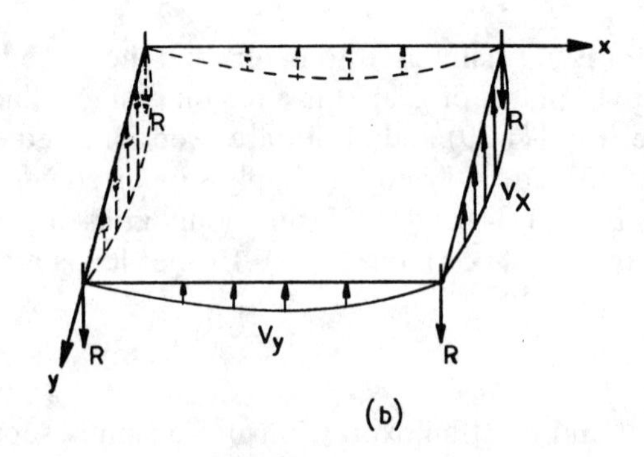

Figure 1-13a and b.

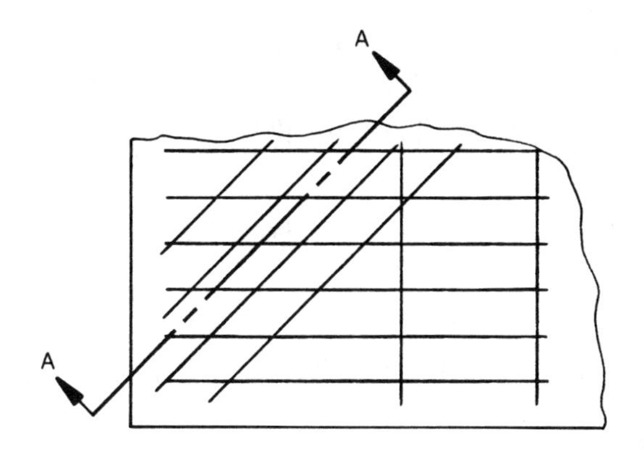

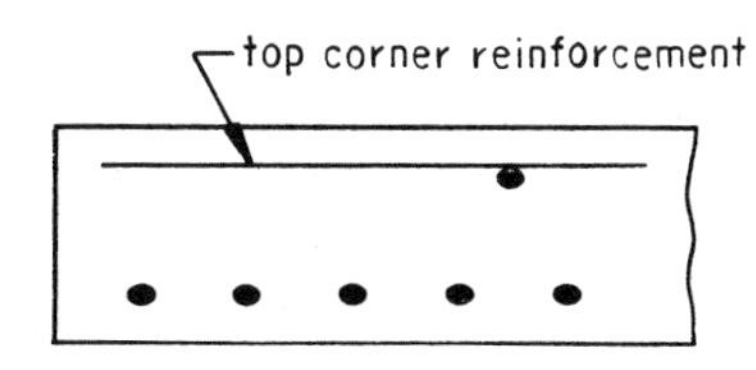

Figure 1-13c. Corner reinforcement.

load given by

$$p = p_o \sin \frac{\pi x}{a} \sin \frac{\pi y}{b}.$$

Solution

The differential Eq. (1-26) is written as

$$\nabla^4 w = \frac{p_o}{D} \sin \frac{\pi x}{a} \sin \frac{\pi y}{b}. \qquad (1)$$

From Fig. 1-13, the boundary conditions are given by

$$w = 0 \quad \text{and} \quad \frac{\partial^2 w}{\partial x^2} = 0 \quad \text{at } x = 0 \quad \text{and} \quad x = a$$

$$w = 0 \quad \text{and} \quad \frac{\partial^2 w}{\partial y^2} = 0 \quad \text{at } y = 0 \quad \text{and} \quad y = b.$$

The assumed expression for the deflection must be of the same general format as that of the applied load in order to solve Eq. (1). It must also satisfy the boundary conditions of the plate. Hence, a deflection of the form

$$w = C \sin \frac{\pi x}{a} \sin \frac{\pi y}{b} \tag{2}$$

satisfies the given boundary conditions.

Substituting Eq. (2) into Eq. (1) gives

$$C = \frac{p_o}{D\pi^4\left(\frac{1}{a^2} + \frac{1}{b^2}\right)^2}$$

and the expression for w becomes

$$w = \frac{p_o}{D\pi^4\left(\frac{1}{a^2} + \frac{1}{b^2}\right)^2} \sin \frac{\pi x}{a} \sin \frac{\pi y}{b}. \tag{3}$$

Substituting this expression into Eq. (1-17) gives

$$M_x = \frac{p_o}{\pi^2\left(\frac{1}{a^2} + \frac{1}{b^2}\right)^2}\left(\frac{1}{a^2} + \frac{\mu}{b^2}\right) \sin \frac{\pi x}{a} \sin \frac{\pi y}{b}$$

$$M_y = \frac{p_o}{\pi^2\left(\frac{1}{a^2} + \frac{1}{b^2}\right)^2}\left(\frac{\mu}{a^2} + \frac{1}{b^2}\right) \sin \frac{\pi x}{a} \sin \frac{\pi y}{b}$$

$$M_{xy} = \frac{p_o(1 - \mu)}{\pi^2\left(\frac{1}{a^2} + \frac{1}{b^2}\right)^2 ab} \cos \frac{\pi x}{a} \cos \frac{\pi y}{b}.$$

The maximum value for moments M_x and M_y occur at $x = a/2$ and $y = b/2$.

To find the reactions, we calculate Q_x and Q_y from Eqs. (1-27) and

(1-28). This gives

$$Q_x = \frac{p_o}{\pi a\left(\dfrac{1}{a^2} + \dfrac{1}{b^2}\right)} \cos\frac{\pi x}{a} \sin\frac{\pi y}{b}$$

$$Q_y = \frac{p_o}{\pi b\left(\dfrac{1}{a^2} + \dfrac{1}{b^2}\right)} \sin\frac{\pi x}{a} \cos\frac{\pi y}{b}$$

for edge $x = a$, the reaction is given by

$$V_x = \left(Q_x - \frac{\partial M_{xy}}{\partial y}\right)\bigg|_{x=a}$$

$$= \frac{-p_o}{\pi a\left(\dfrac{1}{a^2} + \dfrac{1}{b^2}\right)^2}\left(\frac{1}{a^2} + \frac{2-\mu}{b^2}\right) \sin\frac{\pi y}{b} \tag{4}$$

and for edge $y = b$, the reaction is given by

$$V_y = \left(Q_y - \frac{\partial M_{xy}}{\partial x}\right)\bigg|_{y=b}$$

$$= \frac{-p_o}{\pi b\left(\dfrac{1}{a^2} + \dfrac{1}{b^2}\right)^2}\left(\frac{1}{b^2} + \frac{2-\mu}{a^2}\right) \sin\frac{\pi x}{a}. \tag{5}$$

The total reaction around the plate is obtained by integrating Eqs. (4) and (5) from $x = 0$ to $x = a$ and from $y = 0$ to $y = b$ and then multiplying the result by 2 due to symmetry. This gives

$$\text{total reaction} = \frac{4p_o ab}{\pi^2} + \frac{8p_o(1-\mu)}{\pi^2 ab\left(\dfrac{1}{a^2} + \dfrac{1}{b^2}\right)^2}. \tag{6}$$

The first part of this equation can also be obtained by integrating the applied load over the total area, or

$$\int_0^b \int_0^a p \sin\frac{\pi x}{a} \sin\frac{\partial y}{b}\, dx\, dy.$$

The second expression in Eq. (6) is the summation of the four corner reactions given by Eq. (1-38). Hence, at $x = 0$ and $y = 0$ the expression for the reaction is

$$R = 2(M_{xy})\bigg|_{\substack{x=0\\y=0}} = \frac{2p_o(1 - \mu)}{\pi^2 ab\left(\dfrac{1}{a^2} + \dfrac{1}{b^2}\right)^2}.$$

A plot of the shear distribution and reaction is shown in Fig. 1-13b. The positive value of R indicates that the corners have a tendency to lift up and a downward force is needed to keep them in place. This action must be considered when designing cover plates and concrete slabs. An example of the reinforcement at the corners of a concrete slab is shown in Fig. 1-13c.

1-7 Double Series Solution of Simply Supported Plates

The first successful solution of a simply supported rectangular plate subjected to uniform load was made by Navier (Timoshenko 1983) in 1820. He assumed the load p in Eq. (1-26) to be represented by the double Fourier series, Appendix A, of the form

$$p(x, y) = \sum_{m=1}^{\infty}\sum_{n=1}^{\infty} p_{mn} \sin\frac{m\pi x}{a} \sin\frac{n\pi y}{b} \tag{1-39}$$

where p_{mn} is obtained from

$$p_{mn} = \frac{4}{ab}\int_0^b\int_0^a f(x, y) \sin\frac{m\pi x}{a} \sin\frac{n\pi y}{b}\, dx\, dy \tag{1-40}$$

and $f(x,y)$ is the shape of the applied load.

Similarly the deflection w is expressed by

$$w(x, y) = \sum_{m=1}^{\infty}\sum_{n=1}^{\infty} w_{mn} \sin\frac{m\pi x}{a} \sin\frac{n\pi y}{b}. \tag{1-41}$$

This equation automatically satisfies four boundary conditions of a simply supported plate and w_{mn} is a constant that is determined from the differential equation.

The solution of a rectangular plate problem consists of obtaining a load function form Eq. (1-39). The the unknown constant w_{mn} is obtained by substituting Eqs. (1-39) and (1-41) into Eq. (1-26).

Example 1-2

(a) Determine the maximum bending moment of a simply supported plate due to a uniformly applied load.

(b) Let a steel rectangular plate with dimensions $a = 30$ inch and $b = 20$ inch be subjected to a pressure of 15 psi. Determine the maximum bending moment and deflection if $\mu = 0.3$, $E = 30{,}000$ ksi, and $t = 0.38$ inch.

Solution

(a) Let the coordinate system be as shown in Fig. 1-14. Equation (1-40) can be solved by letting $f(x, y)$ equal a constant p_o because the load is uniform over the entire plate. Hence,

$$\begin{aligned} p_{mn} &= \frac{4p_o}{ab}\int_0^b\int_0^a \sin\frac{m\pi x}{a}\sin\frac{n\pi y}{b}\,dx\,dy \\ &= \frac{4p_o}{\pi^2 mn}(\cos m\pi - 1)(\cos n\pi - 1) \\ &= \frac{16p_o}{\pi^2 mn} \end{aligned}$$

where

$m = 1, 3, 5, \ldots$
$n = 1, 3, 5, \ldots$

From Eq. (1-39),

$$p = \frac{16p_o}{\pi^2 mn}\sum_{m=1,3,\ldots}^{\infty}\sum_{n=1,3,\ldots}^{\infty}\sin\frac{m\pi x}{a}\sin\frac{n\pi y}{b}.$$

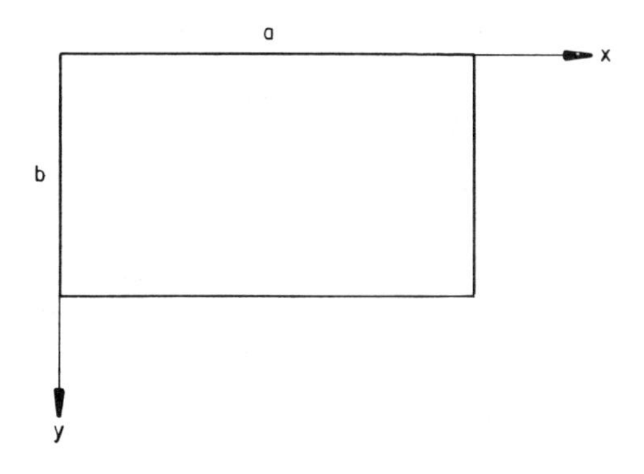

Figure 1-14.

Substituting this equation and Eq. (1-41) into Eq. (1-26) gives

$$w_{mn} = \frac{16p_o}{\pi^6 mnD[(m/a)^2 + (n/b)^2]^2} \qquad \begin{aligned} m &= 1, 3, 5, \ldots \\ n &= 1, 3, 5, \ldots \end{aligned}$$

Hence, the deflection expression becomes

$$w = \frac{16p_o}{\pi^6 D} \sum_{m=1,3,\ldots}^{\infty} \sum_{n=1,3,\ldots}^{\infty} \frac{\sin \dfrac{m\pi x}{a} \sin \dfrac{n\pi y}{b}}{mn[(m/a)^2 + (n/b)^2]^2}. \tag{1}$$

The bending and torsional moment expressions are given by Eq. (1-17) and are expressed as

$$M_x = \frac{16p_o}{\pi^4} \left[\sum_{m=1,3,\ldots}^{\infty} \sum_{n=1,3,\ldots}^{\infty} F_{mn} \sin \frac{m\pi x}{a} \sin \frac{n\pi y}{b} \right] \tag{2}$$

$$M_y = \frac{16p_o}{\pi^4} \left[\sum_{m=1,3,\ldots}^{\infty} \sum_{n=1,3,\ldots}^{\infty} G_{mn} \sin \frac{m\pi x}{a} \sin \frac{n\pi y}{b} \right] \tag{3}$$

$$M_{xy} = \frac{16p_o(1 - \mu)}{\pi^4} \left[\sum_{m=1,3,\ldots}^{\infty} \sum_{n=1,3,\ldots}^{\infty} H_{mn} \cos \frac{m\pi x}{a} \cos \frac{n\pi y}{b} \right] \tag{4}$$

where

$$F_{mn} = \frac{(m/a)^2 + \mu(n/b)^2}{mn[(m/a)^2 + (n/b)^2]^2}$$

$$G_{mn} = \frac{\mu(m/a)^2 + (n/b)^2}{mn[(m/a)^2 + (n/b)^2]^2}$$

$$H_{mn} = \frac{1}{ab[(m/a)^2 + (n/b)^2]^2}.$$

The maximum values of deflection and bending moments occur at $x = a/2$ and $y = b/2$.

A plot of Eqs. (2), (3), and (4) is shown in Fig. 1-15 for a square plate. The figure also shows a plot of M_1 and M_2 obtained from Mohr's circle along the diagonal of the plate. M_1 becomes negative near the corner of the plate. This is due to the uplift tendency at the corners. This uplift is resisted by the reaction R that causes tension at the top portion of the plate near the corners. This tension must be properly reinforced in concrete slabs as shown in Fig. 1-13c.

(b) The maximum values of M_x, M_y, and w are obtained from expression (2), (3), and (4) above. The computer program "DBLSUM" in Table A-2

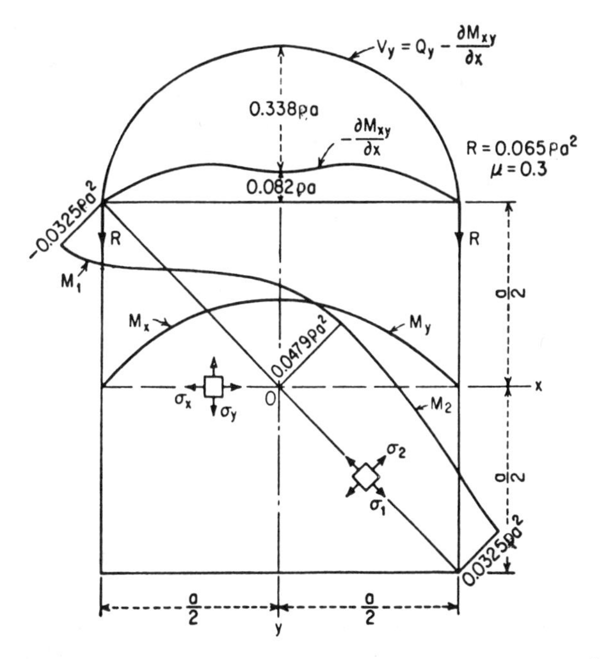

Figure 1-15. (Timoshenko and Woinowsky-Krieger 1959.)

of Appendix A is used to determine the series summation for m and n.

$$M_x = \frac{16p_o b^2}{\pi^4}(0.3035) = 299.1 \text{ inch-lbs/inch}$$

$$M_y = \frac{16p_o b^2}{\pi^4}(0.4941) = 487.0 \text{ inch-lbs/inch}$$

$$w = \frac{16p_o b^4}{\pi^6 D}(0.4647)$$

The value of D is given by Eq. (1-18) as

$$D = \frac{Et^3}{12(1-\mu^2)} = \frac{30{,}000{,}000 \times 0.38^3}{12\,(0.91)} = 150{,}747 \text{ lbs-inch.}$$

the maximum deflection is

$$w = 0.12 \text{ inch.}$$

Example 1-3

Find the deflection expression for the simply supported plate loaded as shown in Fig. 1-16.

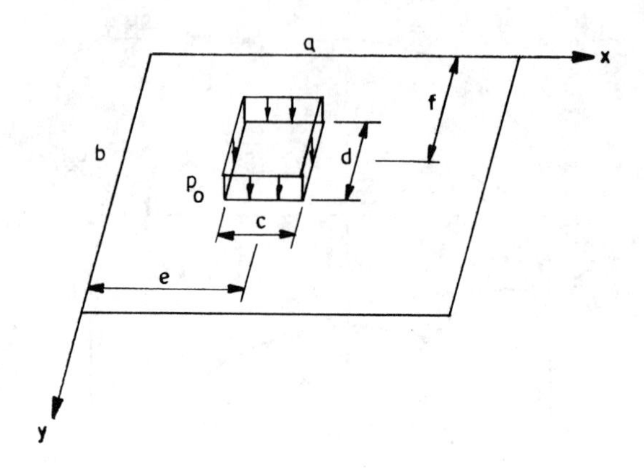

Figure 1-16.

Solution

The Fourier expansion of the load is obtained from Eqs. (1-39) and (1-40) as

$$p_{mn} = \frac{4}{ab\ cd}\int_{f-d/2}^{f+d/2}\int_{e-c/2}^{e+c/2} p_o \sin\frac{m\pi x}{a}\sin\frac{n\pi y}{b}\,dx\,dy$$

$$p_{mn} = \frac{16p_o}{mn\ cd\pi^2}\sin\frac{m\pi e}{a}\sin\frac{m\pi c}{2a}\sin\frac{n\pi f}{b}\sin\frac{n\pi d}{2b}. \tag{1}$$

Substituting this expression and Eq. (1-41) into Eq. (1-26) gives

$$w_{mn} = \frac{16p_o}{\pi^6 cdD}\,\frac{\sin\dfrac{m\pi e}{a}\sin\dfrac{m\pi c}{2a}\sin\dfrac{n\pi f}{b}\sin\dfrac{n\pi d}{2b}}{mn[(m/a)^2 + (n/b)^2]^2} \tag{2}$$

and

$$w = \sum_{m=1,3,\ldots}^{\infty}\sum_{n=1,3,\ldots}^{\infty} w_{mn}\sin\frac{m\pi x}{a}\sin\frac{n\pi y}{b}. \tag{3}$$

This equation reduces to Eq. (1) of Example 1-2 for a uniformly loaded plate when $c = a$, $d = b$, $e = a/2$, and $f = b/2$.

Problems

1-8 Find the expression for moments M_x and M_y in Example 1-3.

1-9 A tabletop is loaded as shown in Fig. 1-16. Find the maximum stress in the table if a = 200 cm, b = 80 cm, c = 30 cm, d = 15 cm, e = 80 cm, f = 40 cm, E = 840 kg/mm^2, p = 0.2 kg/cm^2, and μ = 0.30. Assume the tabletop to be simply supported.

1-10 Show that in Example 1-3 the value of p_{mn} for a concentrated load, p_o, is

$$p_{mn} = \frac{4p_o}{ab} \sin \frac{m\pi e}{a} \sin \frac{m\pi f}{b}.$$

1-8 Single Series Solution of Simply Supported Plates

Levy (Timoshenko 1983) in 1900 developed a method for solving simply supported plates subjected to various loading conditions using single Fourier series. This method is more practical then Navier's solution because it is also applicable to plates with various boundary conditions as discussed in Chapter 2. Levy suggested the solution of Eq. (1-26) to be expressed in terms of homogeneous and particular parts each of which consists of a single Fourier series where the unknown function is determined from the boundary conditions. Thus the solution is expressed as

$$w = w_h + w_p. \tag{1-42}$$

The homogeneous solution is written as

$$w_h = \sum_{m=1}^{\infty} f_m(y) \sin \frac{m\pi x}{a} \tag{1-43}$$

where $f(y)$ indicates that it is a function of y only. This equation also satisfies a simply supported boundary condition at $x = 0$ and $x = a$. Substituting Eq. (1-43) into the differential equation

$$\nabla^4 w = 0$$

gives

$$\left[\left(\frac{m\pi}{a}\right)^4 f_m(y) - 2\left(\frac{m\pi}{a}\right)^2 \frac{d^2 f_m(y)}{dy^2} + \frac{d^4 f_m(y)}{dy^4}\right] \sin \frac{m\pi x}{a} = 0$$

which is satisfied when the bracketed term is equal to zero. Thus,

$$\frac{d^4 f_m(y)}{dy^4} - 2\left(\frac{m\pi}{a}\right)^2 \frac{d^2 f_m(y)}{dy^2} + \left(\frac{m\pi}{a}\right)^4 f_m(y) = 0. \tag{1-44}$$

The solution of this differential equation can be expressed as

$$f_m(y) = F_m e^{R_m y}. \tag{1-45}$$

Substituting Eq. (1-45) into Eq. (1-44) gives

$$R_m^4 - 2\left(\frac{m\pi}{a}\right)^2 R_m^2 + \left(\frac{m\pi}{a}\right)^4 = 0$$

which has the roots

$$R_m = \pm\frac{m\pi}{a}, \quad \pm\frac{m\pi}{a}.$$

Thus, the solution of Eq. (1-44) is

$$f_m(y) = C_{1m}e^{\frac{m\pi y}{a}} + C_{2m}e^{-\frac{m\pi y}{a}} + C_{3m}ye^{\frac{m\pi y}{a}} + C_{4m}ye^{-\frac{m\pi y}{a}}$$

where C_{1m}, C_{2m}, C_{3m}, and C_{4m} are constants. This equation can also be written as

$$f_m(y) = A_m \sinh\frac{m\pi y}{a} + B_m \cosh\frac{m\pi y}{a} + C_m y \sinh\frac{m\pi y}{a} + D_m y \cosh\frac{m\pi y}{a}.$$

Hence, the homogeneous solution given by Eq. (1-43) becomes

$$w_h = \sum_{m=1}^{\infty}\left[A_m \sinh\frac{m\pi y}{a} + B_m \cosh\frac{m\pi y}{a} + C_m y \sinh\frac{m\pi y}{a} + D_m y \cosh\frac{m\pi y}{a}\right]\sin\frac{m\pi x}{a} \qquad \text{(1-46)}$$

where the constants A_m, B_m, C_m, and D_m are obtained from the boundary conditions.

The particular solution, w_p, in Eq. (1-42) can be expressed in a single Fourier series as

$$w_p = \sum_{m=1}^{\infty} k_m(y) \sin\frac{m\pi x}{a}. \qquad \text{(1-47)}$$

The load p is expressed as

$$p(x, y) = \sum_{m=1}^{\infty} p_m(y) \sin\frac{m\pi x}{a} \qquad \text{(1-48)}$$

where

$$p_m(y) = \frac{2}{a}\int_0^a p(x, y) \sin\frac{m\pi x}{a}\, dx. \qquad \text{(1-49)}$$

Substituting Eqs. (1-47) and (1-48) into Eq. (1-26) gives

$$\frac{d^4k_m}{dy^4} - 2\left(\frac{m\pi}{a}\right)^2 \frac{d^2k_m}{dy^2} + \left(\frac{m\pi}{a}\right)^4 k_m = \frac{p_m(y)}{D}. \tag{1-50}$$

Thus, the solution of the differential equation (1-26) consists of solving Eqs. (1-46) and (1-50) as shown in the following example.

Example 1-4

The rectangular titanium plate shown in Fig. 1-17 is subjected to a uniform load p_o. Determine the expression for the deflection.

Solution

From Eq. (1-49),

$$\begin{aligned} p_m(y) &= \frac{2p_o}{a}\int_0^a \sin\frac{m\pi x}{a}\,dx \\ &= \frac{2p_o}{m\pi}(\cos m\pi - 1) = \frac{4p_o}{m\pi} \qquad m = 1, 3, \ldots \end{aligned}$$

Hence, Eq. (1-50) becomes

$$\frac{d^4k_m}{dy^4} - 2\left(\frac{m\pi}{a}\right)^2 \frac{d^2k_m}{dy^2} + \left(\frac{m\pi}{a}\right)^4 k_m = \frac{4p_o}{m\pi D}. \tag{1}$$

The particular solution of this equation can be taken as

$$k_m = C.$$

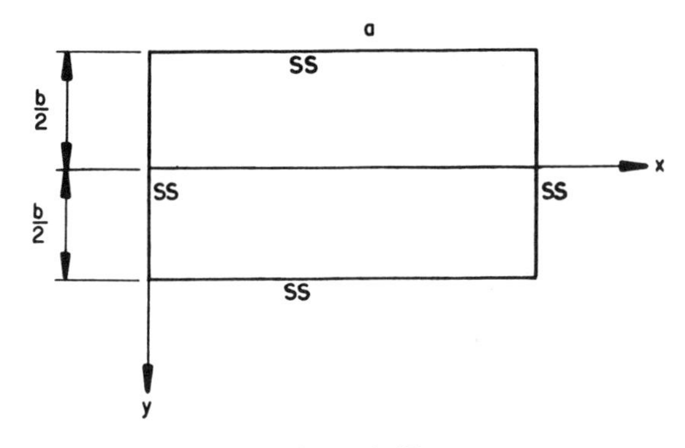

Figure 1-17.

Substituting this expression into Eq. (1) gives

$$k_m = \frac{4a^4 p_o}{m^5 \pi^5 D} \qquad m = 1, 3, \ldots$$

And Eq. (1-47) for the particular solution becomes

$$w_p = \frac{4a^4 p_o}{\pi^5 D} \sum_{m=1,3,\ldots}^{\infty} \frac{1}{m^5} \sin \frac{m\pi x}{a}. \tag{2}$$

The homogeneous solution for the deflection is obtained from Eq. (1-46). Referring to Fig. 1-17, the deflection in the y-direction due to uniform load is symmetric about the x-axis. Hence, the constants A_m and D_m must be set to zero since the quantities $\sinh \frac{m\pi y}{a}$ and $y \cosh \frac{m\pi y}{a}$ are odd functions as y varies from positive to negative. Also, m must be set to 1, 3, 5, etc. in order for $\sin \frac{m\pi x}{a}$ to be symmetric around $x = a/2$. Hence,

$$w_h = \sum_{m=1,3,\ldots}^{\infty} \left(B_m \cosh \frac{m\pi y}{a} + C_m y \sinh \frac{m\pi y}{a} \right) \sin \frac{m\pi x}{a}$$

and the total deflection can now be expressed as

$$w = \sum_{m=1,3,\ldots}^{\infty} \left(B_m \cosh \frac{m\pi y}{a} + C_m y \sinh \frac{m\pi y}{a} + \frac{4p_o a^4}{m^5 \pi^5 D} \right) \sin \frac{m\pi x}{a}. \tag{3}$$

The boundary conditions along the y-axis are expressed as

$$w = 0 \quad \text{at } y = \pm\, b/2$$

and

$$\frac{\partial^2 w}{\partial y^2} = 0 \quad \text{at } y = \pm\, b/2.$$

From the first of these boundary conditions we get

$$B_m \cosh \frac{m\pi b}{2a} + C_m \frac{b}{2} \sinh \frac{m\pi b}{2a} + \frac{4a^4 p_o}{m^5 \pi^5 D} = 0$$

and from the second boundary condition we get

$$\left[B_m \left(\frac{m\pi}{a} \right) + bc_m \right] \cosh \frac{m\pi b}{2a} + C_m \left(\frac{m\pi b}{2a} \right) \sinh \frac{m\pi b}{2a} = 0.$$

Solving these two simultaneous equations gives

$$C_m = \frac{2a^3p_o}{m^4\pi^4 D \cosh \dfrac{m\pi b}{2a}}$$

$$B_m = \frac{4a^4p_o + m\pi p_o a^3 b \tanh \dfrac{m\pi b}{2a}}{m^5\pi^5 D \cosh \dfrac{m\pi b}{2a}}.$$

With these two expressions known, Eq. (3) can now be solved for various values of x and y.

The load in the previous few examples was assumed uniform in distribution. Other distributions can be used in Eq. (1-49) as long as they can be expressed in terms of x and y. Thus, if the load in Example 1-4 is triangular in distribution as shown in Fig. 1-18, then it can be expressed as

$$p_m = \frac{p_o x}{a}$$

and Eq. (1-49) becomes

$$\begin{aligned} p_m(y) &= \frac{2}{a}\int_0^a \frac{p_o x}{a} \sin \frac{m\pi x}{a}\, dx \\ &= \frac{2p_o}{m\pi}(-1)^{m+1} \qquad m = 1, 2, \ldots \end{aligned}$$

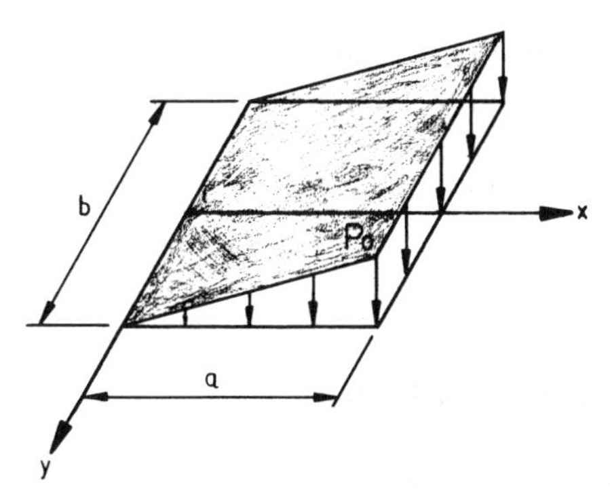

Figure 1-18.

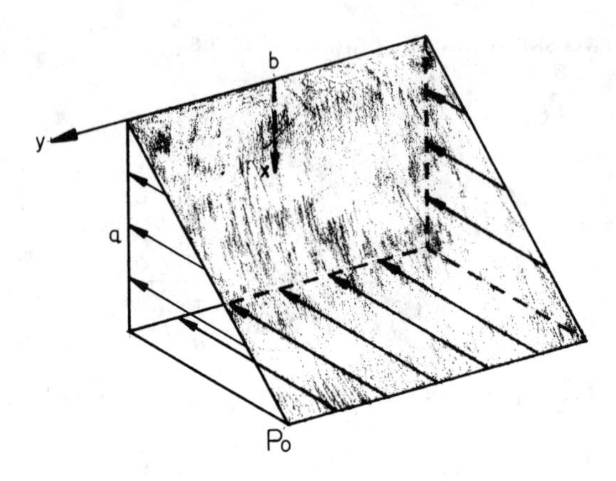

Figure P1-12.

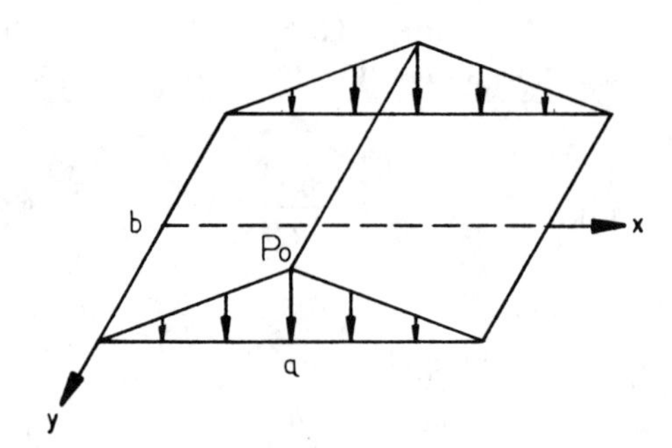

Figure P1-13.

Problems

1-11 Find the expression for M_x, M_y and M_{xy} in Example 1-4.
1-12 A channel weir is approximated as a simply supported rectangular plate subjected to the loading shown in Fig. P1-12. Find the expression for the moments.
1-13 An internal zirconium bulkhead is loaded as shown in Fig. P1-13. Find the expression for the moments assuming the plate to be simply supported.

1-9 Design of Rectangular Plates

The procedure for designing a rectangular plate with a given boundary condition and applied lateral loads is

1. Express the loads in a Fourier series and define a similar expression for the deflection.
2. Solve Eq. (1-26) for the actual deflection, w, by utilizing the boundary conditions in the solution.
3. Determine the maximum moments from Eq. (1-17).
4. For metallic plates, the required thickness is calculated from the expression $\sigma = Mc/I$ where, for a unit width, it reduces to

$$t = \sqrt{6M/\sigma}. \tag{1-51}$$

5. For reinforced concrete slabs where the reinforcement is about the same in the x- and y-axes, the design can be approximated by Eq. (1-26) as discussed above. The design of concrete plates is more complicated than that of metallic plates because the engineer has to determine not only the magnitude of the bending moments, but also their direction and location in order to properly space the reinforcing bars. Standards such as the ACI 318 establish minimum requirements for concrete thickness and reinforcement spacing throughout the slab.
6. Orthotropic plates and reinforced concrete slabs where the reinforcement is not the same in the x- and y-directions are analyzed in accordance with the orthotropic plate theory discussed in Chapter 5.

Allowable stress values at various temperatures of various materials are published in many international codes. The ASME VIII-1 code publishes allowable stresses for over 500 different steels and nonferrous alloys. These stresses are based on the smaller of two-thirds of the yield stress or one-fourth of the tensile stress of the material at a given temperature. Table 1-2 lists allowable stress values for a few materials at temperatures below the creep and rupture values. Allowable stresses at elevated temperatures are discussed in Chapter 2. Allowable stress values for reinforced concrete are given in various standards such as ACI 318.

Table 1-2. ASME VIII-1 allowable stress values, ksi

		Temperature, °F			
Material	ASME Designation	Room Temperature	300	500	700
Carbon steel	SA 516-70	17.5	17.5	17.5	16.6
Stainless steel	SA 240-304	18.8	16.6	15.9	15.9
Aluminum	SB 209-6061 T4	7.5	6.9		
Copper alloy	SB 171-715 70/30	12.5	10.4	10.4	10.4
Nickel alloy	SB 575-276	25.0	25.0	23.9	23.1
Titanium alloy	SB 265-Gr 3	16.3	11.7	7.5	

Table 1-3. Simply supported rectangular plates

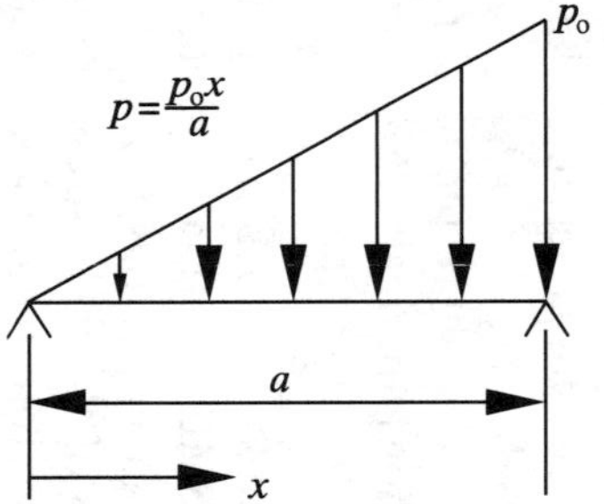

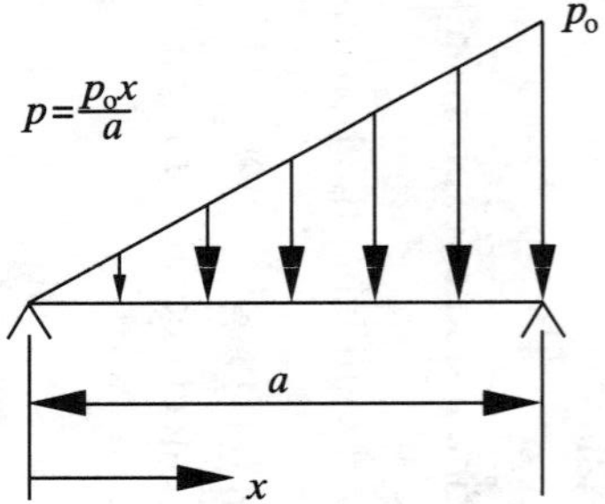

	Uniform pressure, p_o, over total surface		$p = \frac{p_o x}{a}$ (linearly varying, 0 to p_o over a)			$p = \frac{p_o x}{a}$ (linearly varying, 0 to p_o over a)	
a/b	K_1	K_2	K_1	K_2	b/a	K_1	K_2
1.0	0.2873	0.04436	0.1619	0.02243	1.0	0.1619	0.02243
1.1	0.3291	0.05317	0.1710	0.02697	1.1	0.1838	0.02682
1.2	0.3761	0.06170	0.1938	0.03142	1.2	0.2045	0.03108
1.3	0.4163	0.06980	0.2160	0.03570	1.3	0.2239	0.03511
1.4	0.4533	0.07737	0.2376	0.03978	1.4	0.2417	0.03888
1.5	0.4872	0.08435	0.2568	0.04362	1.5	0.2578	0.04215
1.6	0.5174	0.09073	0.2742	0.04723	1.6	0.2725	0.04555
1.8	0.5691	0.1017	0.3072	0.05375	1.8	0.2976	0.05103
2.0	0.6101	0.1106	0.3358	0.05943	2.0	0.3174	0.05546
2.5	0.6776	0.1255	0.3922	0.07077	2.5	0.3498	0.06291
3.0	0.7134	0.1336	0.4341	0.07920	3.0	0.3670	0.06693
3.5	0.7314	0.1378	0.4664	0.08570	3.5	0.3759	0.06904
4.0	0.7410	0.1400	0.4922	0.09089	4.0	0.3804	0.07012
6.0	0.7493	0.1421	0.5588	0.1042	6.0	0.3847	0.07116

Notation: Maximum stress $S = K_1 p_o b^2/t^2$; maximum deflection $w = K_2 p_o b^4/Et^3$. a = plate dimension; b = plate dimension; E = modulus of elasticity; K_1 = stress factor ($\mu = 0.3$); K_2 = deflection factor ($\mu = 0.3$); p_o = maximum pressure; S = maximum stress; t = thickness; w = maximum deflection.

Maximum moments and stresses in most of the frequently encountered load cases for simply supported rectangular plates have been tabulated in many references for the convenience of the engineer. Timoshenko (Timoshenko and Woinowsky-Krieger 1959) lists tables and charts for maximum moments and deflections of numerous loading conditions. Roark (Roark and Young 1975) has similar tables and so does Pilkey (Pilkey and Chang 1978). Table 1-3 gives maximum deflection and stress values for simply supported plates with two commonly encountered loading conditions.

Loading conditions not found in published references must be solved by developing a Fourier series for the loads and deflection and then satisfying the boundary conditions as discussed in this chapter.

The exact analysis of perforated rectangular plates, which are used in boilers and pressure vessels, is difficult to obtain. However, various approximations can be made to obtain a solution. One such approximation is given by the ASME boiler code, Section I, and consists of using the

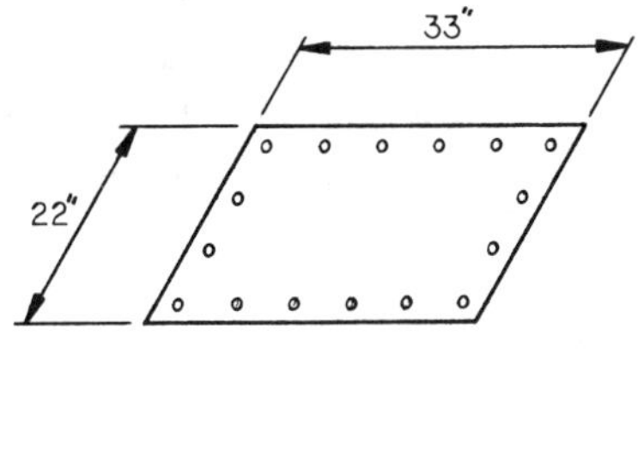

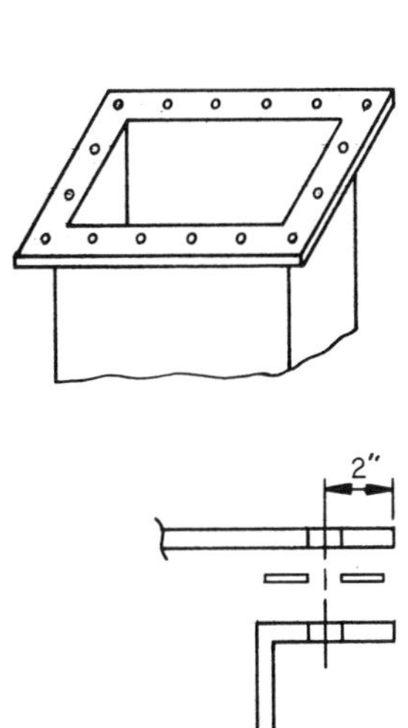

Figure P1-14.

classical solution of a solid plate and then modifying it by using ligament efficiency factors to account for the effect of the perforations on the plate stress.

Problems

1-14 What is the required thickness of the rectangular cover for the opening shown in Fig. P1-14? The cover is made of aluminum SA209-6061 T4 material. The temperature is 300°F and the applied pressure is 100 psi. Calculate the required number of bolts if they are made of the same material. Do the corner bolts have to be larger than the rest of the bolts?

1-15 What is the required thickness of the internal partition *ABCD* of the holding tank shown in Fig. P1-15? The bottom of the tank is enclosed by a circular plate while the top is enclosed by a semicircular plate that covers one side of the top. The tank is full of water on one side of the partition. The material is SA 240-304 and the temperature is 100°F. Assume the partition *ABCD* to be simply supported on all four sides.

1-16 In Example 1-2, what modifications must be made to Eq. (1) if the thickness of the plate is variable and is a function of x and y.

1-17 Discuss the modifications that have to be made to the differential equation and boundary conditions if the corners of a simply supported plate are allowed to curl up due to applied pressure.

1-18 In Fig. 1-11, how should the reinforcing bars in a concrete slab be placed to ensure continuity of moments, shears, etc. between the slab and beam?

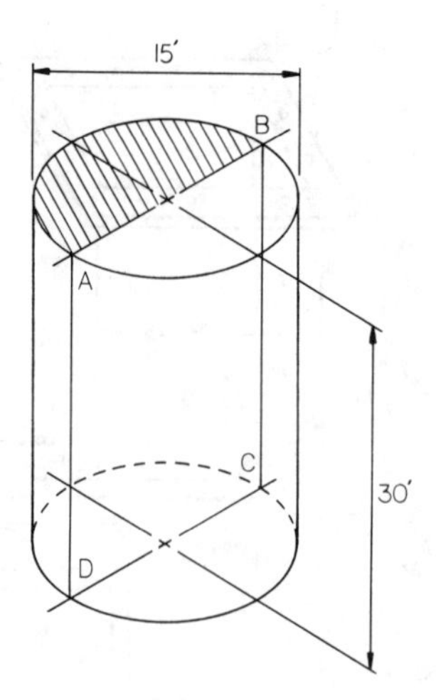

Figure P1-15.

2

Bending of Various Rectangular Plates

2-1 Plates with Various Boundary Conditions

Most rectangular plates in building slabs, ship hulls, aircraft skins (Fig. 2-1), and rectangular holding tanks have boundary conditions other than simply supported. The Levy solution discussed in Chapter 1 can be utilized very effectively in solving rectangular plates with various boundary conditions. Equations (1-46) and (1-50) are readily applicable to plates with two opposite sides simply supported, and the boundary conditions for the other two sides can then be incorporated into the total solution. For plates that do not have two opposite simply supported sides, the solution is more difficult because various cases have to be superimposed to arrive at a solution. The following examples illustrate the general procedure to be followed in solving plates with various boundary conditions.

Example 2-1

Find the expression for the deflection of a uniformly loaded plate having three sides simply supported and the fourth side fixed as shown in Fig. 2-2.

Solution

From Example 1-4, the particular deflection is expressed as

$$w_p = \sum_{m=1,3,\ldots}^{\infty} \frac{4p_o a^4}{m^5 \pi^5 D} \sin \frac{m\pi x}{a}$$

Figure 2-1. F-15 Fighter. (Courtesy of the McDonnell Douglas Corp., St. Louis, MO.)

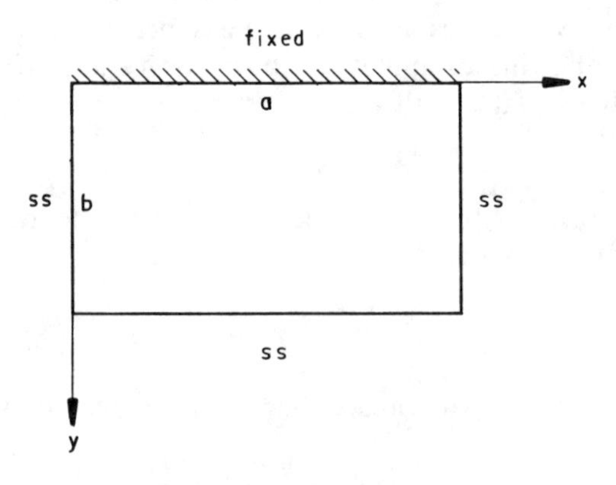

Figure 2-2.

and the total deflection is given by

$$w = \sum_{m=1,2,\ldots}^{\infty} \left(A_m \sinh \frac{m\pi y}{a} + B_m \cosh \frac{m\pi y}{a} + C_m y \sinh \frac{m\pi y}{a} + D_m y \cosh \frac{m\pi y}{a} \right) \sin \frac{m\pi x}{a} + \sum_{m=1,3,\ldots}^{\infty} \frac{4p_o a^4}{m^5 \pi^5 D} \sin \frac{m\pi x}{a}.$$

Since the deflection is symmetric with respect to $x = a/2$, m must be odd and the deflection expression becomes

$$w = \sum_{m=1,3,\ldots}^{\infty} \left(A_m \sinh \frac{m\pi y}{a} + B_m \cosh \frac{m\pi y}{a} + C_m y \sinh \frac{m\pi y}{a} + D_m y \cosh \frac{m\pi y}{a} + \frac{4p_o a^4}{m^5 \pi^5 D} \right) \sin \frac{m\pi x}{a}. \tag{1}$$

The boundary conditions are given by

$$w = 0 \quad \text{and} \quad \frac{\partial w}{\partial y} = 0 \quad \text{at } y = 0$$

$$w = 0 \quad \text{and} \quad \frac{\partial^2 w}{\partial y^2} = 0 \quad \text{at } y = b.$$

Using these four boundary conditions to solve for the unknown constants in Eq. (1) gives

$$A_m = \frac{4p_o a^4}{m^5 \pi^5 D} \frac{F_m}{G_m}$$

$$B_m = \frac{-4p_o a^4}{m^5 \pi^5 D}$$

$$C_m = \frac{4p_o a^4}{m^5 \pi^5 D} \frac{H_m}{G_m}$$

$$D_m = -\frac{m\pi}{a} A_m$$

$$F_m = 2 \cosh^2\left(\frac{m\pi b}{a}\right) - 2 \cosh \frac{m\pi b}{a} - \frac{m\pi b}{a} \sinh \frac{m\pi b}{a}$$

$$G_m = 2 \cosh \frac{m\pi b}{a} \sinh \frac{m\pi b}{a} - 2 \frac{m\pi b}{a}$$

$$H_m = 2 \frac{m\pi}{a} \sinh \frac{m\pi b}{a} \cosh \frac{m\pi b}{a} - \frac{m\pi}{a} \sinh \frac{m\pi b}{a} - \left(\frac{m\pi}{a}\right)^2 b \cosh \frac{m\pi b}{a}.$$

With these constants established, Eq. (1) can be solved for the deflection. The moments throughout the plate can then be obtained from Eq. (1-17).

Example 2-2

For the plate shown in Fig. 2-3, find the maximum bending moment if $p = 10$ psi, $a = 30$ inches, $b = 24$ inches, $E = 30{,}000$ ksi, and $\mu = 0.3$. Use two terms of the series only. What is the required thickness if the allowable stress is 20,000 psi?

Solution

The boundary conditions are

$$(1)\ \text{at } y = 0, \quad w = 0$$

$$(2)\ \text{at } y = 0, \quad \frac{\partial^2 w}{\partial y^2} = 0$$

$$(3)\ \text{at } y = b, \quad \frac{\partial^2 w}{\partial y^2} + \mu \frac{\partial^2 w}{\partial x^2} = 0$$

$$(4)\ \text{at } y = b, \quad \frac{\partial^3 w}{\partial y^3} + (2 - \mu) \frac{\partial^3 w}{\partial x^2 \partial y} = 0.$$

Since the solution is symmetric around $x = a/2$, m must be odd and the general solution is given by

$$w = \sum_{m=1,3,\ldots}^{\infty} \left(A_m \sinh \frac{m\pi y}{a} + B_m \cosh \frac{m\pi y}{a} + C_m y \sinh \frac{m\pi y}{a} + D_m y \cosh \frac{m\pi y}{a} + \frac{4p_o a^4}{m^5 \pi^5 D} \right) \sin \frac{m\pi x}{a}. \quad (1)$$

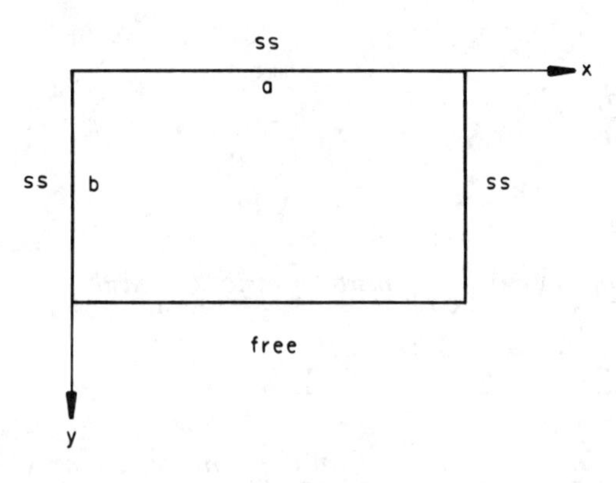

Figure 2-3.

From the first boundary condition, Eq. (1) gives

$$B_m = \frac{-4p_o a^4}{m^5 \pi^5 D}.$$

From the second boundary condition, Eq. (1) gives

$$C_m = \frac{2p_o a^3}{m^4 \pi^4 D}.$$

Solving the third and fourth boundary conditions yields

$$A_m = \frac{K_m G_m - J_m H_m}{I_m G_m - J_m F_m}$$

and

$$D_m = \frac{H_m - A_m F_m}{G_m}$$

where

$$F_m = (1 - \mu)\frac{m^2\pi^2}{a^2}\sinh\frac{m\pi b}{a}$$

$$G_m = \frac{2m\pi}{a}\sinh\frac{m\pi b}{a} + \frac{m^2\pi^2}{a^2}b(1 - \mu)\cosh\frac{m\pi b}{a}$$

$$\begin{aligned} H_m = {} & B_m\frac{m^2\pi^2}{a^2}(\mu - 1)\cosh\frac{m\pi b}{a} \\ & + C_m\frac{m^2\pi^2}{a^2}b(\mu - 1)\sinh\frac{m\pi b}{a} - \frac{2m\pi}{a}C_m\cosh\frac{m\pi b}{a} \\ & + \frac{4\mu p_o a^2}{m^3\pi^3 D} \end{aligned}$$

$$I_m = (\mu - 1)\frac{m^3\pi^3}{a^3}\cosh\frac{m\pi b}{a}$$

$$J_m = (1 + \mu)\frac{m^2\pi^2}{a^2}\cosh\frac{m\pi b}{a} + (\mu - 1)b\frac{m^3\pi^3}{a^3}\sinh\frac{m\pi b}{a}$$

$$\begin{aligned} K_m = {} & (1 - \mu)\frac{m^3\pi^3}{a^3}B_m\sinh\frac{m\pi b}{a} - (1 + \mu)\frac{m^2\pi^2}{a^2}C_m\sinh\frac{m\pi b}{a} \\ & + b\frac{m^3\pi^3}{a^3}C_m(1 - \mu)\cosh\frac{m\pi b}{a}. \end{aligned}$$

The bending moment is obtained from Eq. (1) and the expression

$$M_x = -D\left(\frac{\partial^2 w}{\partial x^2} + \mu\frac{\partial^2 w}{\partial y^2}\right).$$

It has a maximum value at $y = b$ and $x = a/2$. Hence,

$$\begin{aligned}
\frac{M_x}{D} = \sum_{m=1,3,\dots}^{\infty} \Bigg\{ \Bigg(& A_m \sinh\frac{m\pi b}{a} + B_m \cosh\frac{m\pi b}{a} + C_m b \sinh\frac{m\pi b}{a} \\
& + D_m b \cosh\frac{m\pi b}{a} + \frac{4p_o a^4}{m^5\pi^5 D}\Bigg)\frac{m^2\pi^2}{a^2} \\
& - \mu\Bigg[\frac{m\pi}{a}\left(\frac{m\pi}{a}A_m + 2D_m\right)\sinh\frac{m\pi b}{a} \\
& + \frac{m\pi}{a}\left(B_m\frac{m\pi}{a} + 2C_m\right)\cosh\frac{m\pi b}{a} + C_m b\frac{m^2\pi^2}{a^2}\sinh\frac{m\pi b}{a} \\
& + D_m b\frac{m^2\pi^2}{a^2}\cosh\frac{m\pi b}{a}\Bigg]\Bigg\}\sin\frac{m\pi}{2}.
\end{aligned} \tag{2}$$

The stiffness factor D can be deleted from the lefthand side of Eq. (2) and from constants A_m, B_m, C_m, and D_m on the righthand side of the equation. Equation (2) is solved by calculating all constants for $m = 1$ and $m = 3$. This can best be done in tabular form as

Value	$m = 1$	$m = 3$
DB_m	−105,875.54	−435.7
DC_m	5543.63	68.44
F_m	0.04707	64.99
G_m	2.429	2150.94
DH_m	−8078.68	−118,879.48
I_m	−4.995E-3	−20.4184
J_m	−0.02973	−369.34
DK_m	−342.02	16,381.24
DA_m	99,789.12	435.47
DD_m	−5259.82	−68.43
M_x	913	−46

Total M_x = 867 inch-lbs/inch

$$t = \sqrt{6M/\sigma} = \sqrt{6 \times 867/20{,}000}$$
$$= 0.51 \text{ inch}$$

Example 2-3

Find the expression for the deflection of the plate shown in Fig. 2-4a due to a uniform load.

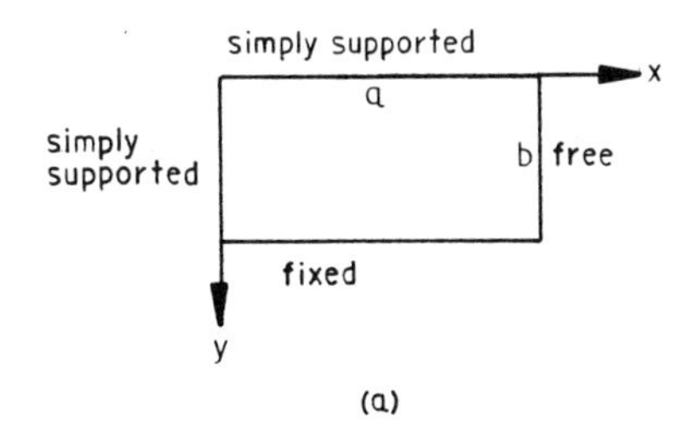

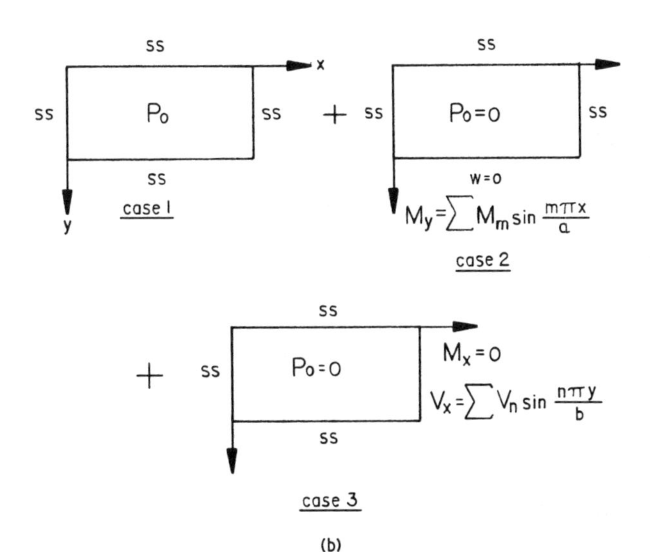

Figure 2-4.

Solution

The solution must be divided into three separate cases as shown in Fig. 2-4b. The differential equations to be solved are

$$\nabla^4 w_1 = p/D, \qquad \nabla^4 w_2 = 0, \qquad \nabla^4 w_3 = 0$$

and

$$w = w_1 + w_2 + w_3. \tag{1}$$

Case 1

The solution of a uniformly loaded, simply supported plate is given in Example 1-4. Using the coordinate system shown in Fig. 2-4b, the solution is expressed as

$$w_1 = \sum_{m=1,3,\ldots}^{\infty} \left(B_m \cosh \frac{m\pi(y - b/2)}{a} + C_m(y - b/2) \sinh \frac{m\pi(y - b/2)}{a} + \frac{4p_o a^4}{m^5 \pi^5 D} \right) \sin \frac{m\pi x}{a} \tag{2}$$

or

$$w_1 = \sum_{m=1,3,\ldots}^{\infty} J_m(y) \sin \frac{m\pi x}{a} \tag{3}$$

where

$$C_m = \frac{2a^3 p_o}{m^4 \pi^4 D \cosh \dfrac{m\pi b}{2a}}$$

$$B_m = -\frac{4a^4 p_o + m\pi p_o a^3 b \tanh \dfrac{m\pi b}{2a}}{m^5 \pi^5 D \cosh \dfrac{m\pi b}{2a}}.$$

Case 2

The boundary conditions are

$$\left.\frac{\partial^2 w_2}{\partial y^2}\right|_{y=0} = w_2|_{y=0} = w_2|_{y=b} = 0$$

and

$$\left. -D\left(\frac{\partial^2 w_2}{\partial y^2} + \mu \frac{\partial^2 w_2}{\partial x^2} \right) \right|_{y=b} = \sum_{m=1,2,\ldots}^{\infty} M_m \sin \frac{m\pi x}{a}.$$

Substituting the function

$$w_2 = \sum_{m=1,2,\ldots}^{\infty} f_m(y) \sin \frac{m\pi x}{a}$$

into the differential equation $\nabla^4 w_2 = 0$ gives

$$f_m(y) = A_m \sinh \frac{m\pi y}{a} + B_m \cosh \frac{m\pi y}{a} + C_m y \sinh \frac{m\pi y}{a} + D_m y \cosh \frac{m\pi y}{a}.$$

Substituting the deflection expression into the first three boundary conditions gives

$$B_m = C_m = 0$$

$$A_m = -D_m b \coth \frac{m\pi b}{a},$$

and from the fourth boundary condition we get

$$D_m = \frac{-M_m}{DT_m}$$

where

$$T_m = \frac{2m\pi}{a} \sinh \frac{m\pi b}{a}.$$

Hence,

$$w_2 = \sum_{m=1,2,\ldots}^{\infty} \frac{M_m}{DT_m} \left(b \coth \frac{m\pi b}{a} \sinh \frac{m\pi y}{a} - y \cosh \frac{m\pi y}{a} \right) \sin \frac{m\pi x}{a}, \tag{4}$$

which can be written as

$$w_2 = \sum_{m=1,2,\ldots}^{\infty} K_m(y) \sin \frac{m\pi x}{a}. \tag{5}$$

Case 3

The boundary conditions are

$$w_3 = \left.\frac{\partial^2 w_3}{\partial x^2}\right|_{x=0} = \left.\left(\frac{\partial^2 w_3}{\partial x^2} + \frac{\partial^2 w_3}{\partial y^2} \right)\right|_{x=a} = 0$$

and

$$\left.\left(\frac{\partial^3 w_3}{\partial x^3} + (2 - \mu) \frac{\partial^3 w_3}{\partial x \, \partial y^2} \right)\right|_{x=a} = \sum_{n=1,2,\ldots}^{\infty} V_n \sin \frac{n\pi y}{b}.$$

Substituting the function

$$w_3 = \sum_{n=1,2,\ldots}^{\infty} g_n(x) \sin \frac{n\pi y}{b}$$

into the differential equation $\nabla^4 w_3 = 0$ gives

$$g_n(x) = A_n \sinh \frac{n\pi x}{b} + B_n \cosh \frac{n\pi x}{b} + C_n x \sinh \frac{n\pi x}{b} + D_n x \cosh \frac{n\pi x}{b}.$$

Substituting this expression into the four boundary conditions results in an expression similar to that of Eq. (5) and can be expressed as

$$w_3 = \sum_{n=1,2,\ldots}^{\infty} L_n(x) \sin \frac{n\pi y}{b}. \tag{6}$$

Substituting Eqs. (3), (5), and (6) into Eq. (1) gives

$$w = \sum_{m=1,3,\ldots}^{\infty} J_m(y) \sin \frac{m\pi x}{a} + \sum_{m=1,2,\ldots}^{\infty} K_m(y) \sin \frac{m\pi x}{a} + \sum_{n=1,2,\ldots}^{\infty} L_n(x) \sin \frac{n\pi y}{b}. \tag{7}$$

This equation has two unknowns. They are M_m in the expression $K_m(y)$, and V_n in the expression $L_n(x)$. The boundary conditions of Fig. 2-4a are

$$\text{at } x = 0, \quad w = 0 \quad \text{and} \quad \frac{\partial^2 w}{\partial x^2} = 0$$

$$\text{at } x = a, \quad M_x = 0 \quad \text{and} \quad V_x = 0$$

$$\text{at } y = 0 \quad w = 0 \quad \text{and} \quad \frac{\partial^2 w}{\partial y^2} = 0$$

$$\text{at } y = b \quad w = 0 \quad \text{and} \quad \frac{\partial w}{\partial y} = 0.$$

Equation (7) satisfies all of the above boundary conditions except

$$\left.\frac{\partial w}{\partial y}\right|_{y=b} = 0 \quad \text{and} \quad V_x|_{x=a} = 0.$$

These two boundary conditions are used to determine the two unknowns in Eq. (7). From the first boundary condition we get

$$\left.\left(\sum_{m=1,3,\ldots}^{\infty} \frac{dJ_m}{dy} \sin \frac{m\pi x}{a} + \sum_{m=1,2,\ldots}^{\infty} \frac{dK_m}{dy} \sin \frac{m\pi x}{a} + \sum_{n=1,2,\ldots}^{\infty} \frac{n\pi}{b} L_n \cos \frac{n\pi y}{b}\right)\right|_{y=b} = 0. \tag{8}$$

In order to solve this expression, the last term needs to be expressed in terms of the quantity $\sin\left(\frac{m\pi x}{a}\right)$. This can be accomplished by letting

$$L_n(x) = \sum_{n=1,2,\ldots}^{\infty} H_{mn} \sin \frac{m\pi x}{a}$$

where

$$H_{mn} = \frac{2}{a}\int_0^a L_n(x) \sin \frac{m\pi x}{a}.$$

Defining

$$\cos \frac{n\pi y}{b} = (-1)^n \quad \text{at } y = b$$

and

$$\delta_m = 0 \quad \text{when } m \text{ is even,}$$

$$\delta_m = 1 \quad \text{when } m \text{ is odd,}$$

eq. (8) becomes

$$\sum_{m=1,2,\ldots}^{\infty} \left[\delta_m \left.\frac{dJ_m}{dy}\right|_{y=b} + \left.\frac{dK_m}{dy}\right|_{y=b} + \sum_{n=1,2,\ldots}^{\infty} \frac{n\pi}{b} H_{mn}(-1)^n \right] \sin \frac{m\pi x}{a} = 0$$

or

$$\delta_m \left.\frac{dJ_m}{dy}\right|_{y=b} + \left.\frac{dK_m}{dy}\right|_{y=b} + \sum_{n=1,2,\ldots}^{\infty} \frac{n\pi}{b} H_{mn}(-1)^n = 0. \tag{9}$$

This equation cannot be solved directly for V_n and M_m. Rather it has infinite solutions of M_m and V_n. Thus, if we truncate the equation after $m = n = 2$, then the equation becomes

$$\left.\frac{dJ_1}{dy}\right|_{y=b} + \left.\frac{dK_1}{dy}\right|_{y=b} - \frac{\pi}{b} H_{11} + \frac{2\pi}{b} H_{12} = 0$$

and

$$\left.\frac{dJ_2}{dy}\right|_{y=b} + \left.\frac{dK_2}{dy}\right|_{y=b} - \frac{\pi}{b} H_{21} + \frac{2\pi}{b} H_{22} = 0.$$

Similarly two other equations can be written to satisfy the boundary condition

$$V_x = 0.$$

From the four simultaneous equations, the unknown quantities M_1, M_2, V_1, and V_2 are determined from the expressions for K_m and L_m. Once these expressions are known, the deflection is obtained from Eq. (7).

Problems

2-1 Use the expressions derived in Example 2-1 to determine the maximum stress in a rectangular plate with $a = 90$ cm, $b = 60$ cm, $p = 1$ kgf/cm^2, $E = 21{,}000$ kgf/mm^2, and $\mu = 0.3$.
2-2 Use the expressions derived in Example 2-2 to determine the maximum stress in a titanium rectangular plate with $a = 40$ inches, $b = 30$ inches, $p = 30$ psi, $E = 15{,}000$ ksi, and $\mu = 0.32$.
2-3 Use the expressions derived in Example 2-2 to determine the maximum stress at a point where $x = 15$ inches and $y = 15$ inches. Let $a = 40$ inches, $b = 50$ inches, $E = 30{,}000$ ksi, $p = 10$ psi, $t = 1.0$ inch, and $\mu = 0.30$.
2-4 A copper internal baffle plate in a reactor has the boundary conditions shown in Fig. 2-4. Find its maximum deflection if $a = 36$ inches, $b = 20$ inches, $p = 100$ psi, $E = 16{,}000$ ksi, and $\mu = 0.33$.
2-5 The top plate of a truck weigh scale is supported by beams such that any portion of the plate can be approximated by Fig. P2-5. Find the expressions for the bending moments at the edges and in the middle due to uniform pressure p.
2-6 Find the expression for the deflection of a rectangular plate shown in Fig. P2-6 due to edge moments given by

$$M_1 = \sum_{m=1}^{\infty} E_m(y) \sin \frac{m\pi x}{a}.$$

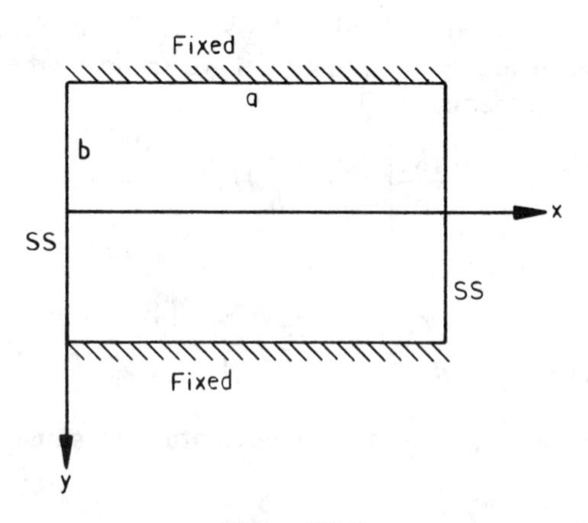

Figure P2-5.

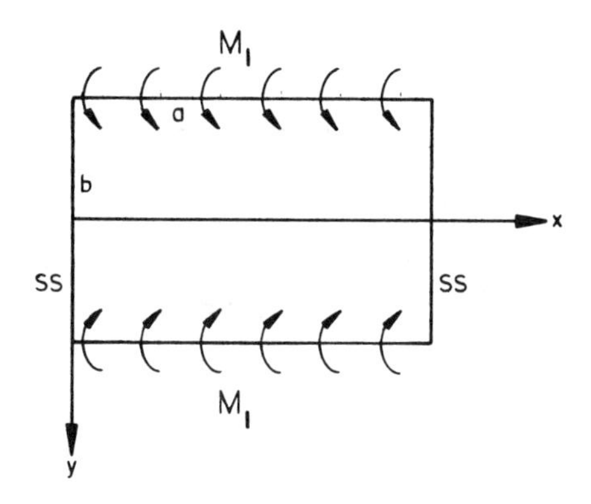

Figure P2-6.

2-7 Find the expression for the deflection of a rectangular plate fixed at all sides.

Hint: Use the results of Problem 2-6 to solve this problem by switching the x- and y-axes.

2-8 Find the moments in the side plates of an oil barge (Fig. P2-8) due to hydrostatic pressure. The pressure distribution can be approximated as triangular in shape with a maximum value at a height of $5'-9''$ and zero at the top and bottom. This is due to the difference in specific gravity between the contents in the inside and water on the outside. Assume the plate panels to be simply supported at the top and bottom and fixed along the sides. Also, assume the length of each side plate panel to be 360 inches which is the spacing of the bulkheads in the barge. This large length disregards the effects of the intermediate vertical stiffeners, between the bulk-

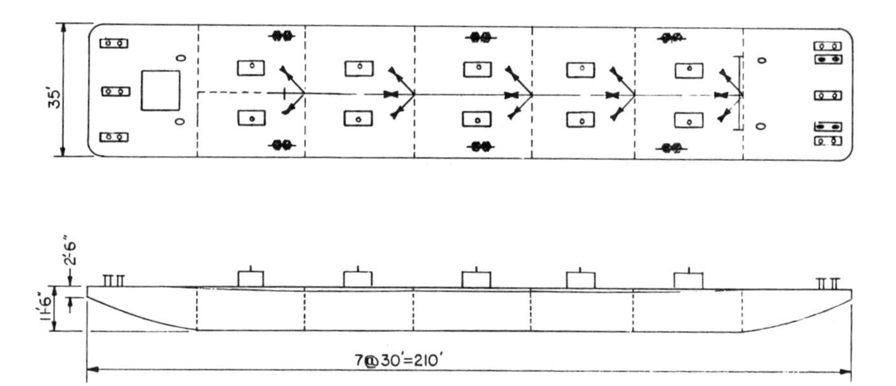

Figure P2-8. Oil barge. (Courtesy of the Mississippi Valley Barge Line.)

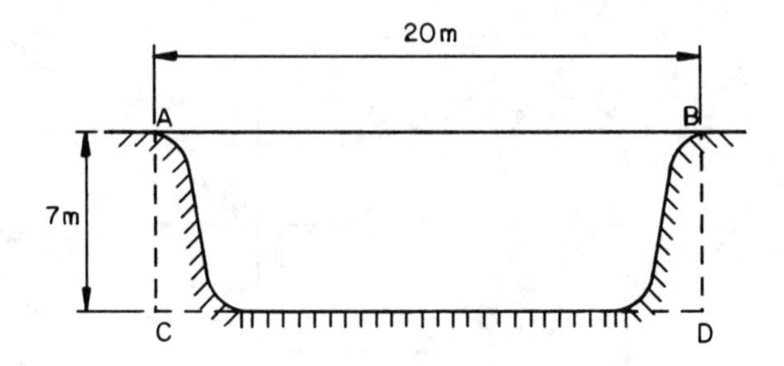

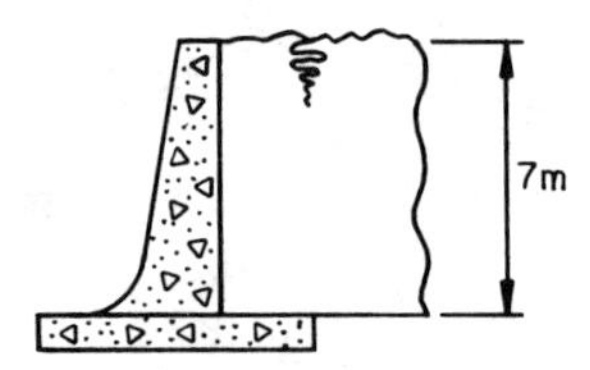

Figure P2-9.

heads, that are welded to the side plates for increased rigidity (see Problem 5-10). Let $E = 29{,}000$ ksi and $\mu = 0.29$.

2-9 Find the maximum moment in the small concrete dam shown in Fig. P2-9 due to hydrostatic pressure. Sides AC and BD are simply supported. Side AB is free and side CD is fixed. Let $E = 2180$ kgf/mm^2 and $\mu = 0$. Assume a uniform thickness.

2-2 Continuous Plates

The classical methods developed so far for solving rectangular plates are also applicable to continuous plates. The boundary conditions of each panel as well as the compatibility of forces or deformations between any two panels across a common boundary, such as *ab* in Fig. 2-5a and *ab* and *ac* in Fig. 2-5b, must be used to determine the constants in the differential equations of each panel. As the number of panels increases, it becomes more tedious to find a solution with the classical plate theory due to the number of simultaneous equations that must be solved to obtain the constants of integration. A more practical approach for solving such plates is to use an approximate solution such as a finite element analysis which is discussed in Chapter 14.

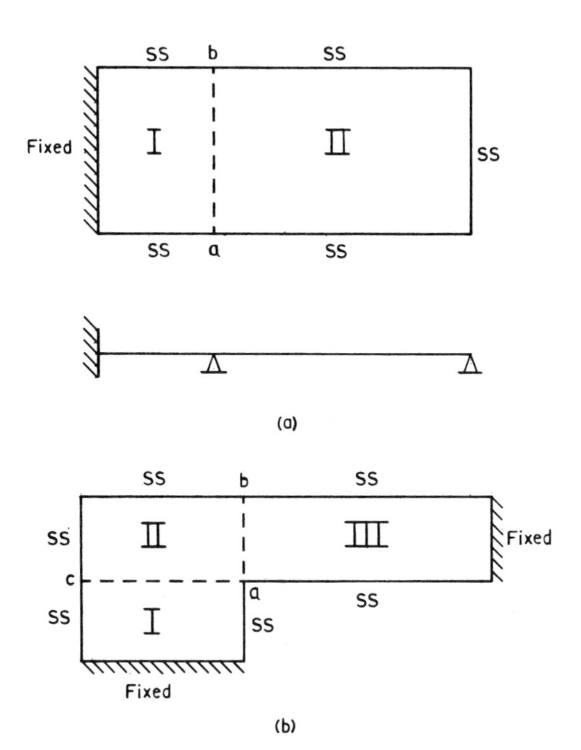

Figure 2-5.

The procedure for solving continuous plates is illustrated in Example 2-4 for a two-panel structure.

Example 2-4

Find the expressions for the deflection in the continuous plate shown in Fig. 2-6a due to a uniform load on panel I only.

Solution

The boundary conditions for panel I are

$$\text{at } y_1 = 0 \qquad w = 0 \tag{1}$$

$$\frac{\partial^2 w}{\partial y_1^2} = M_o \tag{2}$$

$$\text{at } y_1 = b \qquad w = 0 \tag{3}$$

$$\frac{\partial w}{\partial y_1} = 0. \tag{4}$$

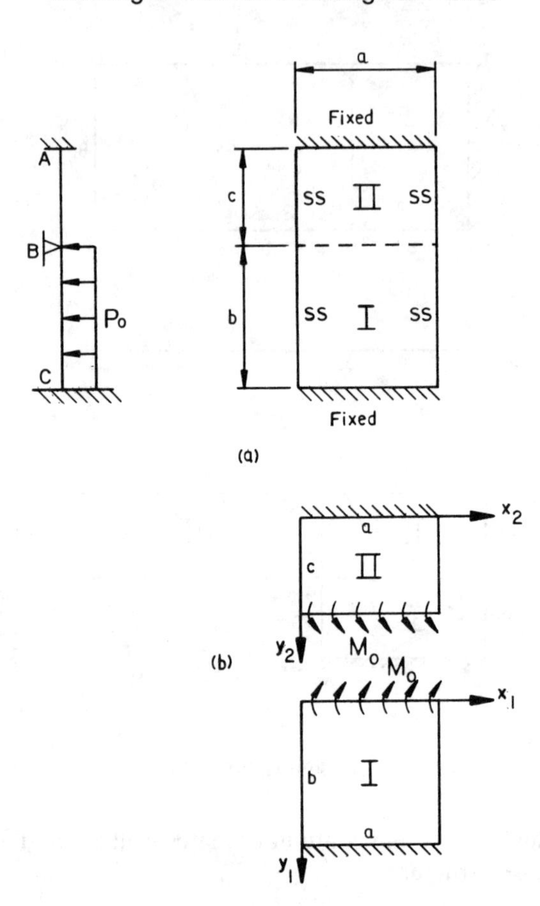

Figure 2-6.

The expression for the deflection is obtained from Example 2-1 as

$$w = \sum_{m=1,3,\dots} \left(A_m \sinh \frac{m\pi y}{a} + B_m \cosh \frac{m\pi y}{a} + C_m y \sinh \frac{m\pi y}{a} + D_m y \cosh \frac{m\pi y}{a} + \frac{4p_o a^4}{m^5 \pi^5 D} \right) \sin \frac{m\pi x}{a}. \tag{5}$$

Let the moment M_o between panels I and II be represented by

$$M_o = \sum_{m=1,3,\dots}^{\infty} E_m(y) \sin \frac{m\pi x}{a}. \tag{6}$$

From the boundary conditions (1) through (4) we get

$$A_m = -B_m\left(\cosh\frac{m\pi b}{a} - 1\right) - C_m b \sinh\frac{m\pi b}{a} \tag{7}$$

$$- D_m b \cosh\frac{m\pi b}{a} \tag{8}$$

$$B_m = -\frac{4p_o a^4}{m^5\pi^5 D} \tag{9}$$

$$C_m = \frac{-E_m}{2\,\dfrac{m\pi}{a}} + \frac{2p_o a^3}{m^4\pi^4 D} \tag{10}$$

$$D_m = \frac{K_1}{K_2}B_m + C_m\frac{\sinh^2\dfrac{m\pi b}{a}}{K_2}$$

where

$$K_1 = \frac{m\pi}{a}\left(\cosh\frac{m\pi b}{a} - 1\right)$$

$$K_2 = \frac{m\pi b}{a} - \sinh\frac{m\pi b}{a}\cosh\frac{m\pi b}{a}.$$

The boundary conditions for panel II are

$$\text{at } y_2 = 0 \qquad w = 0 \tag{11}$$

$$\frac{\partial w}{\partial y_2} = 0 \tag{12}$$

$$\text{at } y_2 = c \qquad w = 0 \tag{13}$$

$$\frac{\partial^2 w}{\partial y_2^2} = M_o \tag{14}$$

The expression for the deflection is written as

$$w = \sum_{m=1,3,\ldots}^{\infty}\left(F_m\sinh\frac{m\pi y}{a} + G_m\cosh\frac{m\pi y}{a} + H_m y\sinh\frac{m\pi y}{a} + I_m y\cosh\frac{m\pi y}{a}\right)\sin\frac{m\pi x}{a}. \tag{15}$$

From the boundary conditions (11) through (14) we get

$$F_m = -\frac{E_m}{K_3 + K_4K_5} \tag{16}$$

$$G_m = 0 \tag{17}$$

$$H_m = \frac{1}{c}\left(1 - \frac{m\pi c}{a}\coth\frac{m\pi c}{a}\right)\frac{E_m}{K_3 + K_4K_5} \tag{18}$$

$$I_m = \frac{m\pi}{a}\frac{E_m}{K_3 + K_4K_5} \tag{19}$$

where

$$K_3 = \frac{m^2\pi^2}{a^2}\left(\sinh\frac{m\pi c}{a} + \frac{m\pi c}{a}\cosh\frac{m\pi c}{a}\right)$$

$$K_4 = \frac{\sinh\dfrac{m\pi c}{a} - \dfrac{m\pi c}{a}\cosh\dfrac{m\pi c}{a}}{c\sinh\dfrac{m\pi c}{a}}$$

$$K_5 = \frac{2m\pi}{a}\cosh\frac{m\pi c}{a} + \frac{m^2\pi^2}{a^2}c\sinh\frac{m\pi c}{a}.$$

The constants A_m through I_m are in terms of the bending moment constant E_m. This constant is obtained by solving the compatibility equation at the common boundary which is

slope in panel I = slope in panel II.

Once E_m is known, then equations (5) and (15) can be solved for the deflections at any location in the plates.

Other applications of the continuous plate concept are for large plates with multiple point supports. Such applications are found in concrete flat slab floors (Winter et al. 1964) with multiple column supports as well as stayed vessels commonly encountered in chemical plants and refineries. These cylindrical vessels consist of inner and outer shells tied together with stays and the annular space between them pressurized. The analysis of both inner and outer shells is based on the theory of plates with multiple-point supports.

Numerous articles have been written on the subject of plates with multiple supports. Some of these articles are listed in Timoshenko (Timoshenko and Woinowsky-Krieger 1959). A large plate under uniform pressure with closely spaced supports (Fig. 2-7a) can be analyzed using the Levy solution. Due to

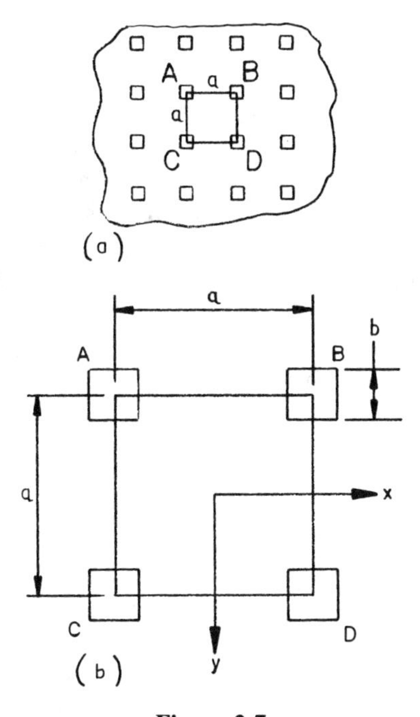

Figure 2-7.

symmetry, the boundary conditions in the plate along $x = \pm a/2$ are

$$\text{slope} = \frac{\partial w}{\partial x} = 0 \tag{2-1}$$

and from Eq. (1-27),

$$\text{shear } Q_x = -D\left(\frac{\partial^3 w}{\partial x^3} + \frac{\partial^3 w}{\partial x\, \partial y^2}\right) = 0. \tag{2-2}$$

Similar boundary conditions exist along the boundary $y = \pm a/2$. The particular solution of the equation

$$\nabla^4 w = q/D \tag{2-3}$$

and the homogeneous solution of

$$\nabla^4 w = 0 \tag{2-4}$$

are accomplished by expressing the deflection by the series

$$w = f_o + \sum_{m=2,4,\ldots}^{\infty} f_m(y) \cos \frac{m\pi x}{a} \tag{2-5}$$

and the load by

$$p = p_o + \sum_{m=2,4,\ldots}^{\infty} p_m(y) \cos \frac{m\pi x}{a}. \tag{2-6}$$

Equation (2-5) satisfies the boundary conditions given by Eqs. (2-1) and (2-2). Substituting Eqs. (2-5) and (2-6) into Eqs. (2-3) and (2-4) and summing the two equations results in an expression for the deflection with three unknown constants. The first is obtained directly from the slope boundary in Eq. (2-1). The second constant is obtained from Eq. (2-2) keeping in mind that the shear, Q, is equal to zero along the unsupported length $(a-b)$ and $pa^2/4$ over the supports b. The third unknown constant is evaluated from the condition that the deflection, w, is equal to zero at the supports.

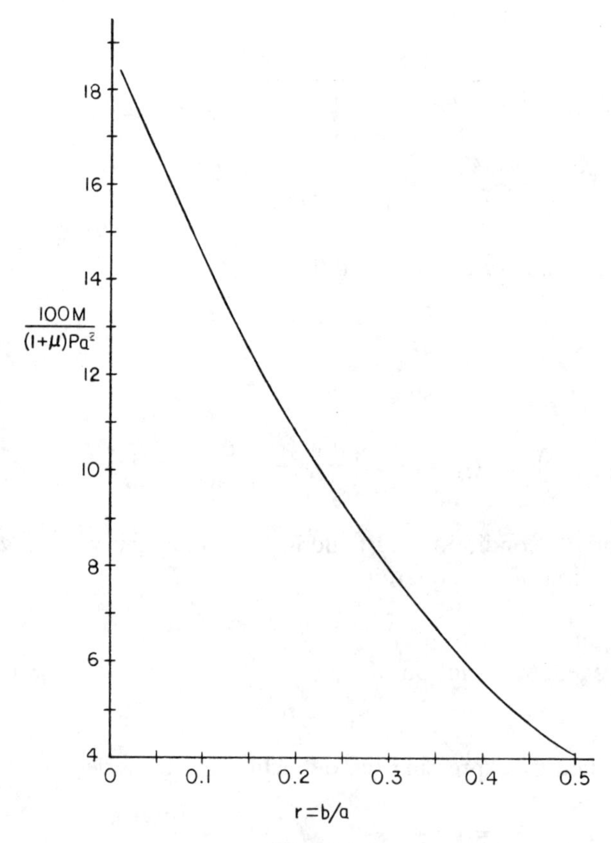

Figure 2-8.

With the deflection known, the bending moments can be determined from Eq. (1-17). The maximum negative bending moment occurs at the supports and is expressed by Woinowsky-Krieger as

$$M_{\max} = (1 + \mu)pa^2\left[\frac{(1 - r)(2 - r)}{48} + \frac{1}{4\pi^3 r^2}\sum_{m=1}^{\infty}\frac{2}{m^3 \sinh m\pi}\sinh\frac{m\pi r}{2}\cosh\frac{m\pi(2 - r)}{2}\sin m\pi r\right] \quad (2\text{-}7)$$

where $r = b/a$. Figure 2-8 shows a plot of this equation for various r ratios.

The analysis expressed by Eq. (2-7) assumes the intermediate supports to be rigid. If the supports deflect due to applied loads, then the moments must be determined from an analysis of a plate on an elastic foundation.

Problems

2-10 Find the expressions for the bending moments in Example 2-4 if $a = 5\ m$, $b = 15\ m$, $C = 10\ m$, $p = 1$ kgf/cm^2, $E = 210{,}000$ MPa, and $\mu = 0$.
2-11 Plot the value of M_x along the length of panels I and II in Fig. 2-6a at $x = a/2$. Compare the result to the bending moments obtained from a beam of unit width and length ABC in Fig. 2-6a.
2-12 A continuous concrete slab is supported by columns as shown in Fig. 2-7. Calculate the maximum bending moment if $a = 20$ ft, $b = 3$ ft, $p = 150$ psf, and $\mu = 0$.

2-3 Plates on an Elastic Foundation

The effective pressure on any point in a plate or slab resting on a continuous foundation such as a concrete road pavement or a rectangular tubesheet in a heat exchanger (Fig. 2-9), is equal to $p - f$ where p is the applied load and f is the resisting pressure of the foundation. If we assume the foundation to be elastic, i.e., its elasticity is defined by a force that causes a unit deflection when applied over a unit area, then we can define

$$f = kw \quad (2\text{-}8)$$

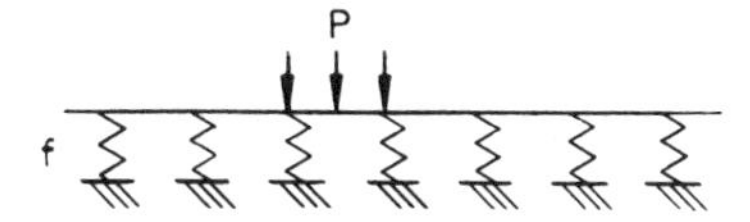

Figure 2-9.

where k is the foundation modulus. Values of k for various soils are given in numerous references such as McFarland (McFarland et al. 1972).

Equation (1-26) for the lateral deflection of a plate can now be modified to

$$\nabla^4 w = \frac{1}{D}(p - kw)$$

or

$$\nabla^4 w + \frac{k}{D} w = p/D \tag{2-9}$$

It must be kept in mind that under certain applied loads and boundary conditions a negative deflection may result somewhere in the plate. This indicates that the foundation must be able to sustain a tensile load at that location. This condition is very common in circular plates used in heat exchangers as discussed in Chapter 3 and the engineer must take appropriate precautions if the foundation cannot undergo a tensile force.

We can use the Levy method to obtain a solution of Eq. (2-9) for plates simply supported at two opposite edges with arbitrary boundary conditions at the other two edges. We proceed by solving the homogeneous and particular parts as in Eq. (1-42). Again expressing the deflection by the Fourier series

$$w_h = \sum_{m=1}^{\infty} f_m(y) \sin \frac{m\pi x}{a}$$

and solving the homogeneous part of Eq. (2-9), we get

$$w_h = \sum_{m=1}^{\infty} [A_m \sinh \alpha_m y \sin \beta_m y + B_m \sinh \alpha_m y \cos \beta_m y$$
$$+ C_m \cosh \alpha_m y \sin \beta_m y + D_m \cosh \alpha_m y \cos \beta_m y] \sin \frac{m\pi x}{a} \tag{2-10}$$

where

$$\alpha_m = \sqrt{\frac{1}{2}\left(\frac{m^2\pi^2}{a^2} + \sqrt{\frac{m^4\pi^4}{a^4} + \frac{k}{D}}\right)} \tag{2-11}$$

and

$$\beta_m = \sqrt{\frac{1}{2}\left(\frac{m^2\pi^2}{a^2} - \sqrt{\frac{m^4\pi^4}{a^4} + \frac{k}{D}}\right)}. \tag{2-12}$$

Similarly, if we express the deflection by

$$w_p = \sum_{m=1}^{\infty} g_m(y) \sin \frac{m\pi x}{a} \tag{2-13}$$

and the applied loads as

$$p = \sum_{m=1}^{\infty} p_m(y) \sin \frac{m\pi x}{a} \tag{2-14}$$

and we substitute these equations into Eq. (2-9) to get

$$\frac{d^4 g_m}{dy^4} - 2\left(\frac{m\pi}{a}\right)^2 \frac{d^2 g_m}{dy^2} + \left(\frac{m^4\pi^4}{a^4} + \frac{k}{D}\right) g_m = \frac{p_m(y)}{D}. \tag{2-15}$$

Equations (2-10) and (2-15) are the two governing expressions for rectangular plates on an elastic foundation with two simply supported edges.

Example 2-5

Find the expression for the deflection of the plate shown in Fig. 2-10 that is resting on a foundation of modulus k_0 and subjected to a uniform pressure p.

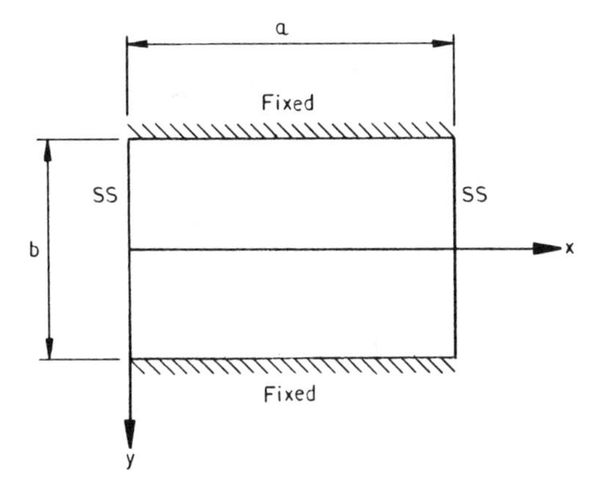

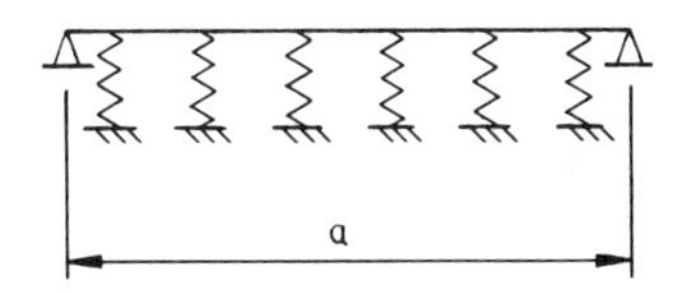

Figure 2-10.

Solution

For a uniform pressure p, the particular solution of Eq. (2-15) gives

$$g_m = \frac{p}{D}\frac{4}{m\pi}\frac{1}{\left(\dfrac{m^4\pi^4}{a^4} + \dfrac{k_o}{D}\right)}. \tag{1}$$

Due to symmetry of the deflection around the x-axis, we can take $B_m = C_m = 0$ in Eq. (2-10) and the total deflection becomes

$$w_h = \sum_{m=1}^{\infty}\left[A_m \sinh \alpha_m y \sin \beta_m y + D_m \cosh \alpha_m y \cos \beta_m y + g_m\right] \sin \frac{m\pi x}{a}. \tag{2}$$

The boundary conditions are

$$\text{at } y = +b/2, \qquad w = 0$$

and

$$\frac{\partial w}{\partial y} = 0.$$

Substituting these boundary conditions into Eq. (2) gives

$$A_m = \frac{g_m}{K_1\left(\cosh \alpha_m \dfrac{b}{2} \cos \beta_m \dfrac{b}{2}\right)}$$

and

$$D_m = -A_m \tanh \alpha_m \frac{b}{2} \tan \beta_m \frac{b}{2} - \frac{g_m}{\cosh \alpha_m \dfrac{b}{2} \cos \beta_m \dfrac{b}{2}}$$

where

$$K_1 = \frac{K_2 + K_3}{K_4 - K_5} - \tanh \alpha_m \frac{b}{2} \tan \beta_m \frac{b}{2}$$

$$K_2 = \alpha_m \cosh \alpha_m \frac{b}{2} \sin \beta_m \frac{b}{2}$$

$$K_3 = \beta_m \sinh \alpha_m \frac{b}{2} \cos \beta_m \frac{b}{2}$$

$$K_4 = \alpha_m \sinh \alpha_m \frac{b}{2} \cos \beta_m \frac{b}{2}$$

$$K_5 = \beta_m \cosh \alpha_m \frac{b}{2} \sin \beta_m \frac{b}{2}.$$

With A_m and D_m known, Eq. (2) can now be solved for the deflection.

Problems

2-13 Find the expression for the deflection of a simply supported plate uniformly loaded and supported on an elastic foundation of modulus k_o.
2-14 Solve Example 1-3 if the plate is resting on an elastic foundation of modulus k_o.

2-4 Thermal Stress

A change in the temperature of a plate may result in a change in the length of the middle surface and a change in the curvature. Accordingly, the basic differential equations derived in Chapter 1 must be modified to consider temperature change. We begin the derivation by defining u and v as the change in length of the middle surface in the x- and y-directions, respectively, of the plate shown in Figs. 1-3 and 1-4. Then Eq. (1-12) becomes

$$\begin{bmatrix} \varepsilon_x \\ \varepsilon_y \\ \gamma_{xy} \end{bmatrix} = \begin{bmatrix} \frac{\partial}{\partial x} & 0 & 0 \\ 0 & \frac{\partial}{\partial y} & 0 \\ \frac{\partial}{\partial y} & \frac{\partial}{\partial x} & 0 \end{bmatrix} \begin{bmatrix} u \\ v \\ 0 \end{bmatrix} - z \begin{bmatrix} 1 & 0 & 0 \\ 0 & 1 & 0 \\ 0 & 0 & 2 \end{bmatrix} \begin{bmatrix} \frac{\partial^2 w}{\partial x^2} \\ \frac{\partial^2 w}{\partial y^2} \\ \frac{\partial^2 w}{\partial x \, \partial y} \end{bmatrix}. \tag{2-16}$$

Next we define the change in strain due to temperature change as $\alpha(\Delta T)$, where α is the coefficient of thermal expansion and ΔT is the temperature change which is a function of z through the thickness of the plate. Values of α for some commonly encountered materials are shown in Table 2-1.

For thin plates with temperature loads, Eq. (1-13) can be expressed as

$$\begin{bmatrix} \varepsilon_x \\ \varepsilon_y \\ \gamma_{xy} \end{bmatrix} = \frac{1}{E} \begin{bmatrix} 1 & -\mu & 0 \\ -\mu & 1 & 0 \\ 0 & 0 & 2(1+\mu) \end{bmatrix} \begin{bmatrix} \sigma_x \\ \sigma_y \\ \tau_{xy} \end{bmatrix} + \alpha(\Delta T) \begin{bmatrix} 1 \\ 1 \\ 0 \end{bmatrix} \tag{2-17}$$

Table 2-1. Coefficients of thermal expansion (multiplied by 10^6)

Material	Temperature, °F					
	Room Temperature	200	400	600	800	1000
Aluminum (6061)	12.6	12.9	13.5			
Brass (Cu-Zn)	9.6	9.7	10.2	10.7	11.2	11.6
Bronze (Cu-Al)				9.0[a]		
Carbon steel	6.5	6.7	7.1	7.4	7.8	
Copper	9.4	9.6	9.8	10.1	10.3	10.5
Cu–Ni (70-30)	8.5	8.5	8.9	9.1		
Nickel alloy C276	6.1	6.3	6.7	7.1	7.3	
Nickel alloy 600	6.9	7.2	7.6	7.8	8.0	
Stainless steel	8.6	8.8	9.2	9.5	9.8	10.1
Titanium (Gr. 1,2)	4.7	4.7	4.8	4.9	5.1	
Zirconium alloys	3.2	3.4	3.7	4.0		
Concrete	5.5					
Wood, hard	2.7[b]					
Wood, soft	3.6[b]					

[a]At 500°F.
[b]Parallel to fibers.

or

$$\begin{bmatrix} \sigma_x \\ \sigma_y \\ \tau_{xy} \end{bmatrix} = \frac{E}{1-\mu^2} \begin{bmatrix} 1 & \mu & 0 \\ \mu & 1 & 0 \\ 0 & 0 & \dfrac{1-\mu}{2} \end{bmatrix} \begin{bmatrix} \varepsilon_x \\ \varepsilon_y \\ \gamma_{xy} \end{bmatrix} - \frac{\alpha \Delta T E}{1-\mu} \begin{bmatrix} 1 \\ 1 \\ 0 \end{bmatrix}. \quad (2\text{-}18)$$

For the case where the plate bends without a change in the length of the middle surface, i.e., $u = v = 0$, we substitute Eqs. (2-16) and (2-18) into Eq. (1-16) and get

$$\begin{bmatrix} M_x \\ M_y \\ M_{xy} \end{bmatrix} = -D \begin{bmatrix} 1 & \mu & 0 \\ \mu & 1 & 0 \\ 0 & 0 & -(1-\mu) \end{bmatrix} \begin{bmatrix} \dfrac{\partial^2 w}{\partial x^2} \\ \dfrac{\partial^2 w}{\partial y^2} \\ \dfrac{\partial^2 w}{\partial x \, \partial y} \end{bmatrix} - \frac{M_o}{1-\mu} \begin{bmatrix} 1 \\ 1 \\ 0 \end{bmatrix} \quad (2\text{-}19)$$

where

$$M_o = \alpha E \int_{-t/2}^{t/2} (\Delta T) z \, dz. \quad (2\text{-}20)$$

Substituting Eqs. (2-19) and (2-20) into the plate equilibrium Eq. (1-24) gives

$$\nabla^4 w = p/D - \frac{1}{D(1-\mu)} \nabla^2 M_o. \quad (2\text{-}21)$$

Equation (2-21) is the governing differential equation for the bending of a rectangular plate due to lateral pressure and thermal loads.

For the case where $p = 0$, Eq. (2-21) reduces to

$$\nabla^2 w = -\frac{M_o}{D(1-\mu)} \quad (2\text{-}22)$$

where

$$\nabla^2 = \frac{\partial^2}{\partial x^2} + \frac{\partial^2}{\partial y^2}.$$

Next we consider the case where the length of the middle surface changes due to temperature variation without any lateral deflection, i.e., $w = 0$. For this case we need, in addition to the forces shown in Fig. 1-6, three

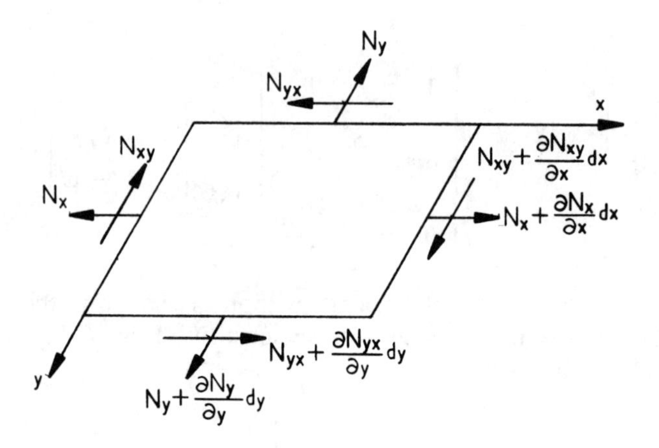

Figure 2-11.

in-plane forces N_x, N_y, and N_{xy} as shown in Fig. 2-11. The stress–force relationship is expressed as

$$\begin{bmatrix} N_x \\ N_y \\ N_{xy} \end{bmatrix} = \int_{-t/2}^{t/2} \begin{bmatrix} \sigma_x \\ \sigma_y \\ \tau_{xy} \end{bmatrix} dz \tag{2-23}$$

Substituting Eq. (2-18) into Eq. (2-23) gives

$$\begin{bmatrix} \varepsilon_x \\ \varepsilon_y \\ \gamma_{xy} \end{bmatrix} = \frac{1}{Et} \begin{bmatrix} 1 & -\mu & 0 \\ -\mu & 1 & 0 \\ 0 & 0 & 2(1+\mu) \end{bmatrix} \begin{bmatrix} N_x \\ N_y \\ N_{xy} \end{bmatrix} + \frac{N_o}{Et} \begin{bmatrix} 1 \\ 1 \\ 0 \end{bmatrix} \tag{2-24}$$

where

$$N_o = \alpha E \int_{-t/2}^{t/2} (\Delta T)\, dz. \tag{2-25}$$

From Eq. (2-16) we observe that each of the three strains is a function of the deflections u, v, and w. Hence, a compatibility equation (Timoshenko and Goodier 1951) that combines the three strains is obtained from Eq. (2-16) and is expressed as

$$\frac{\partial^2 \varepsilon_x}{\partial y^2} + \frac{\partial^2 \varepsilon_y}{\partial x^2} = \frac{\partial^2 \gamma_{xy}}{\partial x\, \partial y}. \tag{2-26}$$

Substituting Eq. (2-24) into this expression yields the differential equation

$$\frac{\partial^2}{\partial y^2}(N_x - \mu N_y + N_o) + \frac{\partial^2}{\partial x^2}(-\mu N_x + N_y + N_o) - 2(1 + \mu)\frac{\partial^2 N_{xy}}{\partial x\, \partial y} = 0. \quad (2\text{-}27)$$

In order to solve this equation, we need to investigate the in-plane forces in the plate. Summing forces (Fig. 2-11) in the x- and y-directions yields

$$\frac{\partial N_x}{\partial x} + \frac{\partial N_{yx}}{\partial y} = 0$$

and (2-28)

$$\frac{\partial N_y}{\partial y} + \frac{\partial N_{xy}}{\partial x} = 0.$$

These equations are satisfied by selecting a stress function $\psi(x, y)$ that is defined by

$$N_x = \frac{\partial^2 \psi}{\partial y^2}, \quad N_y = \frac{\partial^2 \psi}{\partial x^2}, \quad \text{and} \quad N_{xy} = -\frac{\partial^2 \psi}{\partial x\, \partial y}. \quad (2\text{-}29)$$

Substituting Eq. (2-29) into Eq. (2-27) results in the differential equation

$$\nabla^4 \psi + \nabla^2 N_o = 0. \quad (2\text{-}30)$$

Equations (2-21) and (2-30) constitute the general solution of a plate subjected to temperature change. Equation (2-21) is solved by the methods discussed in this book while Eq. (2-30) is solved by methods discussed in the theory of elasticity for plane stress problems which are beyond the scope of this book.

Example 2-6

Find the deflection in a simply supported plate due to decrease in temperature of the top surface of T_o and increase of the bottom surface by T_o.

Solution

$$T = T_o(2z/t)$$

and Eq. (2-20) becomes

$$M_o = \alpha E T_o t^2/6. \quad (1)$$

Let

$$w = \sum_{m=1}^{\infty}\sum_{n=1}^{\infty} w_{mn} \sin\frac{m\pi x}{a} \sin\frac{n\pi y}{b} \tag{2}$$

and

$$M_o = \sum_{m=1}^{\infty}\sum_{n=1}^{\infty} T_{mn} \sin\frac{m\pi x}{a} \sin\frac{n\pi y}{b} \tag{3}$$

where

$$T_{mn} = \frac{4}{ab}\int_0^b\int_0^a M_o \sin\frac{m\pi x}{a} \sin\frac{n\pi y}{b}\, dx\, dy. \tag{4}$$

Substituting Eqs. (2) and (3) into Eq. (2-22) results in

$$w_{mn} = \frac{T_{mn}}{D(1-\mu)\pi^2\left(\frac{m^2}{a^2}+\frac{n^2}{b^2}\right)}. \tag{5}$$

Substituting Eq. (1) and (4) into Eq. (5) gives

$$w_{mn} = \frac{8\alpha E T_o t^2}{3mnD(1-\mu)\pi^4\left(\frac{m^2}{a^2}+\frac{n^2}{b^2}\right)} \tag{6}$$

where m and n are odd.

Since the temperature variation does not affect the middle surface, i.e., $u = v = 0$, Eq. (2-30) is redundant and need not be considered.

Problems

2-15 Calculate the maximum moment in Example 2-6.
2-16 Find the expression for the bending moment in the plate shown in Fig. P2-5 due to the temperature variation given by

$$T = T_o(2z/t).$$

Hint: Use the solution of Problem P2-6 to satisfy the boundary conditions.

2-5 Design of Various Rectangular Plates

The maximum bending and deflection of rectangular plates with various boundary conditions have been solved and tabulated in many references.

Szilard (1974) as well as the references given in Section 1-9 tabulate maximum moments and deflections for rectangular plates with some commonly encountered boundary conditions. Moody (1970) lists numerous moment tables for rectangular plates free at one edge with various boundary conditions at the remaining edges and subjected to various loading conditions. Continuous plates are designed in accordance with the equations developed in Section 2-2.

The ASME VIII-1 code contains rules for the design of jacketed shells with large r/t ratios (Fig. 2-12) that are based on the stayed plate theory and Eq. (2-7). Letting $\mu = 0.3$ and using a b/a ratio of 0.4, Eq. (2-7) can be reduced to

$$t = a\sqrt{\frac{p}{2.28\sigma}}. \tag{2-31}$$

The value of 2.28 in the denominator is varied by the ASME from 2.1 to 3.5 depending on the type of construction and method of weld attachment.

For continuous concrete slabs supported by columns without intermediate beams, the reinforcement cutoffs are also based on Eq. (2-7) and are detailed in the ACI 318 code.

Allowable stress values for some materials were given in Table 1-1 at temperatures below the creep and rupture range as defined by ASME VIII-1. For high-temperature applications, the ASME criteria in the creep and rupture range are based on limiting the allowable stress to the lower of the following values:

1. 100% of the average stress for a creep rate of 0.01%/100 hr.

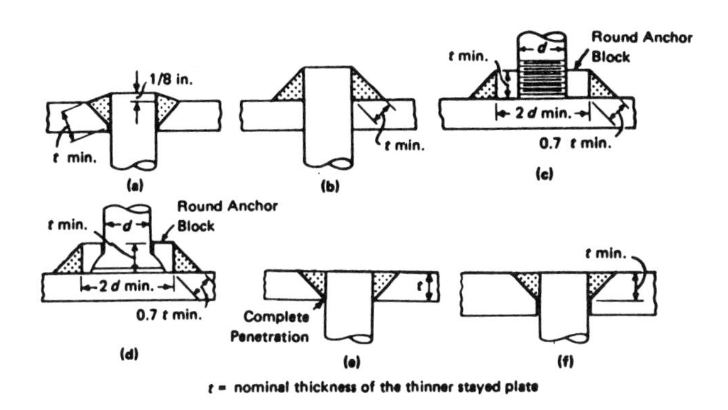

Figure 2-12. Welded staybolts. (Courtesy of ASME.)

Table 2-2. ASME VIII-1 allowable stress values at elevated temperatures, ksi

Material	ASME Designation	Temperature, °F		
		900	1100	1300
Carbon steel	SA 516-70	6.5		
Stainless steel	SA 240-304	14.7	9.8	3.7
Nickel alloy	SB 575-276	22.3	15.0	

2. 67% of the average stress for rupture at the end of 100,000 hr.
3. 80% of the minimum stress for rupture at the end of 100,000 hr.

Using these criteria, the allowable stress values for the materials listed in Table 1-1 that are permitted at high temperatures are shown in Table 2-2.

Problems

2-17 The inner and outer shells of a pressure vessel are stayed together on a 12-inch stay pitch. The pressure between the cylinders is 50 psi. Use Eq. (2-31) to determine the required thickness of the cylinders. Disregard the hoop stress in the cylinders due to pressure because it is small in most applications. Let the allowable bending stress be equal to 15 ksi.
2-18 Determine the required diameter of the stays in Problem 2-17. Let the allowable tensile stress = 20 ksi. If the stays are attached as shown in Fig. 2-12b, calculate the required size of the fillet welds. The allowable stress in shear = 12 ksi.

3

Bending of Circular Plates

3-1 Plates Subjected to Uniform Loads in the θ-Direction

Circular plates are common in many structures such as nozzle covers, end closures in pressure vessels, and bulkheads in submarines and airplanes. The derivation of the classical equations for lateral bending of circular plates dates back to 1828 and is accredited to Poisson (Timoshenko 1983). He used polar coordinates to transfer the differential equations for the bending of a rectangular plate to circular plates. The first rigorous solution of the differential equation of circular plates for various loading and boundary conditions was made around 1900 and is credited to A. E. H. Love (Love 1944).

The five basic assumptions made in deriving the differential equations for lateral bending of rectangular plates in Section 1-1 are also applicable to circular plates. The differential equations for the lateral bending of circular plates subjected to uniform loads in the θ-direction are derived from Fig. 3-1. For sign convention it will be assumed that downward deflections and clockwise rotations are positive. Hence, if a flat plate undergoes a small deflection as shown in Fig. 3-1, then the radius of curvature r at point B is given by

$$\sin(\phi) \approx \phi = r/r_\theta$$

or

$$\frac{1}{r_\theta} = \phi/r = \frac{-1}{r}\frac{dw}{dr}. \qquad (3\text{-}1)$$

The quantity r_θ represents a radius that forms a cone as it rotates around the z-axis (in and out of the plane of the paper). The second radius of curvature is denoted by r_r. The origin of r_r does not necessarily fall on the

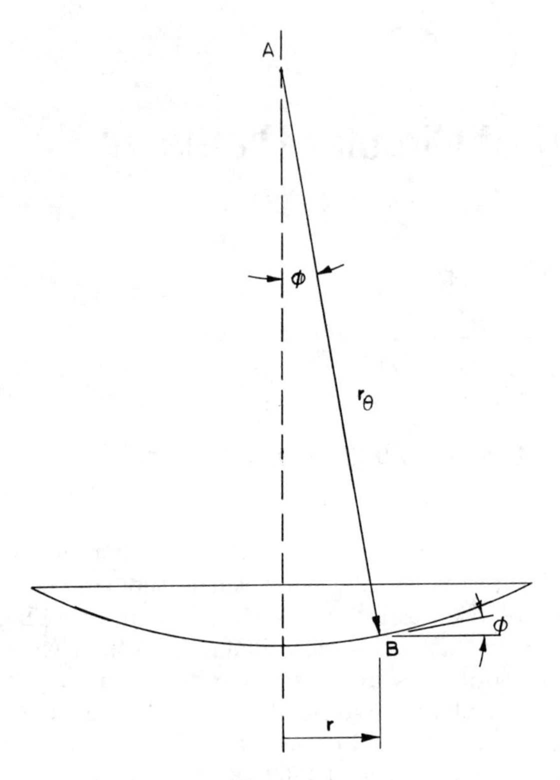

Figure 3-1.

axis of symmetry although, for any point B, the radii r_r and r_θ coincide with each other. The value of r_r is obtained from Eq. (1-6) as

$$\chi = \frac{1}{r_r} = -\frac{d^2w}{dr^2}$$

or

$$\chi = \frac{1}{r_r} = -\frac{d^2w}{dr^2} = \frac{d\phi}{dr}. \tag{3-2}$$

The M_x and M_y expressions in Eq. (1-17) can be written in terms of the radial and tangential directions as

$$\begin{bmatrix} M_r \\ M_t \end{bmatrix} = D \begin{bmatrix} 1 & \mu \\ \mu & 1 \end{bmatrix} \begin{bmatrix} \dfrac{1}{r_r} \\ \dfrac{1}{r_\theta} \end{bmatrix} \tag{3-3}$$

or

$$\begin{bmatrix} M_r \\ M_t \end{bmatrix} = D \begin{bmatrix} 1 & \frac{\mu}{r} \\ \mu & \frac{1}{r} \end{bmatrix} \begin{bmatrix} \frac{d\phi}{dr} \\ \phi \end{bmatrix} \tag{3-4}$$

or

$$\begin{bmatrix} M_r \\ M_t \end{bmatrix} = -D \begin{bmatrix} 1 & \frac{\mu}{r} \\ \mu & \frac{1}{r} \end{bmatrix} \begin{bmatrix} \frac{d^2w}{dr^2} \\ \frac{dw}{dr} \end{bmatrix} \tag{3-5}$$

where

$$D = \frac{Et^3}{12(1 - \mu^2)}.$$

The classical theory of the lateral bending of circular plates discussed in this section is based on the assumption that the loads on the plate are uniformly distributed in the θ direction. In this case the torsional moment $M_{r\theta}$ is zero and the other forces are as shown in Fig. 3-2a. Summing moments around line a–a gives

$$(M_r r d\theta) - \left(M_r + \frac{dM_r}{dr}\,dr\right)(r + dr)\,d\theta + 2(M_t\,dr\,d\theta/2)$$
$$- Qr\,d\theta\,dr/2 - \left(Q + \frac{dQ}{dr}\,dr\right)(r + dr)\,d\theta\,dr/2 = 0. \tag{3-6}$$

The quantity $M_t\,dr\,d\theta/2$ is the component of M_t perpendicular to axis a–a as shown in Fig. 3-2b. Equation (3-6) can be reduced to

$$M_r + \frac{dM_r}{dr}r - M_t + Qr = 0. \tag{3-7}$$

Substituting Eq. (3-5) into this equation gives

$$\frac{d}{dr}\left[\frac{1}{r}\frac{d}{dr}\left(r\frac{dw}{dr}\right)\right] = Q/D. \tag{3-8}$$

Or, since

$$2\pi r Q = \int p 2\pi r\,dr,$$

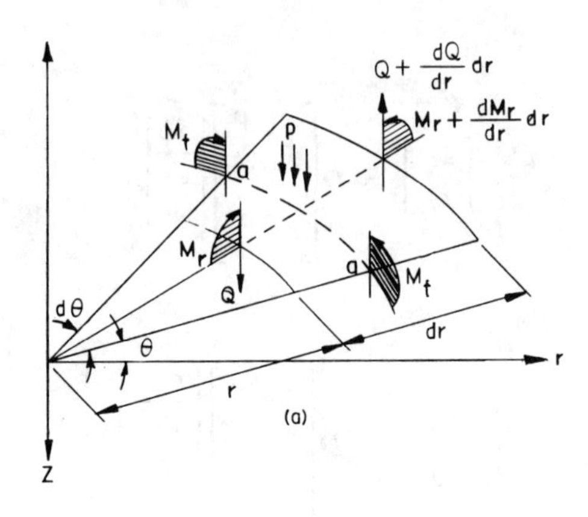

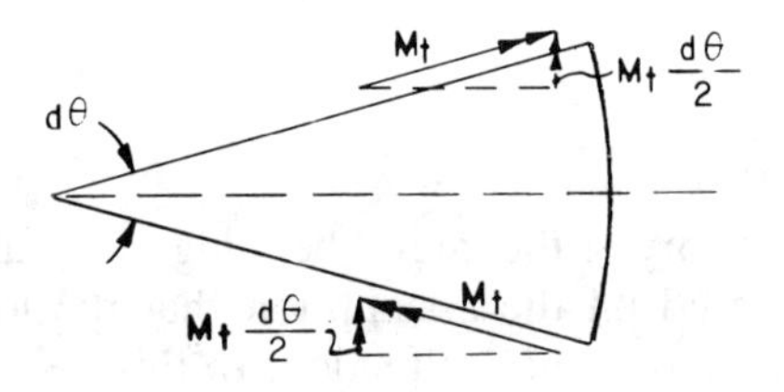

Figure 3-2.

Eq. (3-8) can be written in a different form as

$$\frac{1}{r}\frac{d}{dr}\left\{r\frac{d}{dr}\left[\frac{1}{r}\frac{d}{dr}\left(r\frac{dw}{dr}\right)\right]\right\} = \frac{p}{D} \tag{3-9}$$

where p is a function of r.

The analysis of circular plates with uniform thickness subjected to symmetric lateral loads consists of solving the differential equation for the deflection as given by Eq. (3-8) or (3-9). The bending moments are then calculated from Eq. (3-5). The shearing force is calculated from Eqs. (3-7) and (3-5) as

$$Q = D\left(\frac{d^3w}{dr^3} + \frac{1}{r}\frac{d^2w}{dr^2} - \frac{1}{r^2}\frac{dw}{dr}\right) \tag{3-10}$$

or, from Eqs. (3-7) and (3-4), as

$$Q = -D\left(\frac{d^2\phi}{dr^2} + \frac{1}{r}\frac{d\phi}{dr} - \frac{\phi}{r^2}\right). \tag{3-11}$$

Example 3-1

(a) Find the expression for the maximum moment and deflection of a uniformly loaded circular plate with simply supported edges.
(b) Find the required thickness of a steel plate if the allowable stress = 15,000 psi, $p = 5$ psi, $a = 20$ inches, $E = 30{,}000{,}000$ psi, and $\mu = 0.3$. What is the maximum deflection?
(c) For a concrete plate, μ is usually taken as zero. What are the moment expressions at $r = 0$, $r = a/2$, and $r = a$?

Solution

(a) From Fig. 3-3, the shearing force Q at any radius r is given by

$$2\pi r Q = \pi r^2 p$$

or

$$Q = pr/2.$$

From Eq. (3-8)

$$\frac{d}{dr}\left[\frac{1}{r}\frac{d}{dr}\left(r\frac{dw}{dr}\right)\right] = \frac{pr}{2D}.$$

Integrating this equation gives

$$\text{slope} = \frac{dw}{dr} = \frac{pr^3}{16D} + \frac{C_1 r}{2} + \frac{C_2}{r} \qquad (1)$$

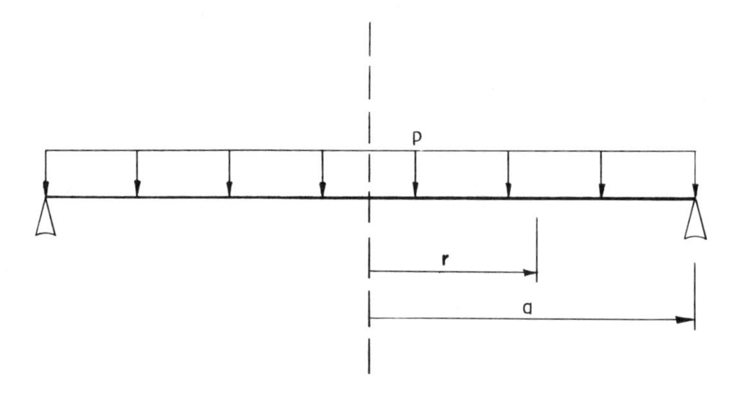

Figure 3-3.

$$\text{deflection} = w = \frac{pr^4}{64D} + \frac{C_1 r^2}{4} + C_2 \ln r + C_3. \tag{2}$$

At $r = 0$ the slope is equal to zero due to symmetry. Hence, from Eq. (1), C_2 must be set to zero. At $r = a$, the moment is zero and

$$-D\left(\frac{d^2w}{dr^2} + \frac{\mu}{r}\frac{dw}{dr}\right) = 0$$

or

$$C_1 = \frac{-(3 + \mu)}{(1 + \mu)} \frac{pa^2}{8D}.$$

At $r = a$, the deflection is zero and Eq. (2) gives

$$C_3 = \frac{pa^4}{64D}\left(\frac{6 + 2\mu}{1 + \mu} - 1\right).$$

The expression for deflection becomes

$$w = \frac{p}{64D}(a^2 - r^2)\left(\frac{5 + \mu}{1 + \mu}a^2 - r^2\right).$$

Substituting this expression into Eq. (3-5) gives

$$M_r = \frac{p}{16}(3 + \mu)(a^2 - r^2)$$

$$M_t = \frac{p}{16}[a^2(3 + \mu) - r^2(1 + 3\mu)].$$

The maximum deflection is at $r = 0$ and is given by

$$\max w = \frac{pa^4}{64D}\left(\frac{5 + \mu}{1 + \mu}\right).$$

A plot of M_r and M_t is shown in Fig. 3-4 for $\mu = 0.3$. The plot shows that the maximum moment is in the center and is equal to

$$M_r = M_t = \frac{pa^2}{16}(3 + \mu).$$

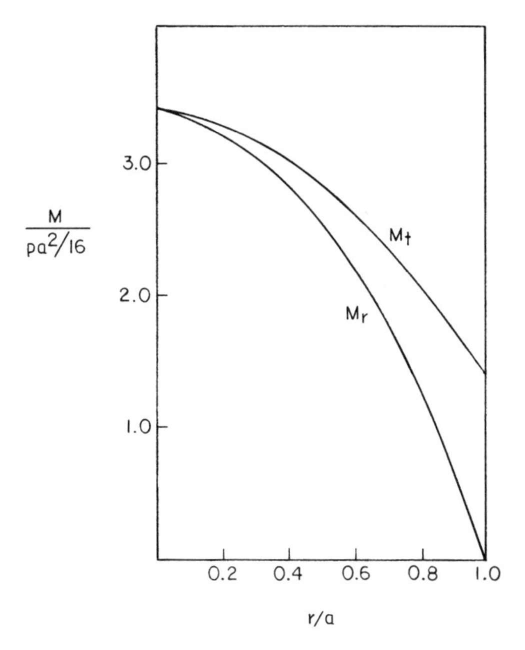

Figure 3-4.

It is of interest to note that M_t is not zero at the edge of the plate. This is important in reinforced concrete plates as reinforcing bars are needed around the perimeter to resist the tension stress caused by M_t.
(b)

$$\text{Maximum } M = \frac{5 \times 20^2}{16}(3.3) = 412.5 \text{ inch-lbs/inch}$$

and

$$t = \sqrt{\frac{6M}{\sigma}}$$

$$t = \sqrt{\frac{6 \times 412.5}{15{,}000}}$$
$$= 0.41 \text{ inch.}$$

$$\text{Maximum } w = \frac{pa^4}{64D}\left(\frac{5 + \mu}{1 + \mu}\right) = 0.27 \text{ inch}$$

(c)

For $\mu = 0$, the moment expressions become

$$M_r = \frac{3p}{16}(a^2 - r^2)$$

$$M_t = \frac{p}{16}(3a^2 - r^2).$$

At $r = 0$,

$$M_r = M_t = \frac{3pa^2}{16}.$$

At $r = a/2$,

$$M_r = \frac{9pa^2}{64}, \qquad M_t = \frac{11pa^2}{64}.$$

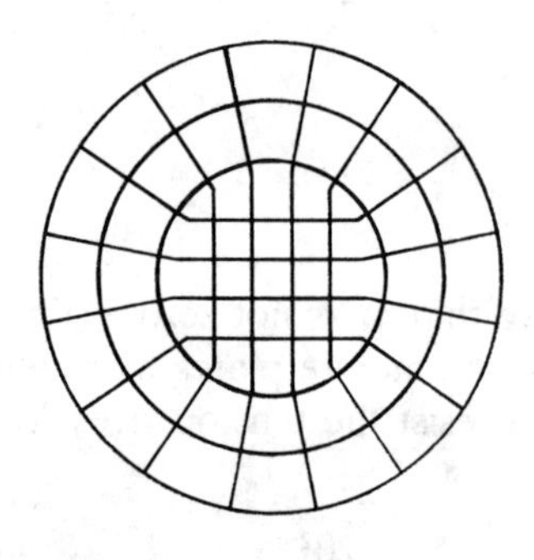

(a) radial and circular pattern

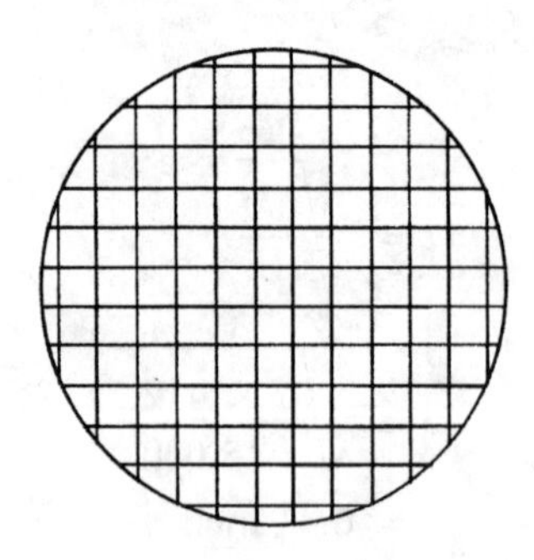

(b) square mesh pattern

Figure 3-5.

At $r = a$,

$$M_r = 0 \qquad M_t = \frac{pa^2}{8}.$$

Figure 3-5 shows a general layout of reinforcing bars in a circular concrete slab.

Example 3-2

Find the stress at $r = a$ and $r = b$ for the plate shown in Fig. 3-6. Let $a = 24$ inches, $b = 12$ inches, $F = 20$ lbs/inch, $t = 0.50$ inch, $E = 30{,}000$ ksi, and $\mu = 0.3$.

Solution

The shearing force at any point is given by

$$Q = bF/r.$$

Substituting this expression into Eq. (3-8) and integrating results in

$$w = \frac{bF}{4D} r^2(\ln r - 1) + \frac{C_1 r^2}{4} + C_2 \ln r + C_3. \tag{1}$$

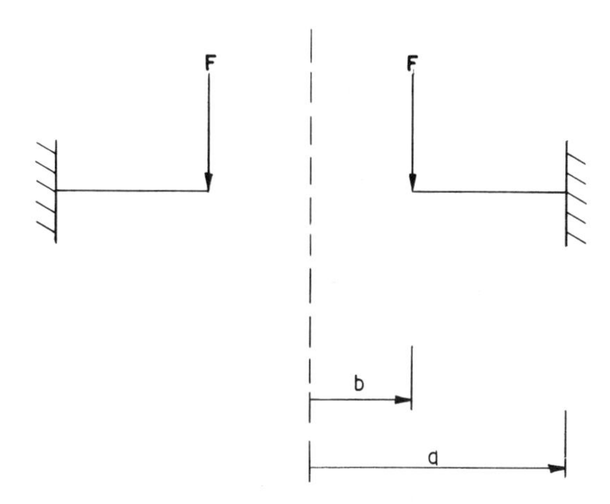

Figure 3-6.

The boundary conditions are

$$w = 0 \quad \text{at } r = a$$
$$dw/dr = 0 \quad \text{at } r = a$$
$$M_r = 0 \quad \text{at } r = b.$$

Evaluating Eq. (1) and its derivatives at the boundary conditions results in

$$C_1 = -648.53/D$$
$$C_2 = 1667.46/D$$
$$C_3 = 12{,}814.46/D.$$

At $r = b$,

$$M_r = 0$$

$$M_t = -D\left(\frac{1}{r}\frac{dw}{dr} + \mu\frac{d^2w}{dr^2}\right)$$

$$= -D\left\{\frac{bF}{2D}\ln r - \frac{bF}{2D}\frac{1}{2} + \mu\left[\frac{bF}{2D}\left(\frac{1}{2} + \ln r\right) + \frac{C_1}{2} - \frac{C_2}{r^2}\right]\right\}$$

or

$$M_t = -67.8 \text{ inch-lbs/inch}$$

and

$$\sigma_t = 6M/t^2 = 1627 \text{ psi}.$$

Similarly, at $r = a$

$$M_t = -34.26 \text{ inch-lbs/inch}$$
$$\sigma_t = 822 \text{ psi}$$
$$M_r = -114.21 \text{ inch-lbs/inch}$$
$$\sigma_r = 2741 \text{ psi}.$$

Example 3-3

Find the expression for the deflection of the plate shown in Fig. 3-7a due to load F.

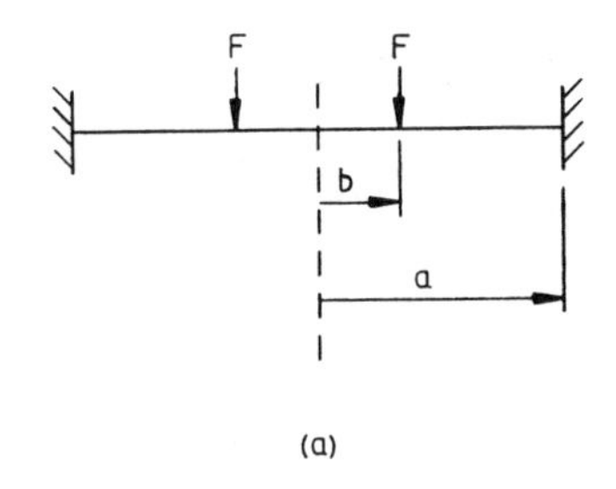

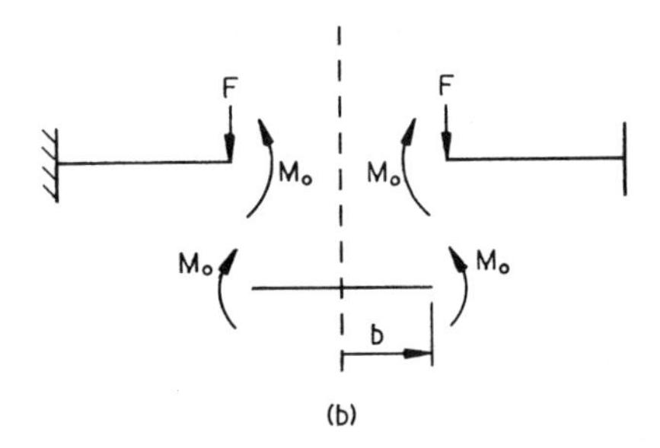

Figure 3-7.

Solution

The plate can be separated into two components (Fig. 3-7b). Continuity between the inner and the outer plate is maintained by applying an unknown moment, M_o, as shown in Fig. 3-7b. The deflection of the inner plate due to M_o is obtained from Eq. (3-8) with $Q = 0$ as

$$w = \frac{C_1 r^2}{4} + C_2 \ln r + C_3. \tag{1}$$

The slope is

$$\frac{dw}{dr} = \frac{C_1 r}{2} + \frac{C_2}{r}, \tag{2}$$

and M_r is obtained from Eq. (3-5) as

$$M_r = -D\left[\frac{C_1}{2}(1 + \mu) - \frac{C_2}{r^2}(1 - \mu)\right]. \tag{3}$$

At $r = 0$, the slope is zero and from Eq. (2) we get $C_2 = 0$.

At $r = b$, $M_r = M_o$ and Eq. (3) yields

$$C_1 = \frac{-2M_o}{D(1 + \mu)}. \tag{4}$$

Equations (1) and (2) can be expressed as

$$w = \frac{-M_o r^2}{2D(1 + \mu)} + C_3 \tag{5}$$

$$\frac{dw}{dr} = \frac{-rM_o}{D(1 + \mu)}. \tag{6}$$

The deflection of the outer plate is obtained from Example 3-2 as

$$w = \frac{bF}{4D} r^2(\ln r - 1) + \frac{C_4 r^2}{4} + C_5 \ln r + C_6. \tag{7}$$

At $r = a$, the slope is zero and Eq. (7) gives

$$\frac{C_4 a}{2} + \frac{C_5}{a} = \frac{-bFa}{4D}(2 \ln a - 1). \tag{8}$$

At $r = b$, $M_r = M_o$ and from Eqs. (7) and (3-5) we get

$$\frac{-M_o}{D} = \frac{bF}{4D}[(1 - \mu) + 2(1 + \mu) \ln b] + \frac{C_4}{2}(1 + \mu) + \frac{C_5}{b^2}(\mu - 1). \tag{9}$$

At $r = b$, the slope of the outer plate is equal to the slope of the inner plate. Taking the derivatives of Eq. (7) and equating it to Eq. (6) at $r = b$ gives

$$\frac{C_4 b}{2} + \frac{C_5}{b} = \frac{-b^2 F}{4D}(2 \ln b - 1) - \frac{bM_o}{D(1 + \mu)}. \tag{10}$$

Equations (8), (9), and (10) contain three unknowns. They are M_o, C_4, and C_5. Solving these three equations yields

$$M_o = \frac{(1 + \mu)bF}{4a^2}[2a^2 \ln(a/b) - (a^2 - b^2)] \tag{11}$$

$$C_4 = \frac{bF}{D}\left(\frac{a^2 - b^2}{2a^2} - \ln a\right) \tag{12}$$

$$C_5 = \frac{b^3 F}{4D}. \tag{13}$$

With these quantities known, the other constants can readily be obtained. Constant C_1 is determined from Eq. (4). Constant C_6 is solved from Eq. (7) for the boundary conditions $w = 0$ at $r = a$. This gives

$$C_6 = \frac{bF}{8D}(a^2 + b^2 - 2b^2 \ln a). \tag{14}$$

Constant C_3 can now be calculated from the equation

$$w \text{ of inner plate}|_{r=b} = w \text{ of outer plate}|_{r=b}.$$

Equating Eqs. (1) and (7) at $r = b$ gives

$$C_3 = \frac{bF}{8a^2D}[-2a^2b^2 \ln(a/b) + (a^2 - b^2)a^2]. \tag{15}$$

Hence, the deflection of the inner plate is obtained by substituting Eqs. (11) and (15) into Eq. (5) to give

$$w = \frac{bF}{8a^2D}[(a^2 - b^2)(a^2 + r^2) - 2a^2(b^2 + r^2)\ln(a/b)]$$

and the deflection of the outer plate is obtained by substituting Eqs. (12), (13), and (14) into Eq. (7). This gives

$$w = \frac{bF}{8a^2D}[(a^2 + b^2)(a^2 - r^2) + 2a^2(b^2 + r^2)\ln(r/a)].$$

Problems

3-1 The double concrete silo shown in Fig. P3-1 is covered by a concrete flat roof as shown. Find the moments in the roof due to an applied uniform load p, and draw the M_r and M_t diagrams. The attachment of the roof to the cylindrical silos is assumed simply supported. Let $\mu = 0$.

3-2 Stainless steel baffles are attached to a vessel that has an agitator shaft as shown in Fig. P3-2. The attachment of the baffles to the vessel is assumed fixed and the uniform pressure due to agitator rotation is 20 psi. What are the maximum values of M_r and M_t and where do they occur? What is the maximum deflection at point b? Let E = 27,000,000 psi and $\mu = 0.29$. Also, if the baffles are assumed as fixed cantilevered beams, what will the maximum moment be and how does it compare to M_r and M_t?

3-3 The pan shown in Fig. P3-3 is made of aluminum and is full of water. If the edge of the bottom plate is assumed fixed, what is the maximum stress due to the exerted water pressure? Let γ = 62.4 pcf, t = 0.030 inch, E = 10,200 ksi, and $\mu = 0.33$. What is the maximum deflection?

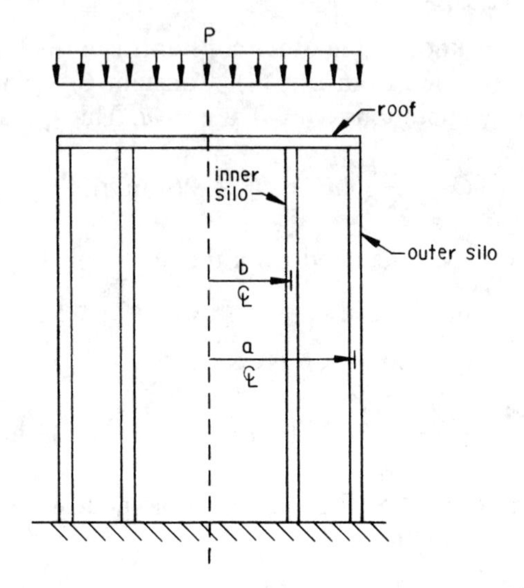

Figure P3-1.

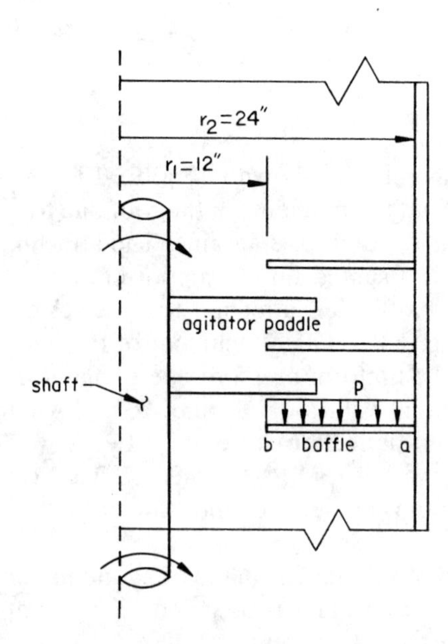

Figure P3-2.

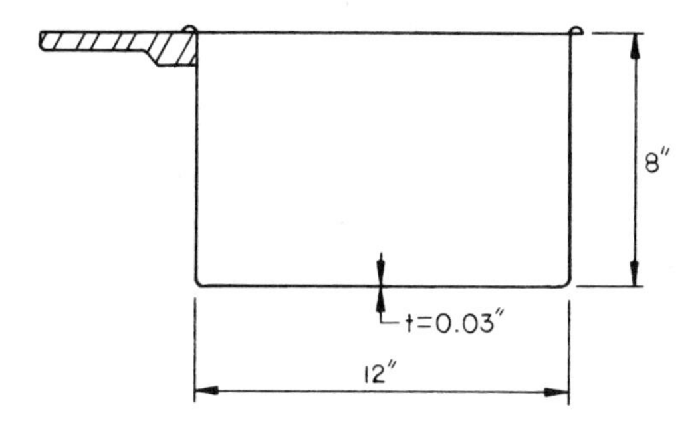

Figure P3-3.

3-4 The pan in Problem 3-3 is empty and is at a temperature of 100°F. What is the thermal stress in the bottom plate if the bottom surface of the bottom plate is subjected to a temperature of 160°F and the top surface is subjected to a temperature of 40°F? Let the coefficient of expansion be 13.5×10^{-6} inches/inch/°F.

3-5 Find the expressions for the moments in the circular plate shown in Fig. P3-5.

3-6 Find the expressions for the moments in the circular plate shown in Fig. P3-6.

3-7 Find the expressions for the moments in the circular plate shown in Fig. P3-7.

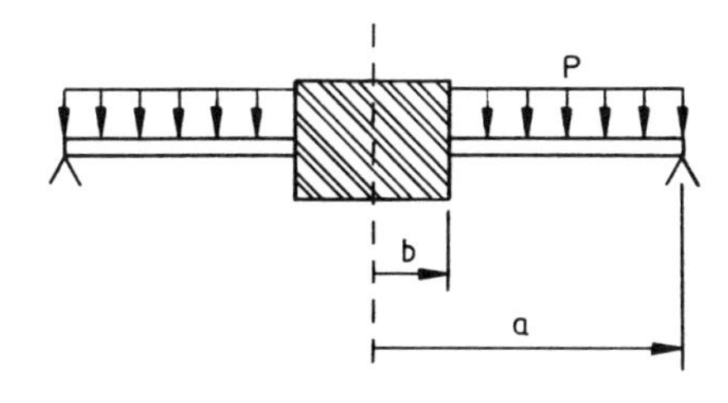

Figure P3-5.

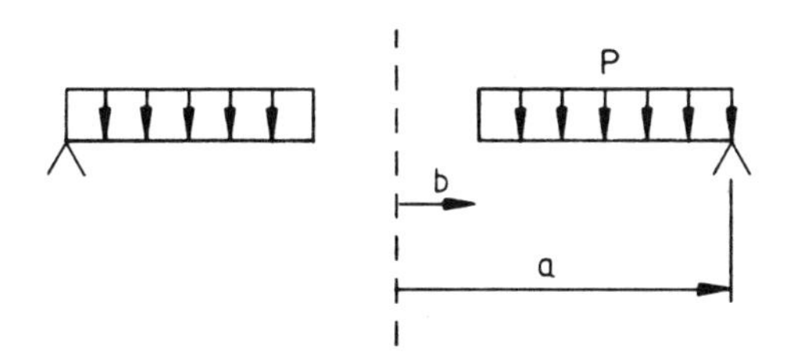

Figure P3-6.

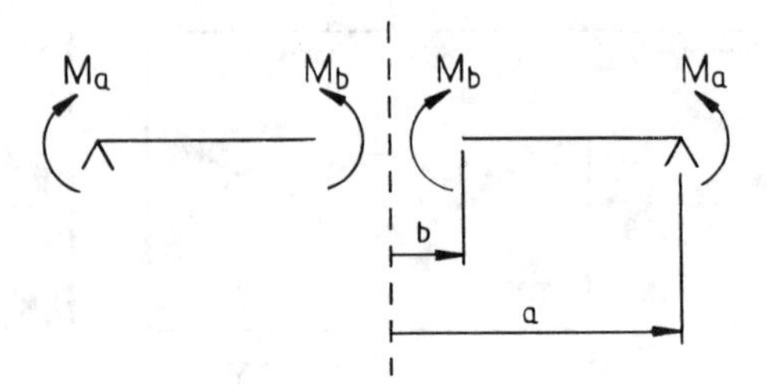

Figure P3-7.

3-2 Plates with Variable Thickness and Subjected to Uniform Loads in the θ-Direction

Circular plates with variable thickness are encountered in many machine parts such as turbine blades, bellows, and springs. The analysis of such plates (Szilard 1974) is similar to that of plates with constant thickness except that the flexural rigidity D is a variable rather than a constant. Substituting Eq. (3-4) into the differential Eq. (3-7) gives

$$D \frac{d}{dr}\left(\frac{d\phi}{dr} + \frac{\phi}{r}\right) + \frac{dD}{dr}\left(\frac{d\phi}{dr} + \mu \frac{\phi}{r}\right) = -Q \qquad (3\text{-}12)$$

or

$$D \frac{d^2\phi}{dr^2} + \left(\frac{D}{r} + \frac{dD}{dr}\right)\frac{d\phi}{dr} + \left(\mu \frac{dD}{dr} - \frac{D}{r}\right)\frac{\phi}{r} = -Q \qquad (3\text{-}13)$$

where

$$\phi = -\frac{dw}{dr}.$$

For a uniformly loaded plate,

$$Q = \frac{1}{2\pi r}\int p(2\pi r)dr = pr/2.$$

Defining

$$\rho = r/a$$

where

a = outer radius of plate.

Eq. (3-13) becomes

$$D \frac{d^2\phi}{d\rho^2} + \left(\frac{D}{\rho} + \frac{dD}{d\rho}\right)\frac{d\phi}{d\rho} + \left(\mu \frac{dD}{d\rho} - \frac{D}{\rho}\right)\frac{\phi}{\rho} = \frac{-p\rho a^3}{2}. \qquad (3\text{-}14)$$

The solution of Eq. (3-14) depends on specifying an expression for the thickness t. A commonly encountered class of plates is shown in Fig. 3-8. Solution of the plate shown in Fig. 3-8a is obtained by defining

$$t = Kr$$

where

$$D = \frac{EK^3r^3}{12(1 - \mu^2)}.$$

Equation (3-14) becomes

$$\rho^3 \frac{d^2\phi}{d\rho^2} + 4\rho^2 \frac{d\phi}{d\rho} - (1 - 3\mu)\rho\phi = \frac{-12Q(1 - \mu^2)}{EK^3a}. \tag{3-15}$$

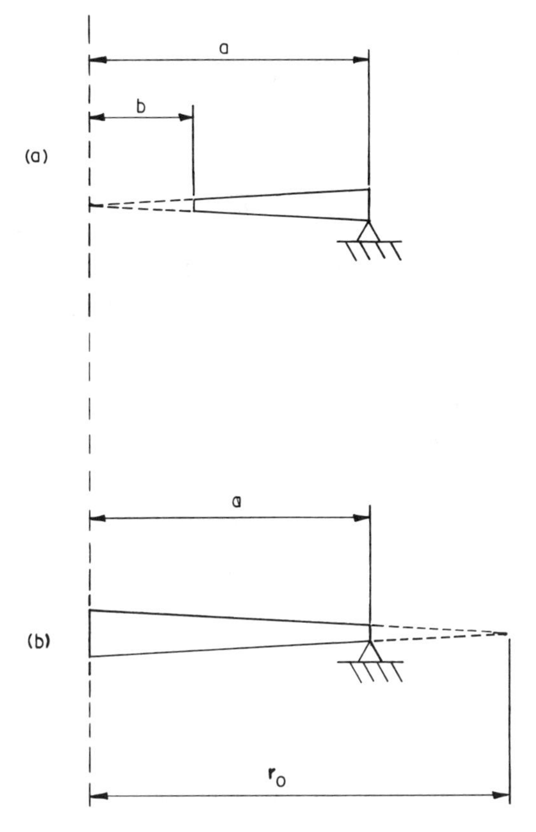

Figure 3-8.

The homogeneous solution of Eq. (3-15) can be expressed as

$$\phi_h = A\rho^a + B\rho^b$$

where

$$a = [-3 + \sqrt{9 - 4(3\mu - 1)}]/2$$

$$b = [-3 - \sqrt{9 - 4(3\mu - 1)}]/2$$

and A and B are constants.

Solution of the plate shown in Fig. 3-8b is obtained by defining

$$t = t_o(1 - r/r_o)$$

and

$$D = D_o(1 - \rho)^3$$

$$D_o = \frac{Et_o^3}{12(1 - \mu^2)} \quad \text{and} \quad \mu = 1/3,$$

and Eq. (3-14) becomes

$$\rho^2(1 - \rho)^3 \frac{d^2\phi}{d\rho^2} + (1 - 4\rho)(1 - \rho)^2 \frac{d\phi}{d\rho} - (1 - \rho)^2\phi = \frac{Qr^2\rho^2}{D_o}. \quad (3\text{-}16)$$

The homogeneous solution of Eq. (3-16) is given by

$$\phi_h = A\left(\frac{1 + 2\rho}{\rho}\right) + B\left(\frac{3\rho - 2\rho^2}{(1 - \rho)^2}\right)$$

where A and B are constants.

The particular solution of Eqs. (3-15) and (3-16) is obtained once the applied loads are defined.

Example 3-4

Determine the expression for the deflection in the plate shown in Fig. 3-9. Let $\mu = 1/3$.

Solution

Let

$$t = Cr.$$

The homogeneous solution of Eq. (3-15) becomes

$$\phi_h = A + B\rho^{-3}.$$

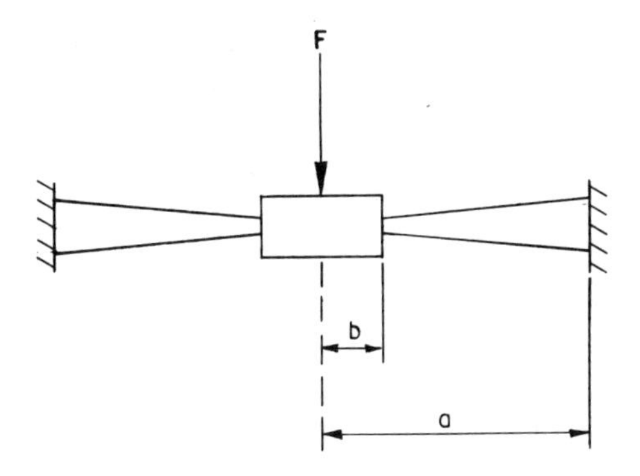

Figure 3-9.

To solve for a particular solution, let $Q = \dfrac{F}{2\pi a\rho}$

and

$$\phi_p = \frac{G}{\rho^2}.$$

Substituting into Eq. (3-15) gives

$$G = \frac{3(1 - \mu^2)F}{\pi E C^3 a^2}$$

and the total solution becomes

$$\phi = A + B\rho^{-3} + \frac{G}{\rho^2}. \tag{1}$$

The boundary conditions are

$$\text{at } r = b, \quad \rho = b/a \quad \text{and} \quad \phi = 0$$

and

$$\text{at } r = a, \quad \rho = 1 \quad \text{and} \quad \phi = 0.$$

Hence, from Eq. (1)

$$A = G\,\frac{a^2(b - a)}{a^3 - b^3}$$

$$B = G\,\frac{b(a^2 - b^2)}{a^3 - b^3}.$$

Integrating Eq. (1) and solving for the boundary condition

$$w = 0 \quad \text{at } \rho = 1$$

results in the following expression for the deflection:

$$w = Aa(\rho - 1) + \frac{Ba}{2}\left(1 - \frac{1}{\rho^2}\right) + Ga\left(1 - \frac{1}{\rho}\right).$$

Another class of problems that is often encountered in machine parts is plates with variable thickness. The variable thickness can be expressed as (Timoshenko and Woinowsky-Krieger 1959)

$$t = t_o e^{-\beta\rho^2/6} \tag{3-17}$$

where

β = factor defining thickness of plate as shown in Fig. 3-10;
t_o = thickness of plate at center.

Substituting Eq. (3-17) into Eq. (3-14) gives

$$\frac{d^2\phi}{d\rho^2} + \left(\frac{1}{\rho} - \beta\rho\right)\frac{d\phi}{d\rho} - \left(\frac{1}{\rho^2} + \mu\beta\right)\phi = -K\rho e^{\beta\rho^2/2} \tag{3-18}$$

where

$$K = \frac{6(1 - \mu^2)a^3 p}{Et_o^3}.$$

The solution of Eq. (3-18) is given by

$$\phi = \phi_h + \phi_p.$$

Let the particular solution be expressed as

$$\phi_p = A\rho e^{\beta\rho^2/2} + Be^{\beta\rho^2/2}.$$

Substituting this into Eq. (3-18) gives

$$\phi_p = \frac{-K}{(3 - \mu)\beta}\rho e^{\beta\rho^2/2}.$$

Pirchler (Timoshenko and Woinowsky-Krieger 1959) suggested a homogeneous solution in terms of a series of the form

$$\phi_h = A_1\left(\rho + \sum_{m=1}^{\infty} \frac{(1 + \mu)(3 + \mu) \ldots (2m - 1 + \mu)}{2.4.4.6.6 \ldots\ldots 2m.2m.(2m + 2)}\rho^{2m+1}\right)$$

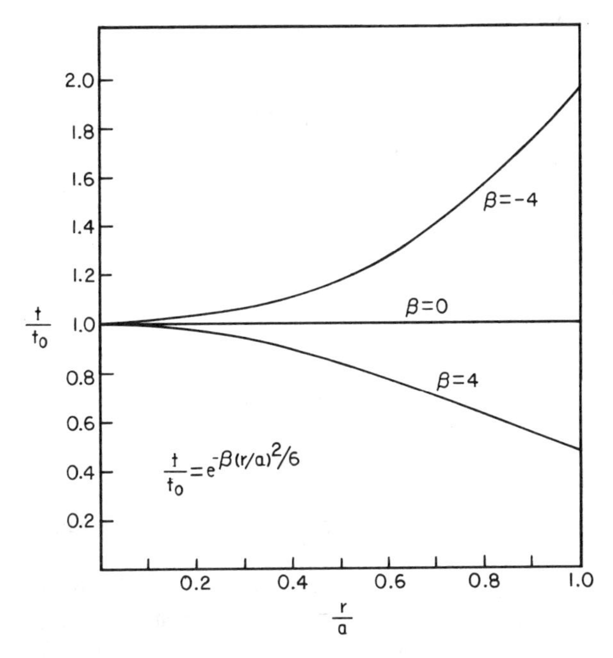

Figure 3-10.

and the total solution is given by

$$\phi = K\left(C\phi_h - \frac{\rho}{(3 - \mu)\beta} e^{\beta\rho^2/2}\right). \tag{3-19}$$

The constant C is obtained from the boundary condition of a solid plate and the maximum moments are determined from Eq. (3-19).

Problems

3-8 Find the expression for the bending moments M_r and M_t for the plate shown in Fig. 3-8. Let $\mu = 1/3$.
3-9 Find the expression for the bending moments M_r and M_t for the plate shown in Fig. P3-9. Let $\mu = 1/3$.
3-10 Find the expression for M_r and M_t for the concrete circular plate shown in Fig. P3-10. Let $p = 1000$ kgf/m^2, $\beta = 4.16$, and $\mu = 0$.

3-3 Plates Subjected to Nonuniform Loads in the θ-Direction

Many structural applications are encountered where the load distribution on the circular plate is variable in the θ-direction. These include stack

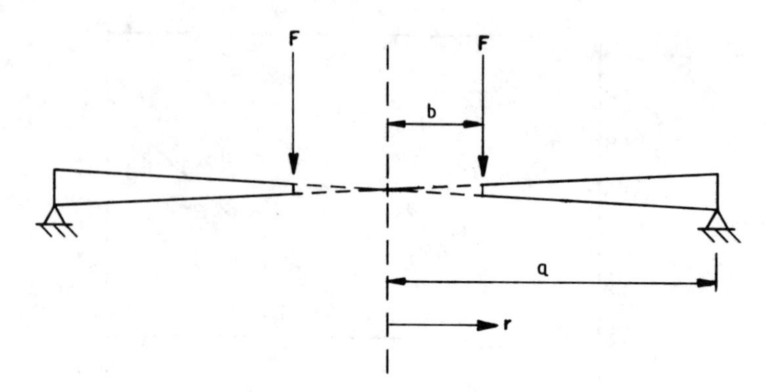

Figure P3-9.

foundations, submerged bulkheads, and nozzle covers subjected to connecting piping loads. The easist method for deriving the governing equation for such problems is from the differential equation of rectangular plates. The reason is that the equation for rectangular plates includes the effects of torsional moments that are ignored in the derivation of the equation for circular plates with uniform loading in the θ-direction and which are needed for the case where the load is variable in the θ-direction. The

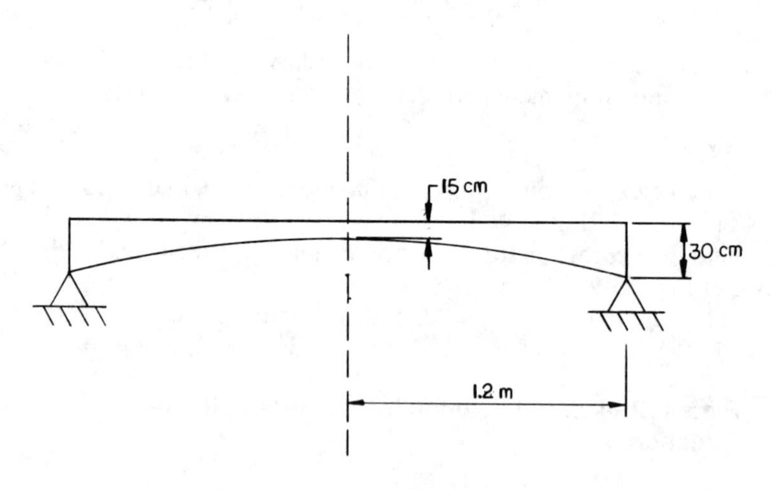

Figure P3-10.

differential expression for rectangular plates is given by Eq. (1-25) as

$$\frac{\partial^4 w}{\partial x^4} + 2\frac{\partial^4 w}{\partial x^2\,\partial y^2} + \frac{\partial^4 w}{\partial y^4} = \frac{p(x, y)}{D}. \tag{3-20}$$

This equation must now be transferred to polar coordinates. Referring to Fig. 3-11,

$$r^2 = x^2 + y^2, \qquad x = r\cos\theta, \qquad y = r\sin\theta, \qquad \tan\theta = y/x.$$

Hence

$$\frac{\partial r}{\partial x} = x/r = \cos\theta \qquad \frac{\partial r}{\partial \theta} = y/r = \sin\theta$$

$$\frac{\partial}{\partial x}\tan\theta = \frac{\partial}{\partial x}(y/x)$$

$$\frac{1}{\cos^2\theta}\frac{\partial\theta}{\partial x} = \frac{\partial}{\partial x}(y/x)$$

$$\frac{\partial\theta}{\partial x} = -y\frac{\cos^2\theta}{x^2} = -y/r^2 = -\sin\theta/r$$

$$\frac{\partial\theta}{\partial y} = x/r^2 = \cos\theta/r.$$

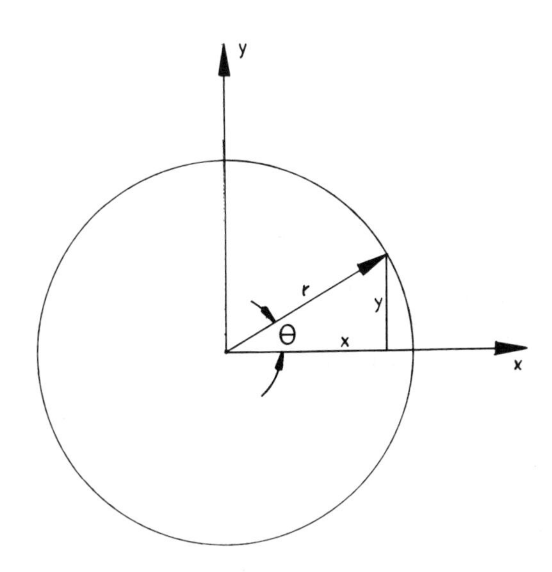

Figure 3-11.

Using the chain rule of partial derivatives,

$$\frac{\partial w}{\partial x} = \frac{\partial w}{\partial r}\frac{\partial r}{\partial x} + \frac{\partial w}{\partial \theta}\frac{\partial \theta}{\partial x}$$

$$= \frac{\partial w}{\partial r}\cos\theta - \frac{1}{r}\frac{\partial w}{\partial \theta}\sin\theta$$

$$\frac{\partial^2 w}{\partial x^2} = \cos\theta\,\frac{\partial}{\partial r}\frac{\partial w}{\partial x} - \frac{1}{r}\sin\theta\,\frac{\partial}{\partial \theta}\frac{\partial w}{\partial x}$$

$$= \frac{\partial^2 w}{\partial r^2}\cos^2\theta - 2\frac{\partial^2 w}{\partial r\,\partial\theta}\frac{\sin\theta\cos\theta}{r} + \frac{\partial w}{\partial r}\frac{\sin^2\theta}{r} + 2\frac{\partial w}{\partial \theta}\frac{\sin\theta\cos\theta}{r^2} + \frac{\partial^2 w}{\partial \theta^2}\frac{\sin^2\theta}{r^2}.$$

Similarly,

$$\frac{\partial^2 w}{\partial y^2} = \frac{\partial^2 w}{\partial r^2}\sin^2\theta + 2\frac{\partial^2 w}{\partial r\,\partial\theta}\frac{\sin\theta\cos\theta}{r} + \frac{\partial w}{\partial r}\frac{\cos^2\theta}{r} - 2\frac{\partial w}{\partial \theta}\frac{\sin\theta\cos\theta}{r^2} + \frac{\partial^2 w}{\partial \theta^2}\frac{\cos^2\theta}{r^2}$$

and

$$\frac{\partial^2 w}{\partial x\,\partial y} = \frac{\partial^2 w}{\partial r^2}\sin\theta\cos\theta + \frac{\partial^2 w}{\partial r\,\partial\theta}\frac{\cos 2\theta}{r} - \frac{\partial w}{\partial \theta}\frac{\cos 2\theta}{r^2} - \frac{\partial w}{\partial r}\frac{\sin\theta\cos\theta}{r} - \frac{\partial^2 w}{\partial \theta^2}\frac{\sin\theta\cos\theta}{r^2}.$$

Hence, Eq. (3-20) becomes

$$\frac{\partial^4 w}{\partial r^4} + \frac{2}{r}\frac{\partial^3 w}{\partial r^3} - \frac{1}{r^2}\frac{\partial^2 w}{\partial r^2} + \frac{1}{r^3}\frac{\partial w}{\partial r} + \frac{2}{r^2}\frac{\partial^4 w}{\partial r^2\partial\theta^2} - \frac{2}{r^3}\frac{\partial^3 w}{\partial\theta^2\,\partial r} + \frac{4}{r^4}\frac{\partial^2 w}{\partial\theta^2} + \frac{1}{r^4}\frac{\partial^4 w}{\partial\theta^4} = p/D. \quad (3\text{-}21)$$

Similarly, the values of M_x, M_y, and M_{xy} become

$$M_r = -D\left(\frac{\partial^2 w}{\partial x^2} + \mu\frac{\partial^2 w}{\partial y^2}\right)$$

$$= -D\left[\frac{\partial^2 w}{\partial r^2} + \mu\left(\frac{1}{r}\frac{\partial w}{\partial r} + \frac{1}{r^2}\frac{\partial^2 w}{\partial\theta^2}\right)\right] \quad (3\text{-}22)$$

$$M_t = -D\left(\frac{1}{r}\frac{\partial w}{\partial r} + \frac{1}{r^2}\frac{\partial^2 w}{\partial\theta^2} + \mu\frac{\partial^2 w}{\partial r^2}\right) \quad (3\text{-}23)$$

$$M_{rt} = (1-\mu)D\left(\frac{1}{r}\frac{\partial^2 w}{\partial r\,\partial\theta} - \frac{1}{r^2}\frac{\partial w}{\partial\theta}\right). \quad (3\text{-}24)$$

The shear forces are expressed as

$$Q_r = -D\frac{\partial}{\partial r}\left(\frac{\partial^2 w}{\partial r^2} + \frac{1}{r}\frac{\partial w}{\partial r} + \frac{1}{r^2}\frac{\partial^2 w}{\partial \theta^2}\right) \quad (3\text{-}25)$$

$$Q_t = -D\frac{1}{r}\frac{\partial}{\partial \theta}\left(\frac{\partial^2 w}{\partial r^2} + \frac{1}{r}\frac{\partial w}{\partial r} + \frac{1}{r^2}\frac{\partial^2 w}{\partial \theta^2}\right) \quad (3\text{-}26)$$

The boundary conditions are:
for simply supported plates

$$w = 0 \quad \text{and} \quad M_r = 0$$

for fixed plates

$$w = 0 \quad \text{and} \quad \frac{\partial w}{\partial r} = 0$$

for plates with free edge

$$M_r = 0 \quad \text{and} \quad V = \left(Q_r - \frac{1}{r}\frac{\partial M_{rt}}{\partial \theta}\right) = 0.$$

Equation (3-21) is solved by letting

$$w = w_h + w_p.$$

The homogeneous solution, w_h, is expressed (McFarland et al. 1972) by the following Fourier series

$$w_h = \sum_{n=0}^{\infty} f_n(r)\cos n\theta + \sum_{n=1}^{\infty} g_n(r)\sin n\theta.$$

Substituting this expression into $\nabla^4 w = 0$ gives

$$\sum_{n=0}^{\infty}\left(\frac{d^4 f_n}{dr^4} + \frac{2}{r}\frac{d^3 f_n}{dr^3} - \frac{1+2n^2}{r^2}\frac{d^2 f_n}{dr^2} + \frac{1+2n^2}{r^3}\frac{df_n}{dr} + \frac{n^2(n^2-4)}{r^4}f_n\right)\cos n\theta$$

$$+ \sum_{n=1}^{\infty}\left(\frac{d^4 g_n}{dr^4} + \frac{2}{r}\frac{d^3 g_n}{dr^3} - \frac{1+2n^2}{r^2}\frac{d^2 g_n}{dr^2} + \frac{1+2n^2}{r^3}\frac{dg_n}{dr} + \frac{n^2(n^2-4)}{r^4}g_n\right)\sin n\theta = 0.$$

This equation is satisfied if

$$\frac{d^4 f_n}{dr^4} + \frac{2}{r}\frac{d^3 f_n}{dr^3} - \frac{1+2n^2}{r^2}\frac{d^2 f_n}{dr^2} + \frac{1+2n^2}{r^3}\frac{df_n}{dr} + \frac{n^2(n^2-4)}{r^4}f_n = 0 \quad (3\text{-}27)$$

and

$$\frac{d^4 g_n}{dr^4} + \frac{2}{r}\frac{d^3 g_n}{dr^3} - \frac{1 + 2n^2}{r^2}\frac{d^2 g_n}{dr^2} + \frac{1 + 2n^2}{r^3}\frac{dg}{dr} + \frac{n^2(n^2 - 4)}{r^4} g_n = 0. \quad (3\text{-}28)$$

Let $f_n(r) = b_n r^m$
and

$$g_n(r) = c_n r^m.$$

Substituting these equations into Eqs. (3-27) and (3-28) gives

$$m(m - 1)(m - 2)(m - 3) + 2m(m - 1)(m - 2) - (1 + 2n^2)m(m - 1) + (1 + 2n^2)m + n^2(n^2 - 4) = 0.$$

The roots of this equation are

$$m_1 = n, \quad m_2 = -n, \quad m_3 = n + 2, \quad m_4 = -n + 2.$$

$$\text{If } n = 0, \quad m_1 = m_2 = 0, \quad m_3 = m_4 = 2$$

and

$$f_o(r) = A_o r^0 + B_o r^2 + C_o r^0 \ln r + D_o r^2 \ln r$$
$$= A_o + B_o r^2 + C_o \ln r + D_o r^2 \ln r.$$

$$\text{If } n = 1, \quad m_1 = m_4 = 1, \quad m_2 = -1, \quad m_3 = 3$$

and

$$f_1 = A_1 r + B_1 r^3 + C_1 r^{-1} + D_1 r \ln r$$
$$g_1 = E_1 r + F_1 r^3 + G_1 r^{-1} + H_1 r \ln r.$$

Similarly,

$$f_n = A_n r^n + B_n r^{-n} + C_n r^{n+2} + D_n r^{-n+2}$$
$$g_n = E_n r^n + F_n r^{-n} + G_n r^{n+2} + H_n r^{-n+2}.$$

Hence, the homogeneous solution, w_h, beomes

$$\begin{aligned} w_h = {} & A_o + B_o r^2 + C_o \ln r + D_o r^2 \ln r \\ & + (A_1 r + B_1 r^3 + C_1 r^{-1} + D_1 r \ln r) \cos \theta \\ & + (E_1 r + F_1 r^3 + G_1 r^{-1} + H_1 r \ln r) \sin \theta \\ & + \sum_{n=2}^{\infty} (A_n r^n + B_n r^{-n} + C_n r^{n+2} + D_n r^{-n+2}) \cos n\theta \\ & + \sum_{n=2}^{\infty} (E_n r^n + F_n r^{-n} + G_n r^{n+2} + H_n r^{-n+2}) \sin n\theta. \end{aligned} \quad (3\text{-}29)$$

The particular solution, w_p, is obtained by letting

$$w_p = I_o(r) + \sum_{n=1}^{\infty} [I_n(r) \cos n\theta + J_n(r) \sin n\theta]$$

and

$$p = p_o(r) + \sum_{n=1}^{\infty} [p_n(r) \cos n\theta + S_n(r) \sin n\theta]$$

where

$$p_n(r) = \frac{1}{\pi} \int_{-\pi}^{\pi} p(r, \theta) \cos n\theta \, d\theta \qquad n = 0, 1, 2, \ldots$$

$$S_n(r) = \frac{1}{\pi} \int_{-\pi}^{\pi} p(r, \theta) \sin n\theta \, d\theta \qquad n = 1, 2, \ldots$$

Substituting w_p into the equation $\nabla^4 w = p/D$ gives

$$\begin{aligned}
&\frac{d^4 I_o}{dr^4} + \frac{2}{r}\frac{d^3 I_o}{dr^3} - \frac{1}{r^2}\frac{d^2 I_o}{dr^2} + \frac{1}{r^3}\frac{dI_o}{dr} + \sum_{n=1}^{\infty} \left[\frac{d^4 I_n}{dr^4} + \frac{2}{r}\frac{d^3 I_n}{dr^3} - \frac{1 + 2n^2}{r^2}\frac{d^2 I_n}{dr^2} \right. \\
&\qquad \left. + \frac{1 + 2n^2}{r^3}\frac{dI_n}{dr} + \frac{n^2(n^2 - 4)}{r^4} I_n \right] \cos n\theta \\
&\qquad + \sum_{n=1}^{\infty} \left[\frac{d^4 J_n}{dr^4} + \frac{2}{r}\frac{d^3 J_n}{dr^3} - \frac{1 + 2n^2}{r^2}\frac{d^2 J_n}{dr^2} \right. \\
&\qquad \left. + \frac{1 + 2n^2}{r^3}\frac{dJ_n}{dr} + \frac{n^2(n^2 - 4)}{r^4} J_n \right] \sin n\theta \\
&\qquad = \frac{p_o(r)}{D} + \frac{1}{D} \sum p_n \cos n\theta + \frac{1}{D} \sum S_n \sin n\theta,
\end{aligned}$$

from which we obtain the following solution:

$$\frac{d^4 I_o}{dr^4} + \frac{2}{r}\frac{d^3 I_o}{dr^3} - \frac{1}{r^2}\frac{d^2 I_o}{dr^2} + \frac{1}{r^3}\frac{dI_o}{dr} = p_o(r)/D \tag{3-30}$$

$$\frac{d^4 I_n}{dr^4} + \frac{2}{r}\frac{d^3 I_n}{dr^3} - \frac{1 + 2n^2}{r^2}\frac{d^2 I_n}{dr^2} + \frac{1 + 2n^2}{r^3}\frac{dI_n}{dr} + \frac{n^2(n^2 - 4)}{r^4} I_n = p_n/D \tag{3-31}$$

$$\frac{d^4 J_n}{dr^4} + \frac{2}{r}\frac{d^3 J_n}{dr^3} - \frac{1 + 2n^2}{r^2}\frac{d^2 J_n}{dr^2} + \frac{1 + 2n^2}{r^3}\frac{dJ_n}{dr} + \frac{n^2(n^2 - 4)}{r^4} J_n = S_n/D. \tag{3-32}$$

Equation (3-21) is the differential equation for the bending of circular plates and is derived from the expression $\nabla^4 w = p/D$. Its solution is given by Eqs. (3-29) through (3-32).

Example 3-5

Find the bending moment in the plate shown in Fig. 3-12. The load distribution on the plate is given by

$$p = p_o \frac{r}{a} \cos \theta.$$

Solution

Since the applied load is a function of θ, all terms in the homogeneous deflection given by Eq. (3-29) are deleted except

$$w_h = (A_{1r} + B_1 r^3 + C_1/r + D_1 r \ln r) \cos \theta. \tag{1}$$

Similarly, Eqs. (3-30) and (3-32) are ignored and Eq. (3-31) is used. The expression for p_n becomes p_1 since the load is a function of θ only. Accordingly,

$$p_1(r) = \frac{1}{\pi} \int_{-\pi}^{\pi} \left(\frac{p_o r}{a} \cos \theta \right) \cos \theta \, d\theta = \frac{p_o r}{a} \int_{-\pi}^{\pi} \cos^2 \theta \, d\theta$$
$$= p_o r/a.$$

Equation (3-31) becomes

$$\frac{d^4 I_1}{dr^4} + \frac{2}{r} \frac{d^3 I_1}{dr^3} - \frac{3}{r^2} \frac{d^2 I_1}{dr^2} + \frac{3}{r^3} \frac{dI_1}{dr} - \frac{3}{r^4} I_1 = \frac{p_o r}{aD}. \tag{2}$$

Let $I_1 = C_1 r^5 + C_2 r^4 + C_3 r^3 + C_4 r^2 + C_5 r + C_6$.

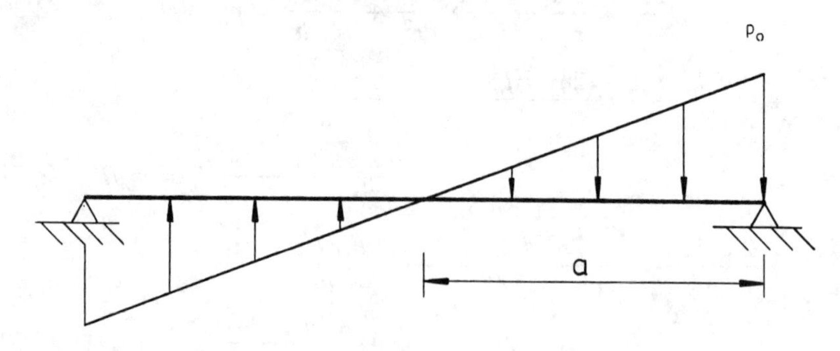

Figure 3-12.

Substituting this expression into Eq. (2) and solving for the constants C_1 through C_6 gives

$$C_1 = \frac{p_o}{192aD}$$

$$C_2 = C_3 = C_4 = C_5 = C_6 = 0$$

and

$$I_1 = \frac{p_o r^5}{192aD}.$$

The solution for w_p is expressed as

$$w_p = \frac{p_o r^5}{192aD} \cos\theta.$$

combining this expression with Eq. (1) gives the total solution for the deflection.

$$w = \left(A_1 r + B_1 r^3 + C_1/r + D_1 r \ln r + \frac{p_o r^5}{192aD}\right) \cos\theta.$$

Since θ and M_r are finite as $r \to 0$, constants C_1 and D_1 must be set to zero. The deflection expression then becomes

$$w = \left(A_1 r + B_1 r^3 + \frac{p_o r^5}{192aD}\right) \cos\theta. \qquad (3)$$

At $r = a$, $M_r = 0$. Equation (3) gives

$$B_1 = -\frac{2(5 + \mu)}{(3 + \mu)} \frac{p_o a}{192D}.$$

At $r = a$, $w = 0$ and Eq. (3) results in

$$A_1 = \frac{(7 + \mu)}{(3 + \mu)} \frac{p_o a^3}{192D}.$$

The final expression for the deflection can now be written as

$$w = \frac{p_o}{192D} \left(\frac{r^5}{a} - \frac{2(5 + \mu)}{(3 + \mu)} a r^3 + \frac{(7 + \mu)}{(3 + \mu)} a^3 r\right) \cos\theta.$$

The equations for M_r and M_t can now be obtained and are expressed as

$$M_r = \frac{p_o a^2 (5 + \mu)}{48} (r/a) \left(1 - \frac{r^2}{a^2}\right) \cos\theta$$

$$M_t = \frac{4 p_o a^2}{192} \left[\left(\frac{r}{a}\right) \frac{(5 + \mu)(1 + 3\mu)}{(3 + \mu)} - \frac{r^3}{a^3}(1 + 5\mu)\right] \cos\theta.$$

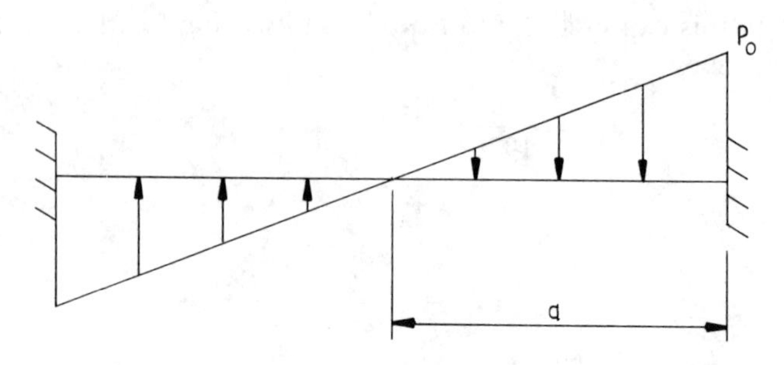

Figure P3-11.

Problems

3-11 Find the maximum bending moment in the plate shown in Fig. P3-11. The edge of the plate is fixed and the applied load is expressed as

$$p = \frac{p_o r}{a} \cos \theta.$$

3-12 Find the expression for the bending moments in the plate shown in Fig. P3-12.

3-4 Plates on an Elastic Foundation

Power and petrochemical plants as well as refineries use evaporators, condensors, and heat exchange units as part of their daily operations. These

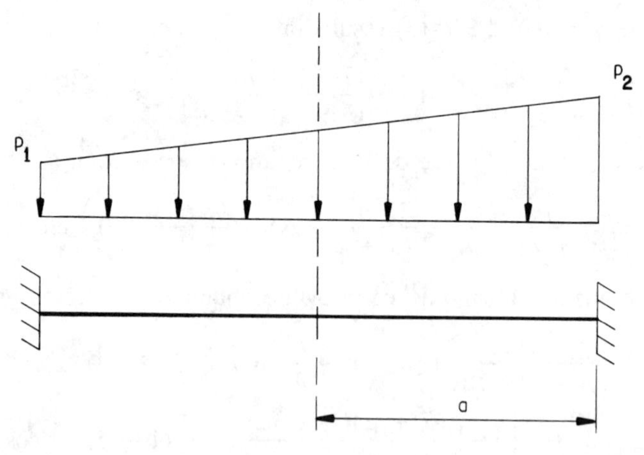

Figure P3-12.

units consist of two perforated circular plates, called tubesheets, that are braced by a number of tubes as shown in Fig. 3-13. The tubesheets and tubes are inserted in a vessel consisting of a cylindrical shell and two end closures. Fluid passing around the outside surface of the tubes exchanges heat with a different fluid passing through the tubes. The tubesheets are assumed to be supported by both the shell and tubes and are analyzed as circular plates on an elastic foundation. Referring to Figs. 3-2 and 3-14, it is seen that the foundation pressure f acts opposite the applied pressure p. Hence Eq. (3-9) can be expressed as (Hetenyi 1964)

$$\frac{1}{r}\frac{d}{dr}\left\{r\frac{d}{dr}\left[\frac{1}{r}\frac{d}{dr}\left(r\frac{dw}{dr}\right)\right]\right\} = \frac{p_r - f_r}{D} \tag{3-33}$$

or

$$\left(\frac{d^2}{dr^2} + \frac{1}{r}\frac{d}{dr}\right)\left(\frac{d^2}{dr^2} + \frac{1}{r}\frac{d}{dr}\right) w = \frac{p_r - kw}{D} \tag{3-34}$$

where

f_r = load exerted by the elastic foundation;
f_r = $k_o w$;
k_o = foundation modulus defined as the modulus of elasticity of foundation divided by the depth of foundation, psi per inch.

A plate on an elastic foundation that is subjected to uniform pressure will settle uniformly without developing any bending moments. If a support is placed at the edge of such a plate, then bending moments are developed because of the nonuniform settlement caused by the boundary condition.

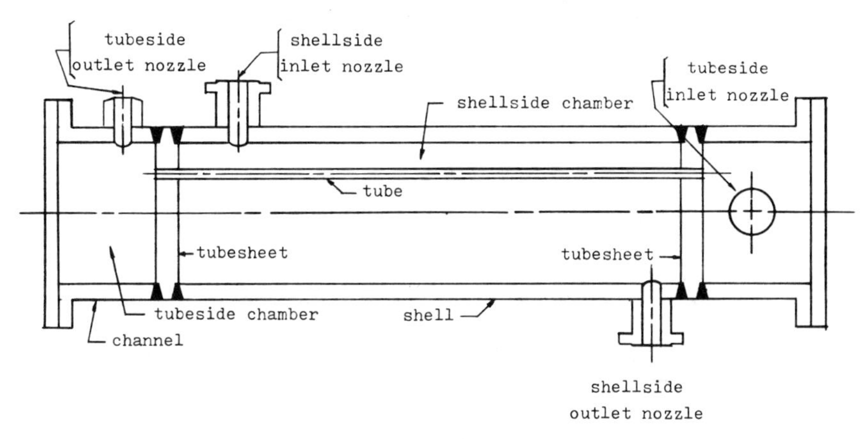

Figure 3-13.

Accordingly, we can investigate the effects of the various boundary conditions on the plate stress by allowing the applied pressure to be set to zero. Letting

$$\alpha = \sqrt[4]{k_o/D}$$

the differential equation becomes

$$\left(\frac{d^2}{dr^2} + \frac{1}{r}\frac{d}{dr}\right)\left(\frac{d^2}{dr^2} + \frac{1}{r}\frac{d}{dr}\right) w + \alpha^4 w = 0.$$

Let $\alpha r = \sqrt[4]{-1}\rho$.
Then the differential equation becomes

$$\nabla^4 w - w = 0 \tag{3-35}$$

where

$$\nabla^4 = \left(\frac{d^2}{d\rho} + \frac{1}{\rho}\frac{d}{d\rho}\right)^2.$$

Equation (3-35) can be written either as

$$\nabla^2(\nabla^2 w + w) - (\nabla^2 w + w) = 0$$

or

$$\nabla^2(\nabla^2 w - w) + (\nabla^2 w - w) = 0.$$

Hence, the solution is a combination of

$$\nabla^2 w + w = 0$$

and

$$\nabla^2 w - w = 0.$$

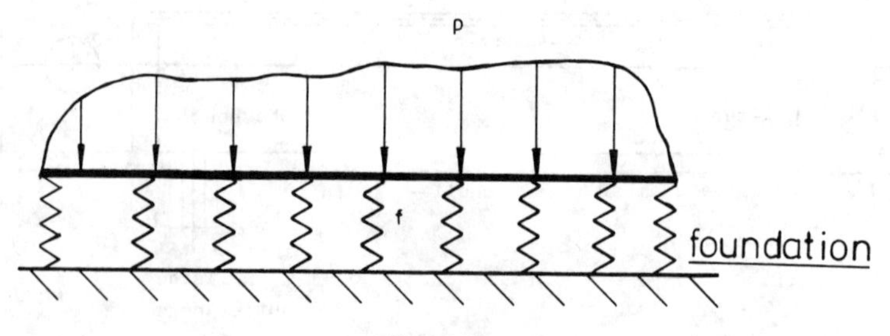

Figure 3-14.

The first equation can be written as

$$\frac{d^2w}{d\rho^2} + \frac{1}{\rho}\frac{dw}{d\rho} + w = 0$$

and the solution is expressed in terms of Bessel function, see Appendix B, as

$$w = A_1J_o(\rho) + A_2Y_o(\rho).$$

The second equation has a solution in the form of

$$w = A_3J_o(i\rho) + A_4Y_o(i\rho).$$

Hence, the total solution is written as

$$w = A_1J_o(\pm\alpha r\sqrt{i}) + A_2Y_o(\pm\alpha r\sqrt{i}) + A_3J_o(\pm\alpha r\sqrt{-i}) + A_4Y_o(\pm\alpha r\sqrt{-i}).$$

This equation can be written as (Hetenyi 1964)

$$w = C_1Z_1(\alpha r) + C_2Z_2(\alpha r) + C_3Z_3(\alpha r) + C_4Z_4(\alpha r) \qquad (3\text{-}36)$$

where the functions Z_1 to Z_4 are modified Bessel functions given in Appendix B.

Example 3-6

A tubesheet in a heat exchanger (Fig. 3-15) is subjected to edge load, Q_o, caused by the difference in expansion between the supporting tubes and the cylindrical shell. Find the expression for the deflection of the tubesheet due to force Q_o if it is assumed simply supported at the edges.

Solution

From Eq. (3-36),

$$w = C_1Z_1(\alpha r) + C_2Z_2(\alpha r) + C_3Z_3(\alpha r) + C_4Z_4(\alpha r). \qquad (1)$$

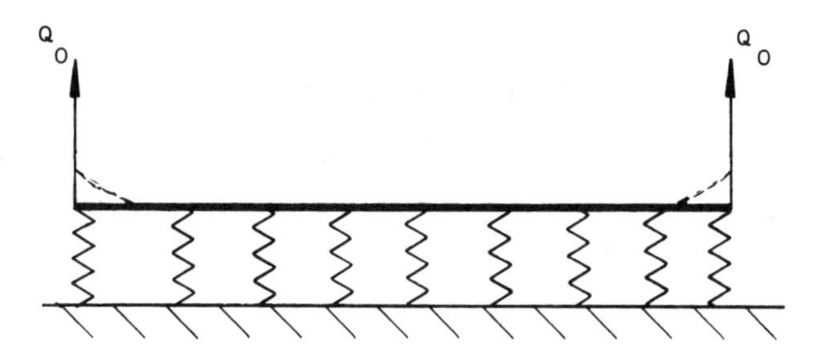

Figure 3-15.

The first constant is determined from the boundary condition at $r = 0$, where the slope dw/dr is equal to zero due to symmetry. Hence,

$$\left.\frac{dw}{dr}\right|_{r=0} = C_1\alpha Z_1'(\alpha r) + C_2\alpha Z_2'(\alpha r) + C_3\alpha Z_3'(\alpha r) + C_4\alpha Z_4'(\alpha r).$$

From Fig. B-3 of Appendix B, the quantity $Z_4'(0)$ approaches infinity as r approaches zero. Hence, C_4 must be set to zero. The second constant is determined from the boundary at $r = 0$ where the shearing force, Q, is zero due to symmetry. The shearing force is expressed as

$$Q = D\left(\frac{d^3w}{dr^3} + \frac{1}{r}\frac{d^2w}{dr^2} - \frac{1}{r^2}\frac{dw}{dr}\right). \tag{2}$$

The derivatives of the first term in Eq. (1) are

$$\frac{dw}{dr} = C_1\alpha Z_1'(\alpha r)$$

$$\frac{d^2w}{dr^2} = C_1\alpha^2 Z_1''(\alpha r)$$

or, from Appendix B,

$$\frac{d^2w}{dr^2} = C_1\left[\alpha^2 Z_2(\alpha r) - \frac{\alpha}{r} Z_1'(\alpha r)\right].$$

The third derivative is

$$\frac{d^3w}{dr^3} = C_1\left[\alpha^3 Z_2'(\alpha r) - \frac{\alpha^2}{r} Z_1''(\alpha r) + \frac{\alpha}{r^2} Z_1'(\alpha r)\right]$$

or, from Appendix B,

$$\frac{d^3w}{dr^3} = C_1\left\{\alpha^3 Z_2'(\alpha r) - \frac{1}{r}\left[\alpha^2 Z_2(\alpha r) - \frac{\alpha}{r} Z_1'(\alpha r)\right] + \frac{\alpha}{r^2} Z_1'(\alpha r)\right\}.$$

Substituting these expressions into Eq. (2) yields

$$Q = D[C_1\alpha^3 Z_2'(\alpha r)].$$

The derivatives of Z_2 and Z_3 in Eq. (1) are similar to those for Z_1. Thus, the total expression for Q in Eq. (2) becomes

$$Q = D[C_1\alpha^3 Z_2'(\alpha r) - C_2\alpha^3 Z_1'(\alpha r) + C_3\alpha^3 Z_4'(\alpha r)].$$

At $r = 0$, $Q = 0$ due to symmetry. From Fig. 8-3 of Appendix B, Z_1' and

Z_2' have a finite value at $r = 0$ while Z_4' approaches ∞. Hence, C_3 must be set to zero. At the boundary condition $r = a$ we have

$$M_r|_{r=a} = 0 = -D\left(\frac{d^2w}{dr^2} + \frac{\mu}{r}\frac{dw}{dr}\right) \tag{3}$$

and

$$Q|_{r=a} = Q_o = D\left(\frac{d^3w}{dr^3} + \frac{1}{r}\frac{d^2w}{dr^2} - \frac{1}{r^2}\frac{dw}{dr}\right). \tag{4}$$

Substituting Eq. (1) into Eq. (3) yields

$$C_2 = C_1 \frac{\alpha Z_2(\alpha a) - \dfrac{1-\mu}{a} Z_1'(\alpha a)}{\alpha Z_1(\alpha a) + \dfrac{1-\mu}{a} Z_2'(\alpha a)}. \tag{5}$$

And from Eqs. (4) and (5) we get

$$C_1 = \frac{-Q_o\alpha}{k_oF}\left[Z_1(\alpha a) + \frac{1-\mu}{\alpha a} Z_2'(\alpha a)\right]$$

where

$$F = Z_1(\alpha a)Z_2'(\alpha a) - Z_1'(\alpha a)Z_2(\alpha a) + \frac{1-\mu}{\alpha a}[Z_1'^2(\alpha a) + Z_2'^2(\alpha a)].$$

Substituting the expression for C_1 into Eq. (5) gives

$$C_2 = \frac{-Q_o\alpha}{k_oF}\left[Z_2(\alpha a) - \frac{1-\mu}{\alpha a} Z_1'(\alpha a)\right].$$

With C_1 and C_2 known, and $C_3 = C_4 = 0$, Eq. (1) can be solved for moments and shears throughout the plate.

Problems

3-13 Show that the maximum deflection of a circular plate on an elastic foundation subjected to a concentrated load, F, in the center is given by the following expression when the radius of the plate is assumed infinitely large:

$$w_{\max} = \frac{F}{8D\alpha^2}.$$

3-14 What are the values of C_1 and C_2 in Example 3-6 if the shear force Q_o is replaced by a bending moment M_o?

3-5 Plates with Variable Boundary Conditions

In many structures such as large oil storage tanks, the surface pressure above the contents causes an uplift force that is transferred to the cylindrical shell as shown in Fig. 3-16a. This force is normally transferred to the foundation through the anchor bolts. Many tanks, however, are not anchored to the foundation, especially in earthquake zones, to avoid damage to the tanks and their attachments. In such cases the uplift force due to surface pressure and earthquake loads is transferred to the base plate as shown in Fig. 3-16b. The edge of the plate tends to lift up and the rest of the plate is kept in place by the pressure of the tank contents. The solution for the deflection of such a plate is obtained from Eqs. (3-29) and (3-30) as

$$w = A + Br^2 + C \ln r + Fr^2 \ln r + \frac{pr^4}{64D} \tag{3-37}$$

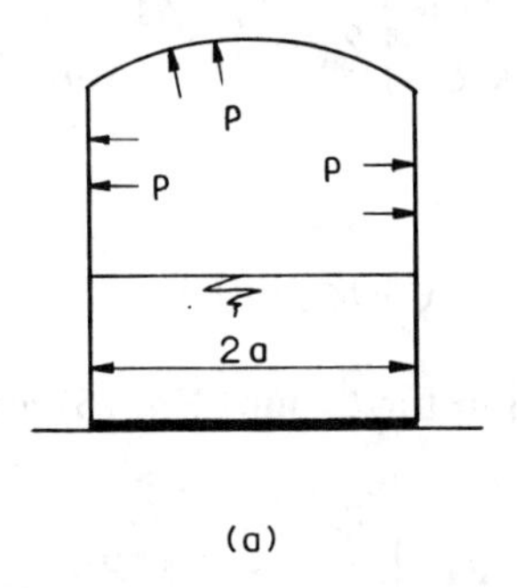

(a)

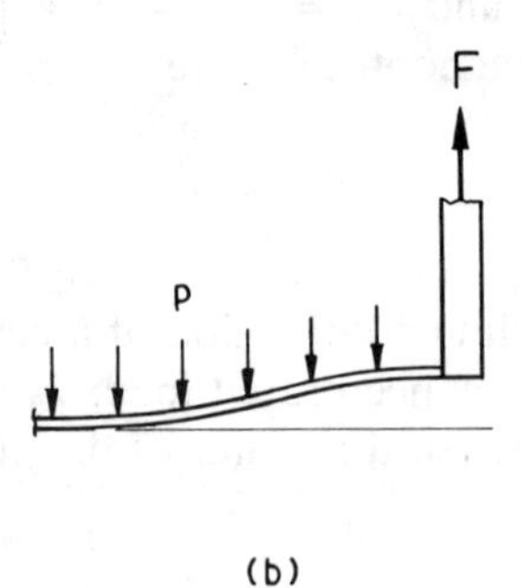

(b)

Figure 3-16.

where the constants A, B, C, and F are determined from the boundary conditions.

At $r = a$, two boundary conditions can be specified. The first is the uplift force Q_o. The other boundary condition is obtained by specifying either M_r or θ. The other boundary conditions are obtained from Fig. 3-17 by assuming an unknown dimension $r = b$ at which the following boundary conditions are satisfied.

$$w = 0, \qquad dw/dr = 0, \qquad M_r = 0.$$

These three boundary conditions plus the two at $r = a$ are used to solve the unknowns b, A, B, C, and F. As the constants A, B, C, and F are a functions of b, the five equations obtained from the boundary conditions cannot be solved directly. A practical solution, however, can easily be obtained by writing a small computer program that increments various values of b until a solution is obtained.

Example 3-7

The tank shown in Fig. 3-18 is subjected to an earthquake motion that results in an upward force at the cylinder-to-plate junction of 234 lbs/inch. Determine the maximum stress in the bottom plate and the maximum uplift. Assume the shell-to-plate junction to have zero rotation as the cylindrical shell is substantially thicker than the base plate. Let $\mu = 0.3$ and $E = 29{,}000$ ksi.

Solution

At $r = b$ the deflection, w, is zero and Eq (3-37) becomes

$$A + Bb^2 + C \ln b + Fb^2 \ln b = \frac{-pb^4}{64D}. \tag{1}$$

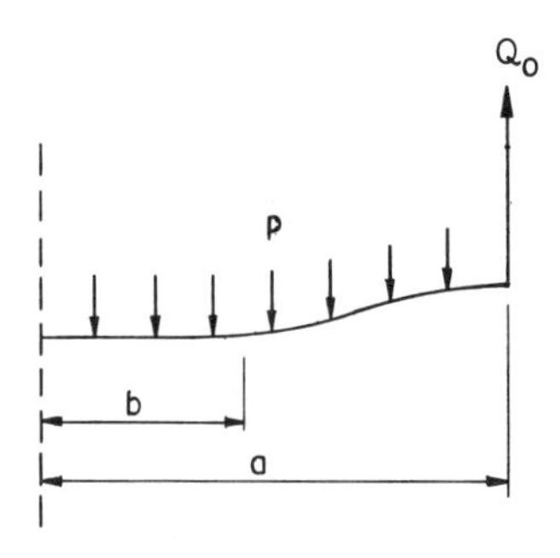

Figure 3-17.

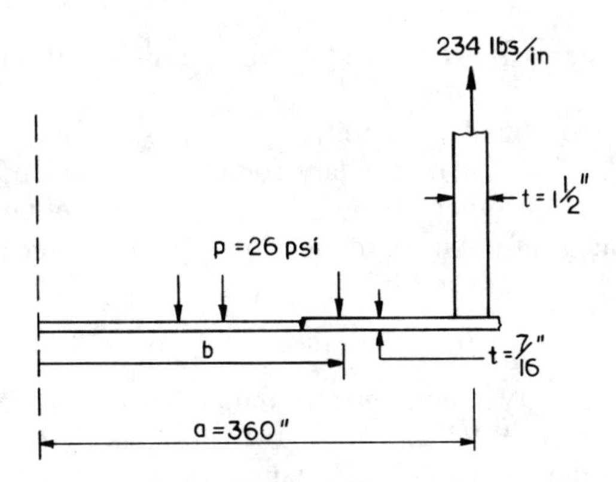

Figure 3-18.

At $r = b$ the slope $dw/dr = 0$. Equation (3-37) gives

$$2Bb + C/b + Fb(2 \ln b + 1) = \frac{-4pb^3}{64D}. \tag{2}$$

At $r = b$, $M_r = 0$ and

$$M_r = -D\left(\frac{d^2w}{dr^2} + \frac{\mu}{r}\frac{dw}{dr}\right) = 0.$$

Substituting w into this equation gives

$$3.7143Bb^2 - C - Fb^2(3.7143 \ln b + 4.7143) + \frac{4.7143pb^4}{16D} = 0. \tag{3}$$

at $r = a$, $dw/dr = 0$ and

$$2Ba + C/a + Fa(2 \ln a + 1) = \frac{-4pa^3}{64D}. \tag{4}$$

At $r = a$, $Q = Q_o$ and

$$Q_o = D\left(\frac{d^3w}{dr^3} + \frac{1}{r}\frac{d^2w}{dr^2} - \frac{1}{r^2}\frac{dw}{dr}\right),$$

which gives

$$F = \frac{Q_o a}{4D} - \frac{pa^2}{8D}. \tag{5}$$

Combining Eqs. (2), (3), (4), and (5) results in an equation that relates the unknown quantity b to the known loads Q_o and p. By placing all terms on one side of the equation, a computer program can be written to increment the quantity b until a solution is found that satisfies this equation. Using such a program results in a value of b of 346.2 inches. With this dimension known, the constants A, B, C, and F can now be determined and are given by

$$A = 1.1347 \times 10^{11} \qquad B = 2.5450 \times 10^{6}$$
$$C = -2.4615 \times 10^{10} \qquad F = -4.0014 \times 10^{5}.$$

The maximum moment, M_r, is found to be 787.9 inch-lbs/inch and the maximum deflection at the edge as 0.02 inch.

$$\text{maximum stress in bottom plate} = 6M/t^2 = 6 \times 787.9/0.4375^2$$
$$= 24{,}700 \text{ psi}.$$

Problem

3-15 Find the stress in the bottom floor plate of the oil tank shown in Fig. P3-15. Let $E = 30{,}000$ ksi and $\mu = 0.30$.

3-6 Design of Circular Plates

The references cited in Sections 1-9 and 2-5 for rectangular plates also contain numerous tables for calculating maximum stress and deflection in circular plates of uniform thickness subjected to various loading and boundary conditions. For concrete slabs, extra precaution must be given to placement of reinforcing bars as discussed in Example 3-1.

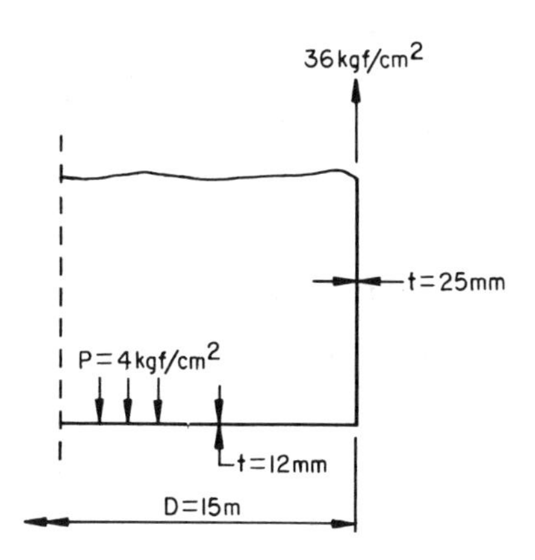

Figure P3-15.

Table 3-1. Circular plates of uniform thickness

Case Number	Maximum Values

1.

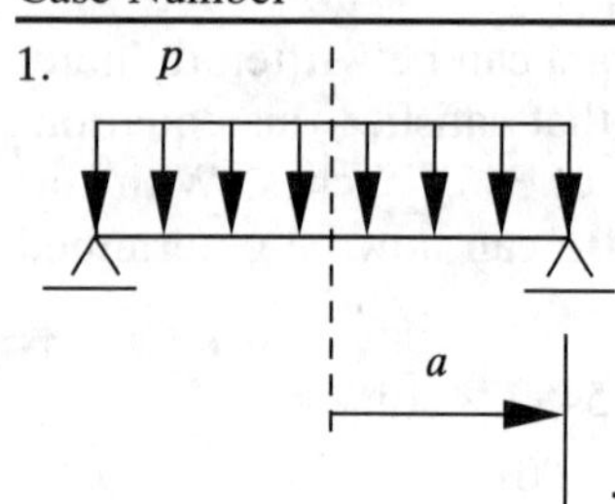

$$S_r = S_t = \frac{3pa^2}{8t^2}(3 + \mu) \quad \text{at center}$$

$$w = \frac{3pa^4(1 - \mu)(5 + \mu)}{16Et^3} \quad \text{at center}$$

$$\theta = -\frac{3pa^3(1 - \mu)}{2Et^3} \quad \text{at edge}$$

For $\mu = 0.3$,

$$S_r = S_t = \frac{1.238pa^2}{t^2} \quad \text{at center}$$

$$w = \frac{0.696pa^4}{Et^3} \quad \text{at center}$$

$$\theta = -\frac{1.050pa^3}{Et^3} \quad \text{at edge}$$

2.

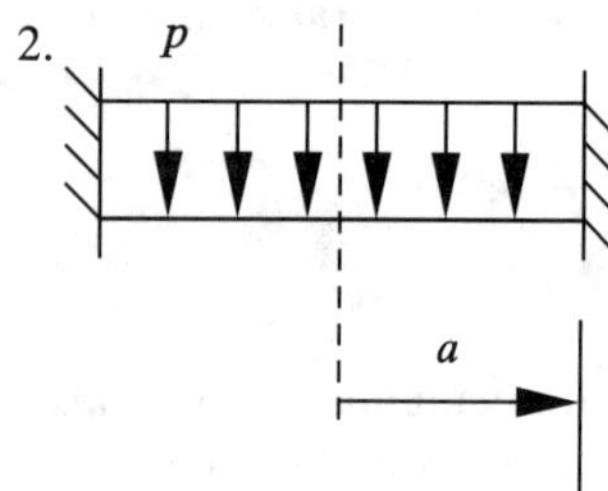

$$S_r = -\frac{3pa^2}{4t^2} \quad \text{at edge}$$

$$w = \frac{3pa^4(1 - \mu^2)}{16Et^3} \quad \text{at center}$$

For $\mu = 0.3$

$$S_r = -\frac{3pa^2}{4t^2} \quad \text{at edge}$$

$$w = \frac{0.171pa^4}{Et^3} \quad \text{at center}$$

3.

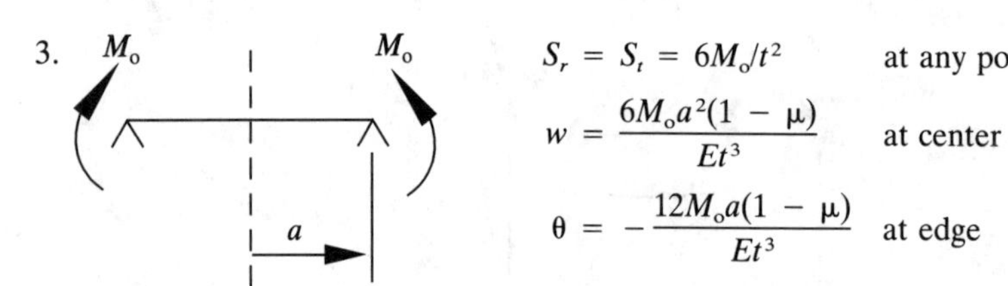

$$S_r = S_t = 6M_o/t^2 \quad \text{at any point}$$

$$w = \frac{6M_oa^2(1 - \mu)}{Et^3} \quad \text{at center}$$

$$\theta = -\frac{12M_oa(1 - \mu)}{Et^3} \quad \text{at edge}$$

For $\mu = 0.3$

$$S_r = S_t = 6M_o/t^2 \quad \text{at any point}$$

$$w = \frac{4.20M_oa^2}{Et^3} \quad \text{at center}$$

$$\theta = -\frac{8.40M_oa}{Et^3} \quad \text{at edge}$$

Notation: a = outside radius of plate; b = radius of applied load; E = modulus of elasticity; p = applied load; r = radius; S_t = tangential stress; S_r = radial stress; t = thickness; w = deflection; θ = rotation, radians; μ = Poisson's ratio.

Circular plates are used as end closures in many shell structures such as reactors, heat exchangers, and distillation towers. Discussion of the interaction of circular plates with various shells will be discussed in later chapters. Table 3-1 lists a few loading conditions that will be utilized later when the interaction of plate and shell components is considered.

The deflection and stress in perforated circular plates are obtained from the theoretical analysis of solid plates modified to take into consideration the effect of the perforations. One procedure that is commonly used is given in the ASME Code, Section VIII-2, Appendix 4. The code uses equivalent values of Poisson's ratio and modulus of elasticity in the theoretical equations for the deflection and stress of solid plates to obtain approximate values for perforated plates. The equivalent values are functions of the pitch and diameter of the perforations. The procedure is based, in part, on O'Donnell's work (O'Donnell and Langer, 1962).

The design of heat exchangers is based on Eq. (3-36) and its solution as shown in Example 3-6. Many codes and standards such as ASME-VIII and TEMA (TEMA 1988) simplify the solution to a set of curves and equations suitable for design purposes.

4

Approximate Analysis of Plates

4-1 The Strain Energy (Ritz) Method

The classical theories discussed in the previous chapters are cumbersome, if not impossible, to solve when the geometry, boundary condition, or load distribution becomes more complicated. Other approximate methods can be utilized to solve such problems. These include the strain energy method and the yield line theory.

The strain energy Method is frequently utilized in solving rectangular plates with intermediate supports and partial loads. We begin the derivation by stating that the strain energy, U, of the infinitesimal element shown in Fig. 4-1a due to applied stress is obtained from the expression

$$\text{strain energy} = \text{force} \times \text{deflection}.$$

The strain energy, U, due to stress σ_x acting on surfaces $ABCD$ and $A'B'C'D'$ in Fig. 4-1a is given by

$$dU = \int_V \sigma_x \, d\left(u + \frac{du}{dx}\,dx\right) dy\,dz - \int_V \sigma_x \, du\,dy\,dz$$

$$dU = \int_V \sigma_x \, d\left(\frac{du}{dx}\right) dx\,dy\,dz$$

where

u = deflection in the x-axis as shown in Fig. 4-1b;
U = strain energy of a solid element.

By defining

$$\varepsilon_x = \frac{du}{dx}$$

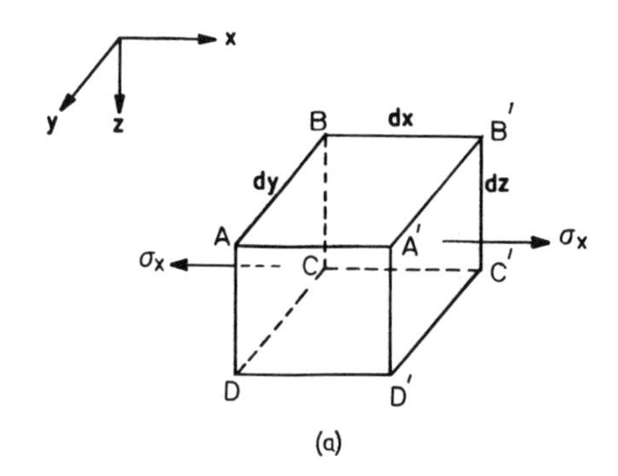

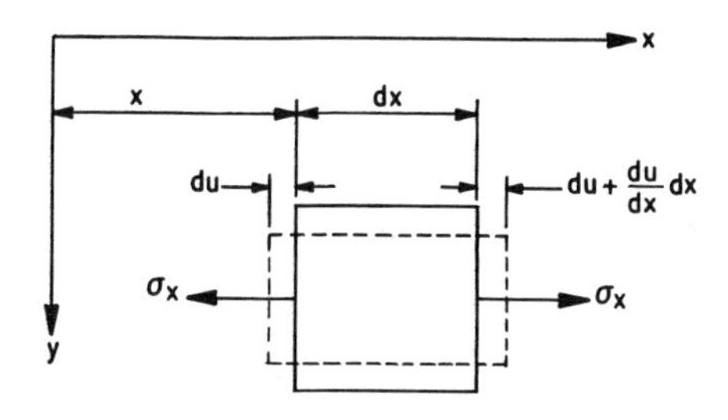

Figure 4-1.

and

$$\varepsilon_x = \frac{\sigma_x}{E}$$

the expression for strain energy becomes

$$dU = \int_V \sigma_x \, d\left(\frac{\sigma_x}{E}\right) dx \, dy \, dz$$

$$dU = \frac{\sigma_x^2}{2E} \, dx \, dy \, dz$$

$$dU = \frac{1}{2} \sigma_x \varepsilon_x \, dx \, dy \, dz.$$

Summation of the strain energy due to σ_x, σ_y, and σ_z results in

$$U = \int_V \frac{1}{2} (\sigma_x \varepsilon_x + \sigma_y \varepsilon_y + \sigma_z \varepsilon_z) \, dx \, dy \, dz.$$

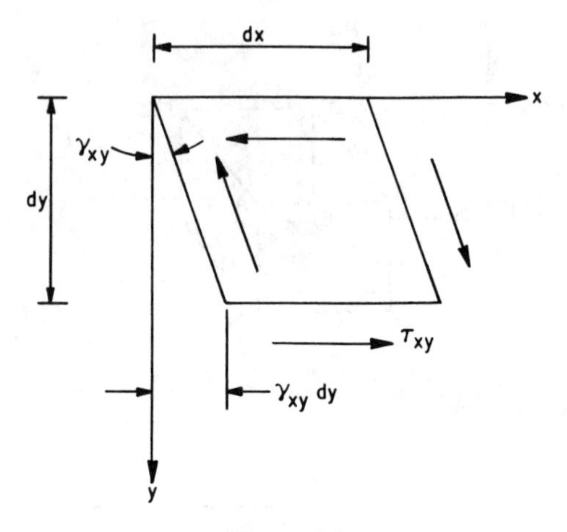

Figure 4-2.

The strain energy due to shearing stress τ_{xy} is obtained from Fig. 4-2 as

$$dU = \frac{1}{2}(\tau_{xy}\, dx\, dz)(\gamma_{xy}\, dy)$$

$$U = \int_V \frac{1}{2}\tau_{xy}\gamma_{xy}\, dx\, dy\, dz.$$

Hence, the total strain energy due to σ_x, σ_y, σ_z, τ_{xy}, τ_{yz}, and τ_{xz} is given by

$$U = \frac{1}{2}\int_V (\sigma_x\varepsilon_x + \sigma_y\varepsilon_y + \sigma_z\varepsilon_z + \tau_{xy}\gamma_{xy}\ \tau_{yz}\gamma_{yz} + \tau_{xz}\gamma_{xz})\, dx\, dy\, dz.$$

For thin plates,

$$\sigma_z = \tau_{yz} = \tau_{xz} = 0$$

and the strain energy expression reduces to

$$U\frac{1}{2}\int_V (\sigma_x\varepsilon_x + \sigma_y\varepsilon_y + \tau_{xy}\gamma_{xy})\, dx\, dy\, dz.$$

Substituting Eqs. (1-12) and (1-15) into this expression results in

$$U = \frac{D}{2}\int\left\{\left(\frac{\partial^2 w}{\partial x^2} + \frac{\partial^2 w}{\partial y^2}\right)^2 - 2(1-\mu)\left[\frac{\partial^2 w}{\partial x^2}\frac{\partial^2 w}{\partial y^2} - \left(\frac{\partial^2 w}{\partial x\, \partial y}\right)^2\right]\right\} dx\, dy. \quad (4\text{-}1)$$

The external work, w, due to applied loads is given by

$$w = \int pw \, dx \, dy. \tag{4-2}$$

The total potential energy of the system is defined as

$$\Pi = U - w \tag{4-3}$$

where Π is the total potential energy that must be minimized in order for the plate to be in stable equilibrium.

Equation (4-3) can be solved by expressing the deflection, w, in a geometric series. Ritz (McFarland et al. 1972) suggested a series of the form

$$w = C_1 f_1(x, y) + C_2 f_2(x, y) + \ldots . \tag{4-4}$$

where the $f(x, y)$ functions represent the deflection of the plate and satisfy the boundary conditions. The constants C are chosen so as to make Eq (4-3) a minimum. Thus

$$\frac{\partial \Pi}{\partial C_1} = 0, \qquad \frac{\partial \Pi}{\partial C_2} = 0, \qquad \text{etc.} \tag{4-5}$$

Example 4-1

Find the deflection of the simply supported plate shown in Fig. 4-3 due to a uniform pressure p.

Solution

Let the deflection given by Eq. (4-4) be represented by an equation of the form

$$w = \sum_{m=1}^{\infty} \sum_{n=1}^{\infty} A_{mn} \sin \frac{m\pi x}{a} \sin \frac{n\pi y}{b} \tag{1}$$

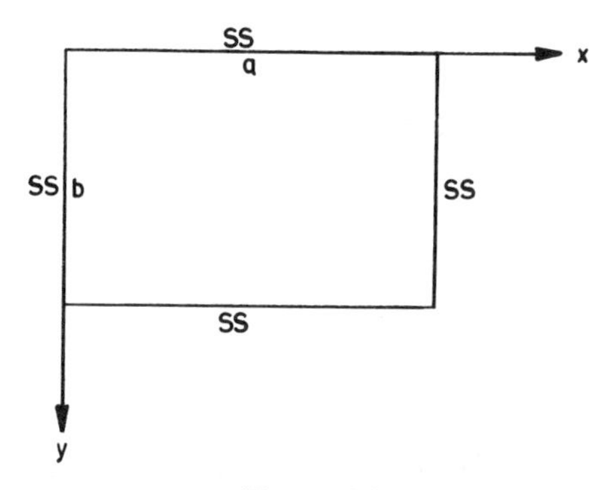

Figure 4-3.

which satisfies the boundary conditions. Substituting this expression into Eq. (4-1) and noting that

$$2(1 - \mu)\left[\frac{\partial^2 w}{\partial x^2}\frac{\partial^2 w}{\partial y^2} - \left(\frac{\partial^2 w}{\partial x\,\partial y}\right)^2\right] = 0$$

gives

$$U = \frac{D}{2}\int_0^b\int_0^a\left[\sum_{m=1}^{\infty}\sum_{n=1}^{\infty} A_{mn}\left(\frac{m^2\pi^2}{a^2} + \frac{n^2\pi^2}{b^2}\right)\sin\frac{m\pi x}{a}\sin\frac{n\pi y}{b}\right]^2 dx\,dy$$

$$U = \frac{ab}{8}D\sum_{m=1}^{\infty}\sum_{n=1}^{\infty} A_{mn}^2\left(\frac{m^2\pi^2}{a^2} + \frac{n^2\pi^2}{b^2}\right)^2. \tag{2}$$

Similarly Eq. (4-2) becomes

$$\begin{aligned} W &= \int_0^b\int_0^a\sum_{m=1}^{\infty}\sum_{n=1}^{\infty} pA_{mn}\sin\frac{m\pi x}{a}\sin\frac{n\pi y}{b}\,dx\,dy \\ &= \sum_{m=1}^{\infty}\sum_{n=1}^{\infty} pA_{mn}\frac{ab}{\pi^2 mn}(\cos m\pi - 1)(\cos n\pi - 1). \end{aligned} \tag{3}$$

Substituting Eqs. (2) and (3) into Eq. (4-3) gives

$$\begin{aligned} \Pi = ab\sum_{m=1}^{\infty}\sum_{n=1}^{\infty}\Bigg[&\frac{D}{8}A_{mn}^2\left(\frac{m^2\pi^2}{a^2} + \frac{n^2\pi^2}{b^2}\right)^2 \\ &- \frac{p}{\pi^2}\frac{A_{mn}}{mn}(\cos m\pi - 1)(\cos n\pi - 1)\Bigg] \end{aligned} \tag{4}$$

From Eq. (4-5),

$$\frac{\partial \Pi}{\partial A_{mn}} = 0$$

and

$$A_{mn} = \frac{4p(\cos m\pi - 1)(\cos n\pi - 1)}{D\pi^6 mn\left(\dfrac{m^2}{a^2} + \dfrac{n^2}{b^2}\right)^2}.$$

Equation (1) for the deflection becomes

$$w = \sum_{m=1}^{\infty}\sum_{n=1}^{\infty}\left(\frac{4p(\cos m\pi - 1)(\cos n\pi - 1)}{D\pi^6 mn\left(\dfrac{m^2}{a^2} + \dfrac{n^2}{b^2}\right)^2}\right)\sin\frac{m\pi x}{a}\sin\frac{n\pi y}{b}$$

or

$$w = \sum_{m=1,3,\ldots}^{\infty} \sum_{n=1,3,\ldots}^{\infty} \frac{16p}{D\pi^6 mn \left(\dfrac{m^2}{a^2} + \dfrac{n^2}{b^2}\right)^2} \sin\frac{m\pi x}{a} \sin\frac{n\pi y}{b}.$$

The solution of rectangular plates by the Ritz method consists of defining an expression in the form of Eq. (4-4) that satisfies all of the boundary conditions. A modification of the Ritz method with Lagrange multipliers, is used to solve plates with various boundary conditions such as intermediate supports where the expression in Eq. (4-4) cannot satisfy all of the boundary conditions. To begin the derivation let us assume that Eq. (4-4) satisfies all of the boundary conditions except two. Let these two boundary conditions be defined as

$$B_1(C_1, C_2, \ldots) = 0$$

and

$$B_2(C_1, C_2, \ldots) = 0.$$

Because B_1 and B_2 are the reaction constraints, they must be added to the total work and Eq (4-3) becomes

$$\Pi = U - w + K_1 B_1 + K_2 B_2 \tag{4-6}$$

where K_1 and K_2 are constants. Minimizing Eq. (4-6) results in the following simultaneous equations

$$\frac{\partial \Pi}{\partial C_1} = 0 = \frac{\partial U}{\partial C_1} - \frac{\partial w}{\partial C_1} + K_1 \frac{\partial B_1}{\partial C_1} + K_2 \frac{\partial B_2}{\partial C_1}$$

$$\frac{\partial \Pi}{\partial C_2} = 0 = \frac{\partial U}{\partial C_2} - \frac{\partial w}{\partial C_2} + K_1 \frac{\partial B_1}{\partial C_2} + K_2 \frac{\partial B_2}{\partial C_2}$$

$$\vdots$$

$$\frac{\partial \Pi}{\partial K_1} = 0 = B_1$$

$$\frac{\partial \Pi}{\partial K_2} = 0 = B_2.$$

Solution of these simultaneous equations yields the expression for the deflection.

Example 4-2

The plate shown in Fig. 4-4 is simply supported at the edges and is also supported by a column at $x = a/2$ and $y = b/2$. Find the deflection due to a uniform pressure p.

Solution

Let the deflection be given by

$$w = \sum_{m=1}^{\infty} \sum_{n=1}^{\infty} A_{mn} \sin \frac{m\pi x}{a} \sin \frac{n\pi y}{b}. \tag{1}$$

Then from Eq. (4) of Example 4-1

$$U - W = ab \sum_{m=1}^{\infty} \sum_{n=1}^{\infty} \left[\frac{D}{8} A_{mn}^2 \left(\frac{m^2\pi^2}{a^2} + \frac{n^2\pi^2}{b^2} \right)^2 - \frac{p}{\pi^2} \frac{A_{mn}}{mn} (\cos m\pi - 1)(\cos n\pi - 1) \right]. \tag{2}$$

As Eq. (1) for the deflection is not satisfied at $x = a/2$ and $y = b/2$, it follows that a constraint equation mut be expressed as

$$\sum_{m=1}^{\infty} \sum_{n=1}^{\infty} A_{mn} \sin \frac{m\pi}{2} \sin \frac{n\pi}{2} = 0$$

and the total energy of the system is

$$\Pi = U - W + K_1 \sum_{m=1}^{\infty} \sum_{n=1}^{\infty} A_{mn} \sin \frac{m\pi}{2} \sin \frac{n\pi}{2}$$

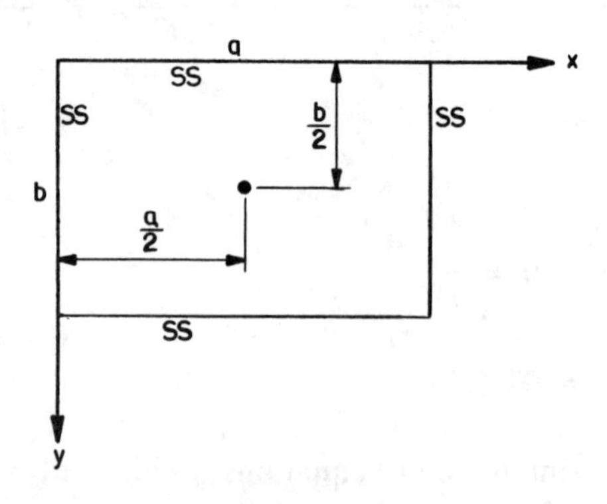

Figure 4-4.

where $U - W$ is given by Eq. (2). Minimizing Π with respect to A_{mn} gives

$$\frac{\partial \Pi}{\partial A_{mn}} = 0 = +\frac{Dab\pi^4}{4}[(m/a)^2 + (n/b)^2]^2 A_{mn} - \frac{pab(\cos m\pi - 1)(\cos n\pi - 1)}{\pi^2 mn} + K_1 \sin\frac{m\pi}{2}\sin\frac{n\pi}{2}$$

$$A_{mn} = \frac{4pa^4(\cos m\pi - 1)(\cos n\pi - 1)}{D\pi^6 mn[m^2 + n^2(a/b)^2]^2} + \frac{4a^3 K_1}{D\pi^4 b[m^2 + n^2(a/b)^2]^2}\sin\frac{m\pi}{2}\sin\frac{n\pi}{2}$$

$$\frac{\partial \Pi}{\partial K_1} = 0 = \sum_{m=1}^{\infty}\sum_{n=1}^{\infty} A_{mn}\sin\frac{m\pi}{2}\sin\frac{n\pi}{2}.$$

Substituting A_{mn} into this equation and solving for K_1 gives

$$K_1 = \frac{pab}{\pi^2}\frac{A_1}{A_2}$$

where

$$A_1 = \sum_{m=1}^{\infty}\sum_{n=1}^{\infty}\frac{(\cos m\pi - 1)(\cos n\pi - 1)}{mn[m^2 + n^2(a/b)^2]^2}\sin\frac{m\pi}{2}\sin\frac{n\pi}{2}$$

$$A_2 = \sum_{m=1}^{\infty}\sum_{n=1}^{\infty}\frac{\sin\frac{m\pi}{2}\sin\frac{n\pi}{2}}{[m^2 + n^2(a/b)^2]^2}$$

and

$$w = \frac{4pa^4}{D\pi^6}\left\{\sum_{m=1}^{\infty}\sum_{n=1}^{\infty}\frac{(\cos m\pi - 1)(\cos n\pi - 1)}{mn[m^2 + n^2(a/b)^2]^2} - \frac{A_1}{A_2}\sum_{m=1}^{\infty}\sum_{n=1}^{\infty}\frac{\sin\frac{m\pi}{2}\sin\frac{n\pi}{2}}{[m^2 + n^2(a/b)^2]^2}\right\}\sin\frac{m\pi x}{a}\sin\frac{n\pi y}{b}.$$

Problems

4-1 Derive Eq. (4-1).

4-2 Find the deflection of the concrete balcony supported as shown in Fig. P4-2. The uniform load is taken as p. Let $a = 72$ inch, $b = 30$ inches,

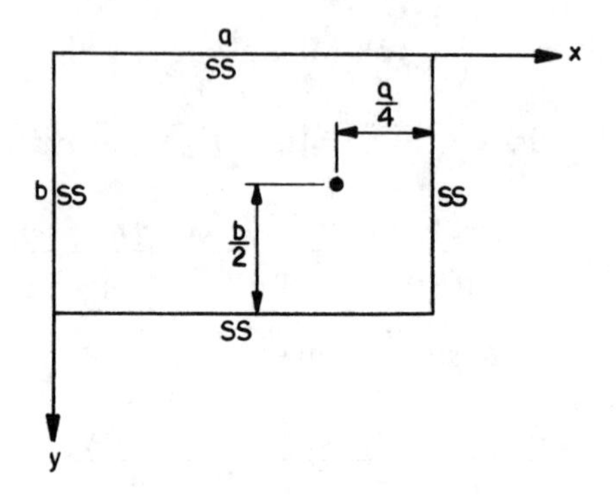

Figure P4-2.

$p = 100$ psf and

$$w = \sum_{m=1}^{\infty} \sum_{n=1}^{\infty} A_{mn} \sin \frac{m\pi x}{a} \sin \frac{n\pi y}{b}.$$

4-3 The steel plate in Fig. P4-3 is part of a shoring box for an earth embankment. Find the deflection due to a uniform earth pressure of 120 psf. Let $a = 60$ inches, $b = 75$ inches, and

$$w = \sum_{m=1}^{\infty} \sum_{n=1}^{\infty} A_{mn} \sin \frac{m\pi x}{a} \sin \frac{n\pi y}{b}.$$

4-2 Yield Line Theory

The yield line theory is a powerful tool for solving many complicated plate problems where an exact elastic solution is impractical to obtain and an

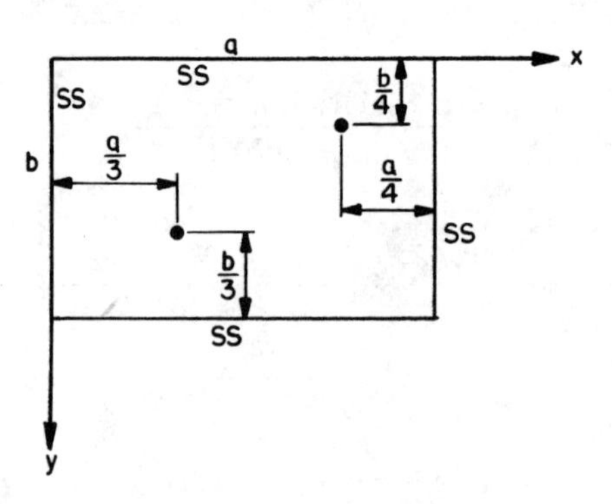

Figure P4-3.

approximate solution is acceptable. It is best suited for plates with free boundary conditions and concentrated loads. The theory is based on the assumption that the stress–strain diagram of the material can be idealized as shown in Fig. 4-5a. At point A on the elastic stress–strain diagrma, the stress distribution in a plate of thickness t is as shown in Fig. 4-5b. For a plate under external moment M, the equilibrium equation for external and internal moments is given by

$$M = (\sigma_y)(t/2)(1/2)(2t/3)$$

or

$$\sigma_y = \frac{6M}{t^2}, \tag{4-7}$$

which is the basic relationship for bending stress of an elastic plate. For design purposes the yield stress is divided by a factor of safety to obtain an allowable stress.

As the load increases, the outer fibers of the plate are strained past the yield point A in Fig. 4-5a. As the strain approaches point B, the stress

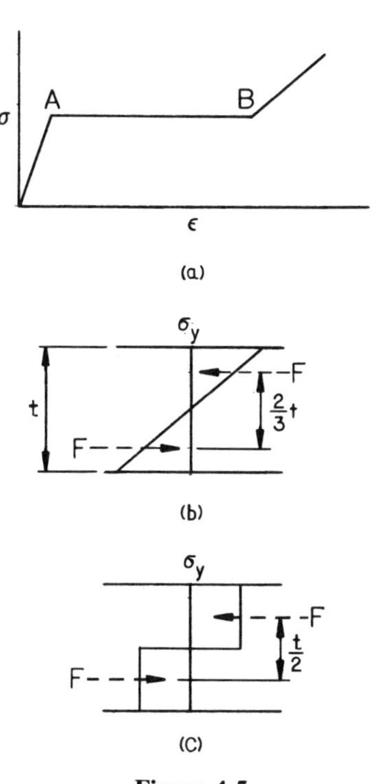

Figure 4-5.

distribution becomes, for all practical purposes, as shown in Fig. 4-5c. Summation of internal and external moments gives

$$M_p = (\sigma_y)(t/2)(t/2)$$

or

$$\sigma_y = \frac{4M_p}{t^2}. \tag{4-8}$$

Equation (4-8) is the basic equation for the bending of a plate in accordance with the plastic theory. For design purposes, the applied loads are multiplied by a load factor to obtain the design loads.

The plastic theory is applied in industry to both metallic plates as well as concrete slabs. Tensile stress in concrete slabs due to plastic moments is resisted by reinforcing bars, and tests conducted on such slabs (Wood 1961) verify the applicability of the plastic theory.

Example 4-3

The maximum bending moment in a circular plate subjected to applied loads is 3000 inch-lbs/inch. Calculate the required thickness using elastic and plastic methods. Use a safety factor of 2.0, a load factor of 2.0, and a yield stress of 36,000 psi.

Solution

(a) Elastic Analysis

$$\sigma_a = \text{allowable stress} = 36{,}000/2.0 = 18{,}000 \text{ psi}$$

Hence,

$$t = \sqrt{\frac{6M}{\sigma_a}} = \sqrt{\frac{6 \times 3000}{18{,}000}} = 1.0 \text{ inch.}$$

(b) Plastic Analysis

$$M_p = 3000 \times 2.0 = 6000 \text{ inch-lbs/inch}$$

and

$$t = \sqrt{\frac{4M_p}{\sigma_y}} = \sqrt{\frac{4 \times 6000}{36{,}000}} = 0.82 \text{ inch.}$$

Hence, a savings of 18% in thickness is obtained by using plastic versus elastic analysis for the same factor of safety.

The yield line theory in plate analysis is very similar to the plastic hinge theory in beam analysis. Application of the plastic hinge theory in beams

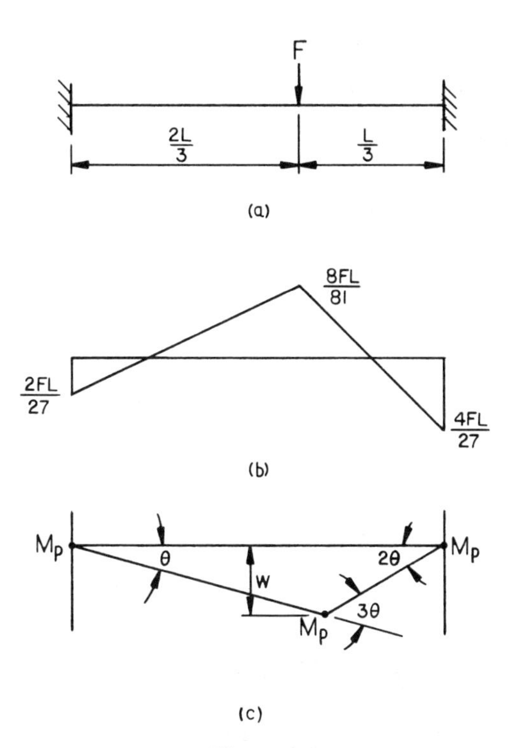

Figure 4-6.

can be illustrated by referring to the fixed beam shown in Fig. 4-6a. The elastic moment diagram due to the concentrated load is shown in Fig. 4-6b. The maximum elastic moment is $4FL/27$ and occurs at the right end support. As the load increases, the moment at the right end reaches M_p and a plastic hinge is developed at that location. However, because the moments at the left end and under the load are below M_p the beam can carry more load because it is still statistically determinate. Eventually the moment under the load reaches M_p and a plastic hinge is developed there also. However, the beam is still stable and more load can be applied to the beam until the moment at the left side reaches M_p and the beam becomes unstable as shown in Fig. 4-6c. At this instance the moment under the load, M_p, is equal to the moment, M_p, at the ends of the span. The magnitude of the moments can be determined by letting the external work equal the internal work. The amount of external work can be expressed as

$$E.W. = (F)(w)$$

while the internal work is expressed as

$$I.W. = M_p(\theta) + M_p(3\theta) + M_p(2\theta).$$

θ can be expressed as

$$\tan\theta = \theta = \frac{w}{(2/3)L}$$

or

$$\theta = \frac{3w}{2L}.$$

Hence, the internal work is given by

$$I.W. = 6M_p\frac{3w}{2L}.$$

Equating external and internal work gives

$$M_p = \frac{FL}{9}.$$

The ratio of M_p to M_e is 0.75. This coupled with the fact that $t_e = \sqrt{6M_e/S}$ while $t_p = \sqrt{4M_p/S}$ results in a net ratio of t_p to t_e of 0.7. Thus, a 30% savings in thickness is achieved by plastic analysis of this beam.

For plate analysis, the plastic hinges become yield lines. Also, axes of rotation develop in plates rather than points of rotation. Some of the properties of yield lines and axes of rotation are

1. In general, yield lines are straight.
2. Axes of rotation of a plate lie along lines of support.
3. Axes of rotation pass over columns.
4. A yield line passes through the intersection of axes of rotation of adjacent plate segments.

Some illustrations of plates with various geometries, supports, and yield lines are shown in Fig. 4-7. It must be noted that the failure mechanism method described here is an upper bound solution and all failure mechanism patterns must be investigated in order to obtain a safe solution. However, the failure mechanisms for the class of problems discussed here have been verified experimentally and can thus be used for design purposes.

Example 4-4

Find the maximum plastic moment in a simply supported square plate subjected to a uniform load of intensity p.

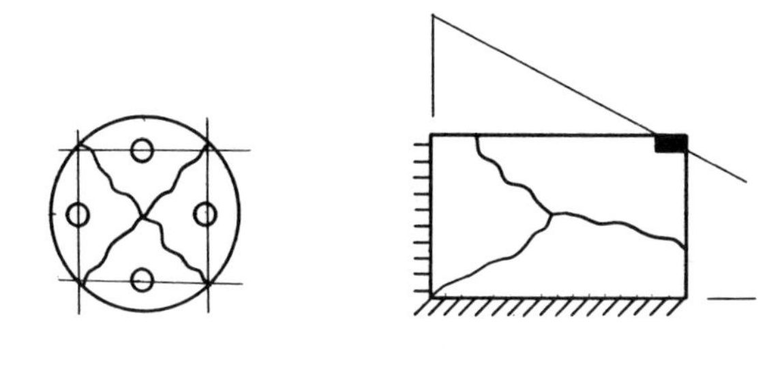

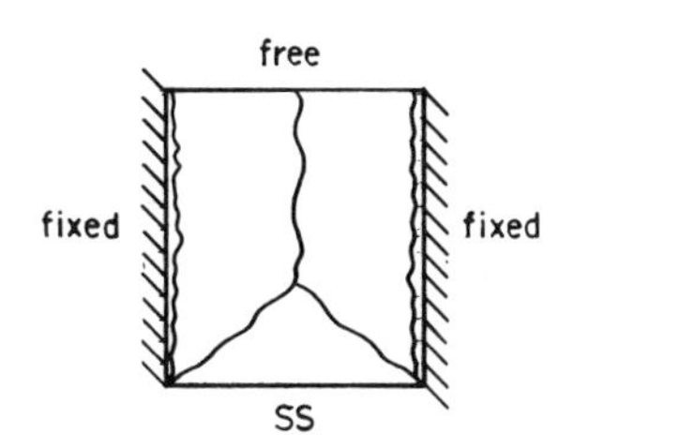

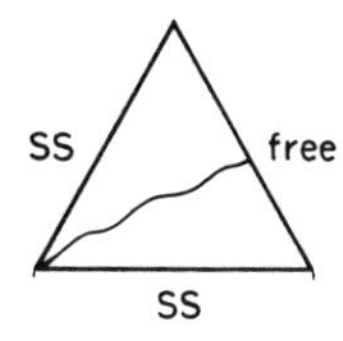

Figure 4-7.

Solution

The collapse mechanism of the square plate (Fig. 4-8a) consists of four yield lines. Section *AA* through the diagonal of the plate (Fig. 4-8b) details the rotation of one of the yield lines. The plastic moment, M_p, at this yield line undergoes a rotation of 2α over the length *od*.

$$\text{external work} = \text{internal work}$$

$$(p)\ (\text{volume of pyramid}) = (M_b)\ (2\alpha)\ (\text{length})\ (4 \text{ yield lines})$$

$$(p)(L^2 \times w/3) = (M_p)(2\alpha)\left(\frac{\sqrt{2}L}{2}\right)(4).$$

For small deflection w, we let

$$\tan \alpha = \alpha = \frac{w}{\sqrt{2}L/2}.$$

We then get

$$M_p = \frac{pL^2}{24}.$$

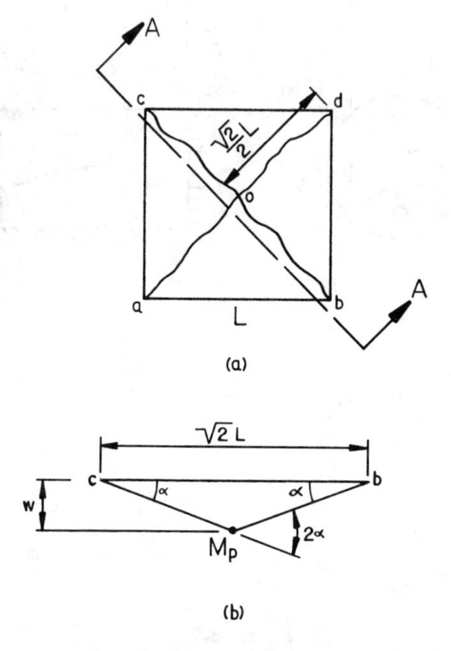

Figure 4-8.

The method discussed in Example 4-4 becomes cumbersome for complicated geometries. A more efficient method of formulating the expression for internal work is to use vector designations. We illustrate the method by referring to the square plate with the failure mechanism shown in Fig. 4-9a. The applied moments in one panel (Fig. 4-9b) can be designated by the vectors given in Fig. 4-9c. The horizontal components of the vector moment (Fig. 4-9d) cancel each other while the vertical vectors are additive.

If we use the approach discussed in Example 4-4, then from sketches (d) and (e) of Fig. 4-9 we get

$$\begin{aligned} I.W. \text{ for one panel} &= 2(M_p^v)\left(\frac{L/2}{\cos\phi}\right)(\theta) \\ &= 2(M_p \cos\phi)\left(\frac{L/2}{\cos\phi}\right)\left(\frac{w}{L/2}\right) \\ &= 2(M_p)w. \end{aligned}$$

This same result can be obtained more efficiently by observing in sketches (d) and (e) of Fig. 4-9 that the product of the quantity M_p times the slanted length is of the same magnitude as the product of the moment M_p applied

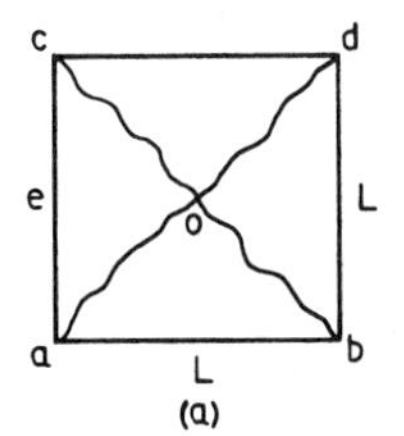

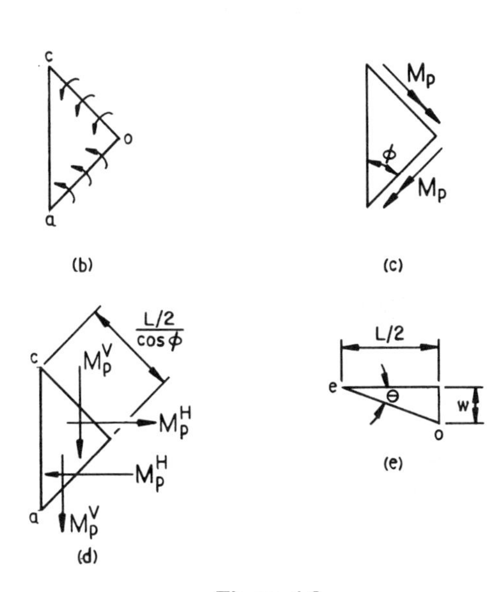

Figure 4-9.

along the edge L. Hence,

$$I.W. \text{ for one panel} = (M_p)(L)(\theta)$$

$$= (M_p)(L)\left(\frac{w}{L/2}\right)$$

$$= 2(M_p)w.$$

As the rotation of the outside edge is obtained more easily than the rotation of the inner yield lines, the vector approach will be used in all subsequent discussions.

Example 4-5

Find the required thickness of a square plate fixed at the edges and subjected to a uniform load of 12 psi. Use a load factor of 2.0, and yield stress of 36 ksi.

Solution

From Fig. 4-10a,

$$E.W. = I.W.$$

$$\left(\frac{1}{3}\right)(20^2)(2.0 \times 12)w = \text{work due to } M_p \text{ at internal yield lines}$$
$$+ \text{ work due to } M_p \text{ at edges}$$

and from Fig. 4-10b,

$$3200\ w = 4(M_p)(L)\left(\frac{w}{L/2}\right) + 4(M_p)(L)\left(\frac{w}{L/2}\right)$$

or,

$$M_p = 200 \text{ inch-lbs/inch.}$$

$$t = \sqrt{\frac{4M_p}{\sigma_y}} = \sqrt{\frac{4 \times 200}{36{,}000}} = 0.15 \text{ inch.}$$

Example 4-6

Find M_p of the rectangular plate in Fig. 4-11a due to uniform load p. The plate is simply supported.

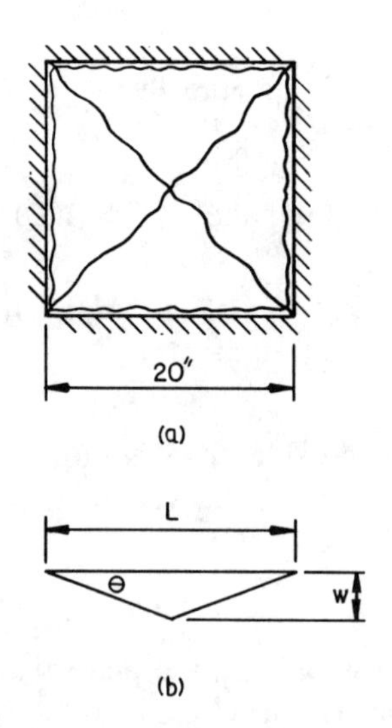

Figure 4-10.

Solution

The yield lines take the shape shown in Fig. 4-11b. Distance x is unknown and must be determined.

$$I.W. = 2(M_p)(a)\left(\frac{w}{x}\right) + 2(M_p)(b)\left(\frac{w}{a/2}\right)$$

From Fig. 4-11c,

$$\begin{aligned} E.W. &= p(E.W.\ \mathrm{I} + E.W.\ \mathrm{II} + E.W.\ \mathrm{III}) \\ &= p\left[2\left(\frac{ax}{2}\right)(1/3) + 2(b - 2x)(a/2)(1/2)\right. \\ &\quad \left. + 4(x)(a/2)(1/2)(1/3)\right]w. \end{aligned}$$

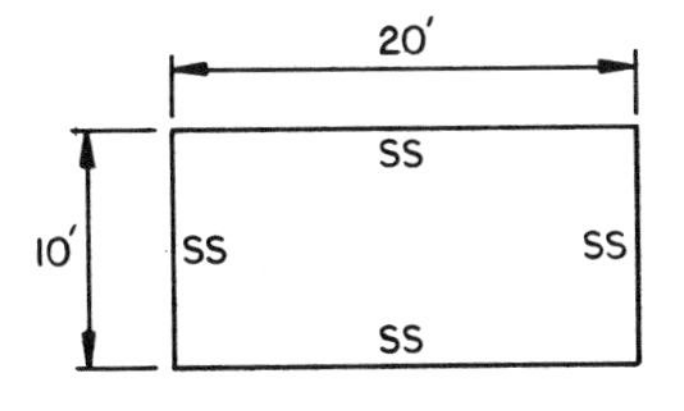

(a)

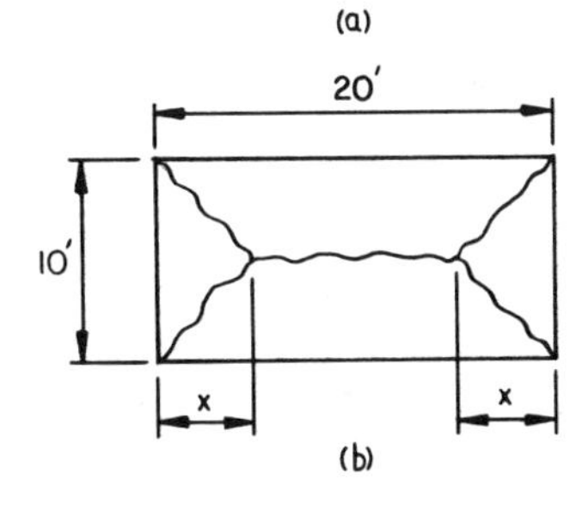

(b)

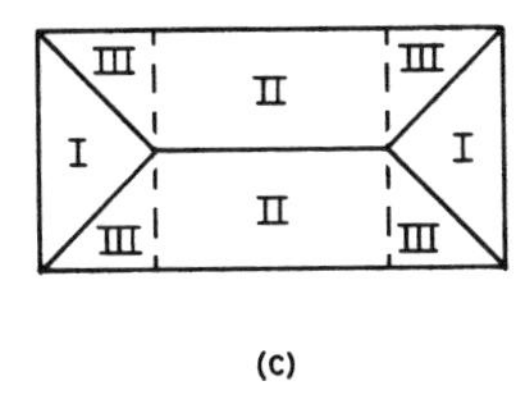

(c)

Figure 4-11.

Equating internal and external work and solving for M_p gives

$$M_p = p\left[\frac{\frac{ab}{2} - \frac{ax}{3}}{2\left(\frac{2b}{a} + \frac{a}{x}\right)}\right]. \tag{1}$$

Minimizing M_p by taking its derivative with respect to x and equating it to zero gives

$$x^2 + \frac{a^2}{b}x - \frac{3a^2}{4} = 0. \tag{2}$$

For $a = 10$ ft, and $b = 20$ ft, Eq. (2) gives $x = 6.51$ ft.
From Eq. (1),

$$\begin{aligned} M_p &= p\left(\frac{14{,}400 - 3124.8}{2(4.00 + 1.54)}\right) \\ &= 1018\ p \text{ inch-lbs/inch.} \end{aligned}$$

Problems

4-4 Find M_p due to a uniform load on the simply supported hexagon plate shown in Fig. P4-4.
4-5 Find M_p in the skewed bridge slab (Fig. P4-5) due to uniform load p.
4-6 Determine M_p in the triangular weir plate shown in Fig. P4-6 due to uniform load p. The plate is fixed at all edges.

4-3 Further Application of the Yield Line Theory

The deflections obtained from the yield line theory are larger than those obtained from the elastic theory due to reduced thicknesses. This should

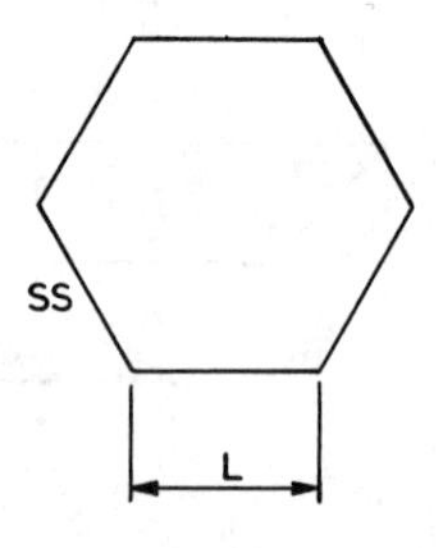

Figure P4-4.

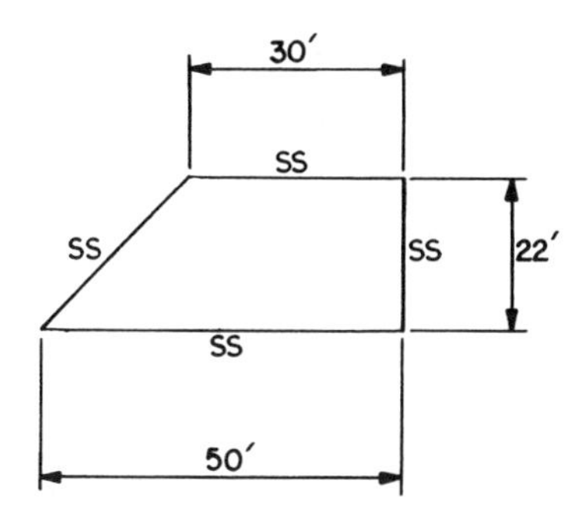

Figure P4-5.

be considered in applications where small deflections are critical to the performance of equipment such as flanges and other sealing components. Also, the yield line theory tends to give an upper bound solution. Accordingly, all possible yield line paths must be investigated in order to obtain a true solution. This is especially important in plates with free edges as illustrated in Example 4-7.

Example 4-7

Find the maximum moment in a triangular section (Fig. 4-12a) subjected to a uniform load p.

Solution

Assume the yield line to be as shown in Fig. 4-12b at an angle a from side A. Then

$$C = a + b$$

$$\begin{aligned} I.W. &= M_p(x \cos a)\frac{w}{x \sin a} + M_p(x \cos b)\frac{w}{x \sin b} \\ &= M_p(\cot a + \cot b)w \\ E.W. &= p\left[\frac{(A)(x \sin a)}{2}\frac{1}{3} + \frac{(B)(x \sin b)}{2}\frac{1}{3}\right]w \\ &= \frac{px}{6}(A \sin a + B \sin b)w. \end{aligned}$$

Equating internal and external work results in

$$M_p = \frac{px}{6}\frac{A \sin a + B \sin b}{\cot a + \cot b}.$$

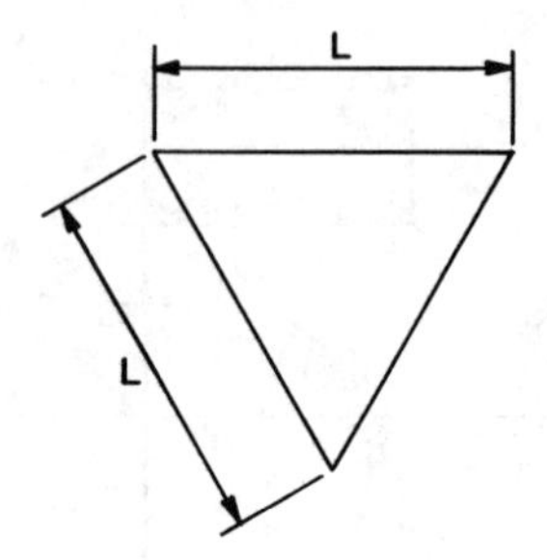

Figure P4-6.

Minimizing M_p with respect to a results in

$$a = C/2$$

which indicates that in any triangle with one edge free, the yield line always bisects the angle between the two simply supported edges. The plastic moment is given by

$$M_p = \frac{px}{6} \frac{A \sin(C/2) + B \sin(C/2)}{\cot(C/2) + \cot(C/2)}.$$

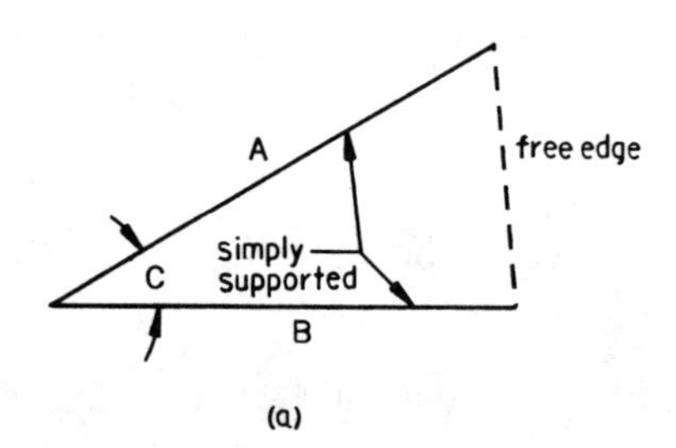

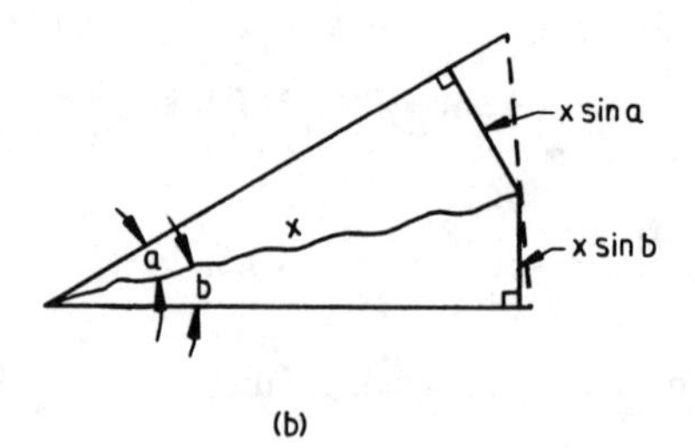

Figure 4-12.

Example 4-8

Find the maximum moment in a uniformly loaded square plate, (Fig. 4-13a) with three sides simply supported and one side free.

Solution

(a) Let the failure pattern be as shown in Fig. 4-13b. Then,

$$I.W. = 2M_p(L)\frac{w}{L/2} + M_p(L)\frac{w}{L-y}$$

$$= \left(4M_p + \frac{M_pL}{L-y}\right)w$$

$$E.W. = p\left[\frac{L(L-y)}{2}\frac{1}{3} + \frac{Ly}{2} + 2(L-y)\frac{L}{2}\frac{1}{2}\frac{1}{3}\right]w$$

$$= p\left[\frac{L(L-y)}{3} + \frac{Ly}{2}\right]w.$$

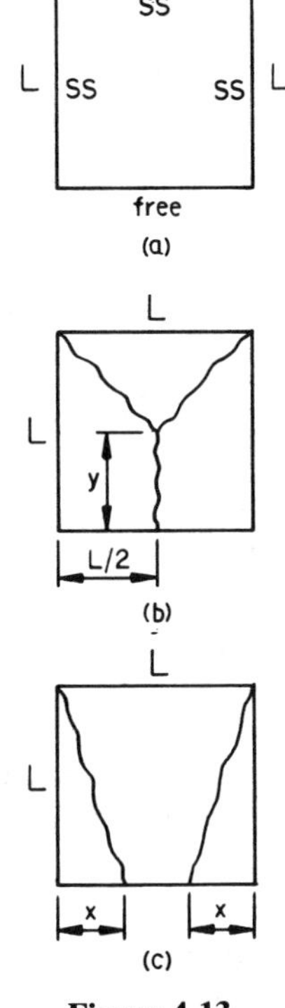

Figure 4-13.

Equating internal and external work gives

$$M_p = \frac{pL}{6}\frac{2L^2 - Ly - y^2}{5L - 4y}.$$

Minimizing M_p with respect to y gives

$$y = 2.15L \quad \text{which is discarded}$$

or

$$y = 0.35L$$

and

$$M_p = \frac{pL^2}{14.14}.$$

(b) Let the failure pattern be as shown in Fig. 4-13c. Then,

$$\begin{aligned} I.W. &= 2M_pL\frac{w}{x} + 2M_px\frac{w}{x} \\ &= \left(\frac{2M_pL}{x} + \frac{2M_px}{L}\right)w \\ E.W. &= p\left[\frac{2Lx}{2}\frac{1}{3} + \frac{2xL}{2}\frac{1}{3} + (L - 2x)L\frac{1}{2}\right]w \\ &= p\left[\frac{2Lx}{3} + \frac{(L - 2x)L}{2}\right]w. \end{aligned}$$

Equating internal and external work results in

$$M_p = \frac{pL}{12}\frac{3L^2x - 2x^2L}{L + x}.$$

Minimizing M_p with respect to x gives

$$x = -1.87L \quad \text{or} \quad x = 0.54L \quad \text{which are impossible.}$$

Thus, the first alternate (a) controls.

The yield line mechanism in circular plates subjected to uniform or concentrated loads have a radial pattern. The failure mode is conical in shape as illustrated in Example 4-9. This circular pattern must also be investigated for rectangular plates under concentrated loads.

Example 4-9

Find the moment in a simply supported circular plate due to (a) uniform load p and (b) concentrated load F in the middle.

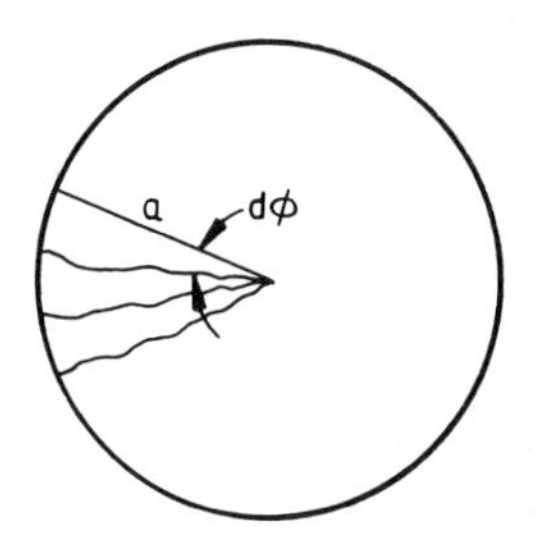

Figure 4-14.

Solution

(a) From Fig. 4-14 using w as the deflection in the middle,

$$I.W. = \int_0^{2\pi} M_p a \, d\phi \, \frac{w}{a}$$
$$= 2\pi M_p w$$
$$E.W. = (\pi a^2 p/3)w$$

and

$$M_p = pa^2/6.$$

(b) Again from Fig. 4-14 using w as the deflection in the middle,

$$E.W. = F(w)$$

and

$$M_p = \frac{F}{2\pi}.$$

It is of interest to note that M_p in this case is independent of the radius.

Example 4-10

Find the moment in a square plate subjected to a concentrated load F in the middle if the plate is (a) simply supported on all sides or (b) fixed on all sides.

Solution

(a) From Figure 4-15a, using a straight line mechanism with a deflection, w, in the middle

$$4(M_p)(a)\frac{w}{a/2} = F(w)$$
$$M_p = 0.125F.$$

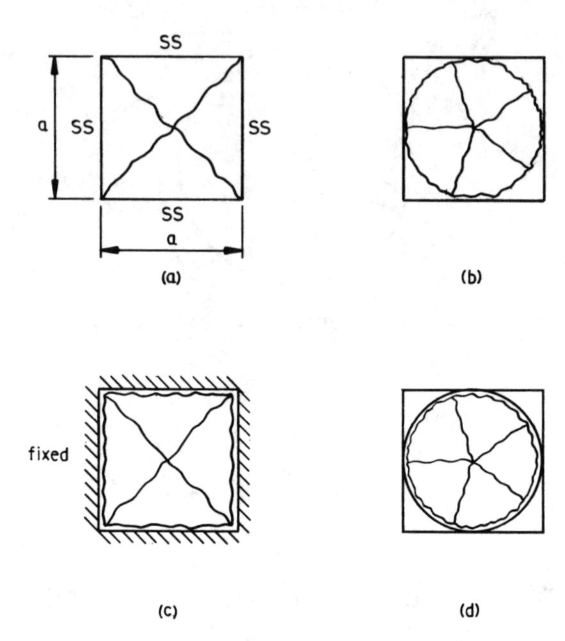

Figure 4-15.

From Figure 4-15b, using a circular mechanism with a deflection, w, in the middle

$$2(2\pi M_p)w = F(w)$$

$$M_p = 0.08F.$$

Hence, the straight line mechanism controls.

(b) From Figure 4-15c, using a straight line mechanism with a deflection, w, in the middle

$$2\left[4(M_p)(a)\frac{1}{a/2}\right]w = F(w)$$

$$M_p = 0.063F.$$

From Figure 4-15d, using a circular mechanism with a deflection, w, in the middle

$$2(2\pi M_p)w = F(w)$$

$$M_p = 0.08F$$

and the circular mechanism controls.

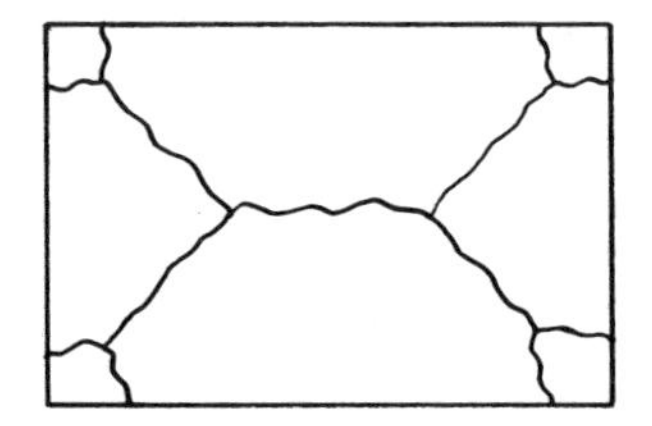

Figure 4-16.

A possible failure mechanism that occurs near corners of reinforced concrete slabs is shown in Fig. 4-16. Experimental (Moy 1981) and theoretical evaluation (Jones 1966) showed that for square slabs without corner reinforcement, this mechanism results in a maximum moment that is about 10% higher than that obtained from a straight line mechanism. For slabs with small acute corner angles and without adequate corner reinforcement, the maximum moment is about 25% higher than that obtained from straight line mechanisms. Thus for reinforced concrete slabs with sharp corners, the plastic moments developed previously must be increased accordingly.

Another failure mechanism that occurs in long narrow plates is shown in Fig. 4-17 and should be investigated together with other failure modes.

Problems

4-7 Find M_p in the square base plate (Fig. P4-7) due to uniform load p.
4-8 Find M_p in Problem 4-7 if a concentrated load F is applied at the middle of the free edge.
4-9 The internal vessel tray (Fig. P4-9) is assumed fixed at the outer edge. Determine M_p due to a concentrated load in the middle.
4-10 Determine M_p in the triangular plate shown in Fig. P4-10 due to uniform load p.
4-11 Find M_p in the rectangular balcony (Fig. P4-11) with two adjacent sides fixed and the opposite corner supported by a column. The balcony is uniformly loaded with 75 psf. How will reinforcing bars be arranged.

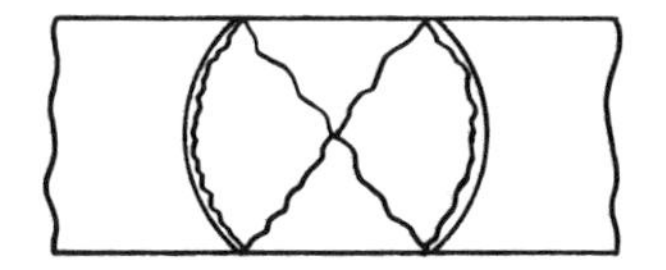

Figure 4-17.

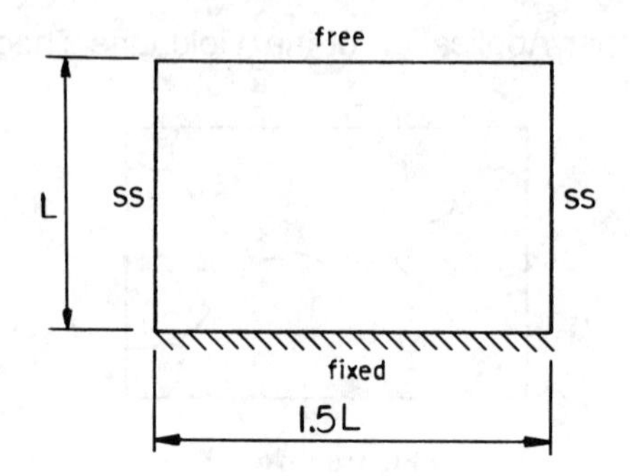

Figure P4-7.

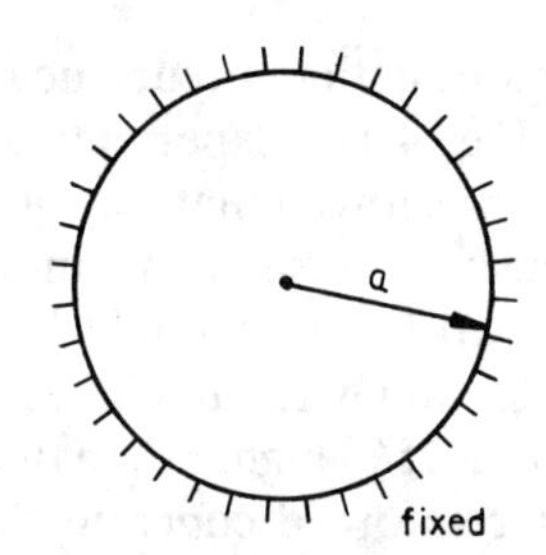

Figure P4-9.

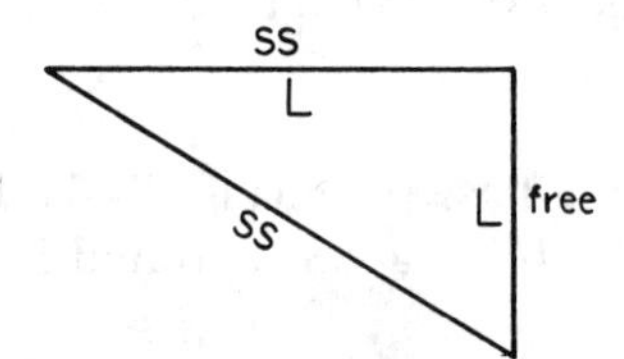

Figure P4-10.

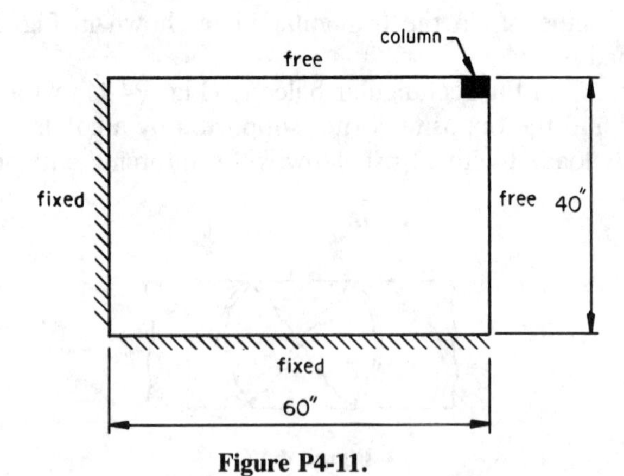

Figure P4-11.

Table 4-1. Plastic bending moments in various plates

Case	Maximum Moment
1. Uniform load, p simply supported edge 	$M_p = \frac{pr^2}{6}$
2. Uniform load, p fixed edge 	$M_p = \frac{pr^2}{12}$
3. Concentrated load, F, in the middle simply supported edge 	$M_p = \frac{F}{2\pi}$
4. Concentrated load, F, in the middle fixed edge 	$M_p = \frac{F}{4\pi}$
5. Uniform load, p simply supported edges 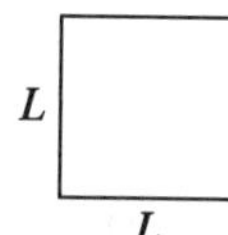	$M_p = \frac{pL^2}{24}$
6. Uniform load, p, fixed edges 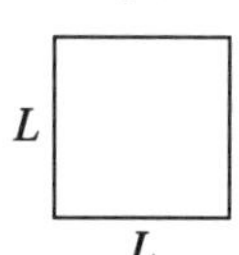	$M_p = \frac{pL^2}{48}$

7. Concentrated load, F, in the middle simply supported edges

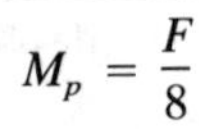

$$M_p = \frac{F}{8}$$

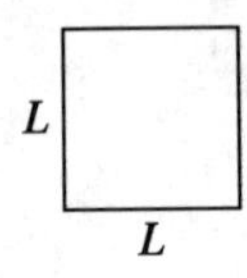

8. Concentrated load, F, in the middle fixed edges

$$M_p = \frac{F}{4\pi}$$

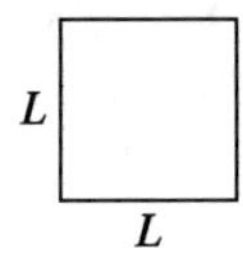

9. Uniform load, p, simply supported edges

$$M_p^a = \frac{pa^2}{72}$$

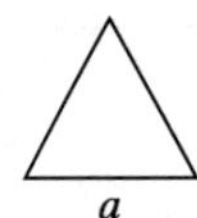

10. Uniform load, p, fixed edges

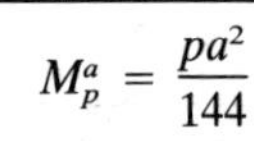

$$M_p^a = \frac{pa^2}{144}$$

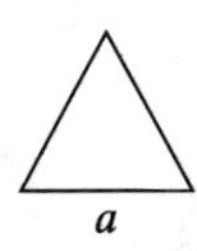

11. Concentrated load, F, at the centroid simply supported edges

$$M_p^a = \frac{F}{6\sqrt{3}}$$

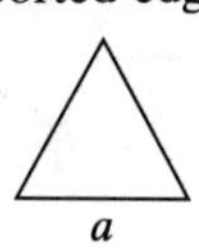

12. Concentrated load, F, at the centroid fixed edges

$$M_p^a = \frac{F}{12\sqrt{3}}$$

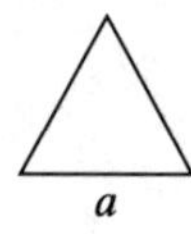

[a]These bending moments are to be increased by a factor of 1.25 to account for sharp angle effect at the corners of reinforced concrete slabs.

4-4 Design Concepts

The plastic theory is used as an approximation for determining maximum moments in plates that cannot be solved by the classical plate theory due to complex geometry, boundary conditions, or applied loading. In many structures this approximation is adequate for design purposes. When a more accurate analysis is needed, the plastic theory is used first as an approximation followed by a more rigorous analysis such as the finite element method. Table 4-1 lists the maximum plastic moment for some frequently encountered plates with various loading and boundary conditions. It should be noted that the deflection due to plastic design is larger than that obtained from the elastic theory due to reduced thickness. Thus, extra precaution must be given to the design of components that cannot tolerate large deflections such as cover plates in flanged openings.

Equation (4-8) for plastic bending of a plate is used by numerous international codes to establish an upper limit on the allowable bending stress values. The ratio obtained by dividing Eq. (4-7) by Eq. (4-8) is 1.5 and is referred to as the shape factor. It indicates that for a given bending moment, plastic analysis of plates results in a stress level that is 50% lower than that determined from the elastic theory for the same factor of safety. Accordingly, many standards such as the ASME Boiler and Pressure Vessel Code use an allowable stress for plates in bending that is 50% higher than the tabulated allowable membrane stress value.

5

Plates of Various Shapes and Properties

5-1 Introduction

Many structures (Fig. 5-1) consist of plates with shapes other than rectangular or circular. In this chapter a brief discussion of elliptic and triangular plates is given. The solution of other shapes is obtained by approximate solutions similar to those discussed in Chapter 4 or by a finite element analysis.

Many structural components such as bridge decks, reinforced concrete slabs, corrugated sheet plates, and composite materials (Fig. 5-1) have physical properties that are different in the x- and y-axes. Accordingly, the equations derived in Chapter 1 cannot be used directly to analyze these components. Rather, a modified theory is needed and is referred to as the orthotropic plate theory. This theory is briefly discussed in this chapter.

5-2 Elliptic Plates

The shape of an elliptic plate is expressed by the equation

$$\frac{x^2}{a^2} + \frac{y^2}{b^2} = 1 \tag{5-1}$$

where a and b are the major and minor axes shown in Fig. 5-2. The boundary conditions for an elliptic plate with a fixed boundary (Fig. 5-2) are given by

$$w = 0 \quad \text{and} \quad \frac{\partial w}{\partial n} = 0$$

Figure 5-1. F117 Fighter (Courtesy of Lockheed Advanced Development Company.)

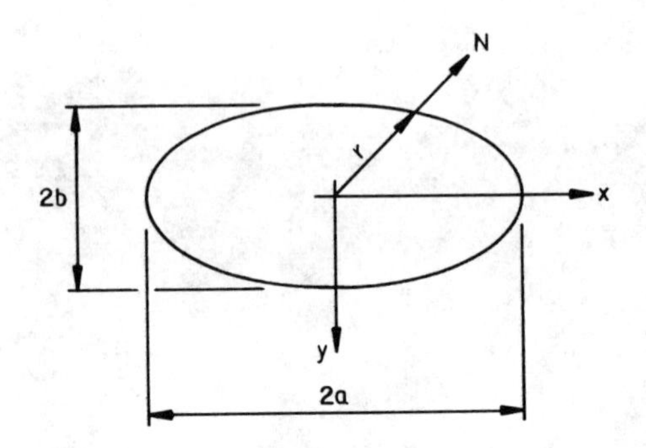

Figure 5-2.

where n is normal to the plate edge.

An expression for the deflection of a uniformly loaded plate that satisfies the boundary conditions is given by

$$w = K\left(\frac{x^2}{a^2} + \frac{y^2}{b^2} - 1\right)^2. \tag{5-2}$$

Substituting Eq. (5-2) into the differential Eq. (1-26) gives

$$K = \frac{p}{8D}\,\frac{a^4b^4}{3a^4 + 3b^4 + 2a^2b^2} \tag{5-3}$$

and the moments are obtained from Eq. (1-17) as

$$M_x = -4DK\left[x^2\left(\frac{3}{a^4} + \frac{\mu}{a^2b^2}\right) + y^2\left(\frac{1}{a^2b^2} + \frac{3\mu}{b^4}\right) - \left(\frac{1}{a^2} + \frac{\mu}{b^2}\right)\right] \tag{5-4}$$

$$M_y = -4DK\left[x^2\left(\frac{3\mu}{a^4} + \frac{1}{a^2b^2}\right) + y^2\left(\frac{\mu}{a^2b^2} + \frac{3}{b^4}\right) - \left(\frac{\mu}{a^2} + \frac{1}{b^2}\right)\right] \tag{5-5}$$

$$M_{xy} = -8DK(1-\mu)\frac{xy}{a^2b^2}. \tag{5-6}$$

For simply supported plates, the expression for deflection is more complicated than that given in Eq. (5-2). The solution has been developed by many authors such as Perry (Perry 1950).

Example 5-1

Determine the required thickness of a fixed elliptic plate with $a = 20$ inches and $b = 15$ inches due to a pressure of 100 psi. Let $\mu = 0.3$ and the allowable stress equal 15,000 psi.

Solution

From Eq. (5-3), $K = 124{,}711/D$, and Eqs. (5-4), (5-5), and (5-6) give

$$M_x = -498{,}844\left(\frac{x^2}{45{,}283} + \frac{y^2}{34{,}615} - \frac{1}{261}\right)$$

$$M_y = -498{,}844\left(\frac{x^2}{59{,}751} + \frac{y^2}{15{,}976} - \frac{1}{192.5}\right)$$

$$M_{xy} = -7.76xy.$$

At $x = 20$ inches and $y = 0$ inches, maximum value of $M_x = -2495$ inch-lbs/inch.

At $x = 0$ inches and $y = 15$ inches, maximum value of $M_y = -4434$ inch-lbs/inch.

At $x = 14.02$ inches and $y = 10.7$ inches, maximum value of $M_{xy} =$ 1164 inch-lbs/inch.

$$t = \sqrt{\frac{6 \times 4434}{15{,}000}}$$

$$= 1.33 \text{ inches.}$$

5-3 Triangular Plates

The solution of a uniformly loaded simply supported isosceles right triangular plate, ABO, of length a is obtained from Fig. 5-3. The plate is assumed to be loaded over a small area of dimension c with a downward load p. In order to find a solution that satisfies the simply supported boundary along line AB, we need to apply a mirror image of the load at a point on the other side of the boundary on a fictitious extension of the plate, ABC, as shown in Fig. 5-3. This fictitious load, p', has an upwards direction. The bending couple due to p' and p results in a zero bending moment and deflection along boundary AB. The deflection due to p is obtained from Example 1-3 with $a = b$ and $c = d$. This gives

$$w_1 = \sum_{m=1}^{\infty}\sum_{n=1}^{\infty} w_{mn} \sin\frac{m\pi x}{a} \sin\frac{n\pi y}{a} \tag{5-7}$$

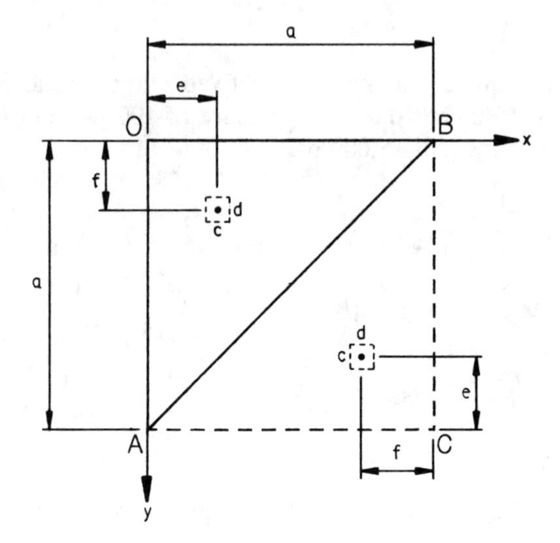

Figure 5-3.

where

$$w_{mn} = \frac{16p}{\pi^6 Dc^2} \frac{\sin \frac{m\pi e}{a} \sin \frac{m\pi c}{2a} \sin \frac{n\pi f}{a} \sin \frac{n\pi c}{2a}}{mn[(m/a)^2 + (n/a)^2]^2}. \tag{5-8}$$

For load p' we substitute in Example 1-3 the quantity $(a - f)$ for e and the quantity $(a - e)$ for f. This gives

$$w_2 = \sum_{m=1}^{\infty} \sum_{n=1}^{\infty} g_{mn} \sin \frac{m\pi x}{a} \sin \frac{n\pi y}{a} \tag{5-9}$$

where

$$g_{mn} = \frac{16p'}{\pi^6 Dc^2} \frac{\sin \frac{m\pi f}{a} \sin \frac{m\pi c}{2a} \sin \frac{n\pi e}{a} \sin \frac{n\pi c}{2a} \cos m\pi \cos n\pi}{mn[(m/a)^2 + (n/a)^2]^2} \tag{5-10}$$

and the total solution of the triangular plate is given by

$$w = w_1 + w_2. \tag{5-11}$$

The deflection of a simply supported equilateral triangular plate of length a is obtained by defining the coordinate system as shown in Fig. 5-4. The

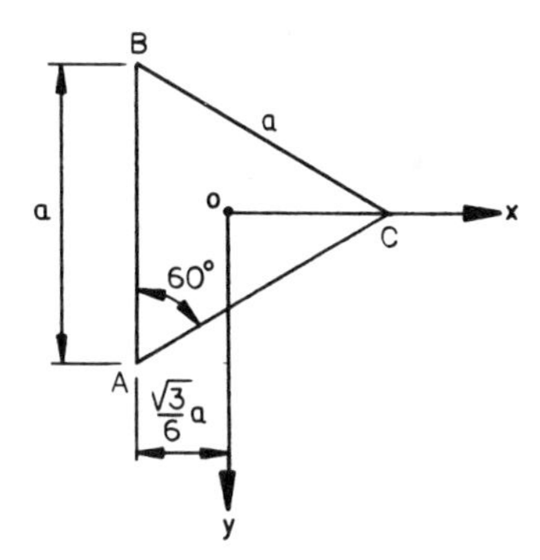

Figure 5-4.

equations for the boundary conditions become

$$x = -\frac{\sqrt{3}}{6}a \quad \text{at boundary } AB$$

$$y = -\frac{1}{\sqrt{3}}x + \frac{a}{3} \quad \text{at boundary } BC$$

$$y = \frac{1}{\sqrt{3}}x - \frac{a}{3} \quad \text{at boundary } AC.$$

Hence, a deflection expression that vanishes at the boundaries is given by

$$w = K\left(x + \frac{\sqrt{3}}{6}a\right)\left(\frac{x}{\sqrt{3}} + y - \frac{a}{3}\right)\left(\frac{x}{\sqrt{3}} - y - \frac{a}{3}\right)$$

or

$$w = \frac{K}{6}\left(2x^3 - \sqrt{3}\,x^2a - 6xy^2 - \sqrt{3}\,ay^2 + \frac{\sqrt{3}}{9}a^3\right). \qquad (5\text{-}12)$$

Woinowsky-Kreiger (Timoshenko and Woinowsky-Kreiger 1959) obtained a value of K that satisfies the simply supported boundary condition in the form of

$$K = \frac{p}{64aD}(a^2 - 3x^2 - 3y^2). \qquad (5\text{-}13)$$

Problems

5-1 What is the required thickness of the fixed elliptic plate (Fig. P5-1) due to a 7 kgf/cm² pressure? Let the allowable stress be 1400 kgf/cm² and $\mu = 0.26$.

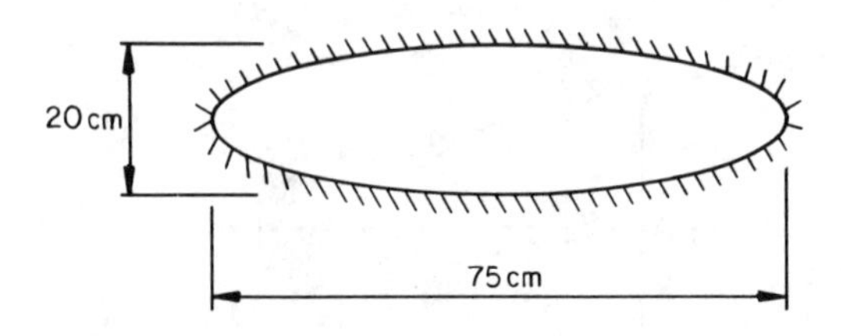

Figure P5-1.

5-2 Determine the expressions for M_x, M_y, and M_{xy} for a simply supported isosceles right triangular plate.
5-3 Compare the maximum M_x value obtained from Problem 5-2 with that obtained from the yield line theory of Chapter 4.
5-4 Determine the expressions for M_x, M_y, and M_{xy} for a simply supported equilateral triangular plate.

5-4 Orthotropic Plate Theory

In our discussion of plates so far, the material was assumed homogeneous and isotropic. In an isotropic material subjected to an axial stress in a principal direction, the major deformation occurs in the direction of applied load. Lateral deformation of smaller magnitude (Fig. 5-5a) occurs in the other principal directions. Also, shearing stress causes only shearing deformation as discussed in Section 1-3. The deformation is dependent on the elastic constants E and μ. Many materials of construction such as steel, aluminum, and titanium fall into this category.

In orthotropic materials stressed in one of the principal directions, the lateral deformation in the other principal directions could be smaller or larger than the deformation in the direction of the applied stress depending on the material properties (Fig. 5-5b). Also, the magnitude of the shearing deformation (Jones 1975) is independent of the elastic constants. Some materials of construction that fall into this category are bridge steel ducts (Troitsky 1987), reinforced concrete, plywood sheets, and composite materials (ASMEc 1992).

In anisotropic plates, or orthogonal plates stressed in other than the principal axes, the applied stress in a given direction causes not only extension in the same direction and deformation in the other two directions, but also shearing deformation (Fig. 5-5c). Similarly shearing stress causes not only shearing deformations, but also axial deformations. The state of stress in an anisotropic plate is very complicated and is beyond the scope of this book.

The development of the plate theory for orthotropic materials is similar

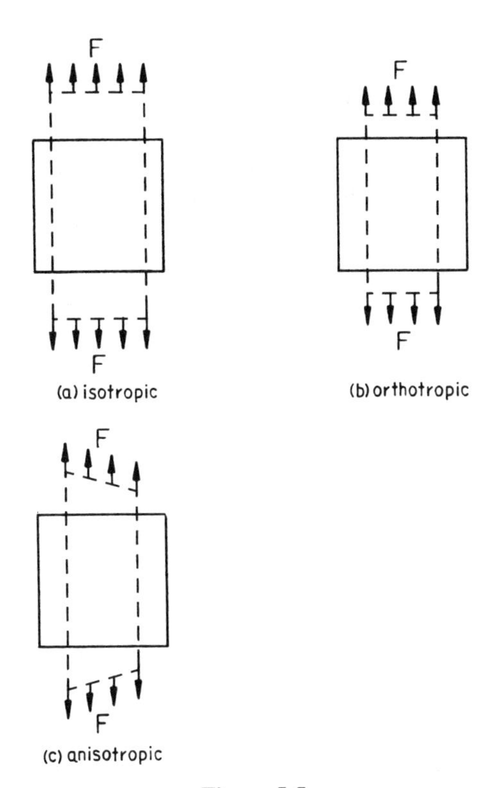

Figure 5-5.

to that of isotropic materials. The stress–strain Eq. (1-14) can be rewritten as (McFarland et al. 1972)

$$\begin{bmatrix} \sigma_x \\ \sigma_y \\ \tau_{xy} \end{bmatrix} = \begin{bmatrix} E_1 & E_{12} & 0 \\ E_{12} & E_2 & 0 \\ 0 & 0 & G \end{bmatrix} \begin{bmatrix} \varepsilon_x \\ \varepsilon_y \\ \gamma_{xy} \end{bmatrix} \tag{5-14}$$

where E_1, E_2, E_{12}, and G are constants defined as

$$E_1 = \frac{E_x}{1 - \mu_x \mu_y}$$

$$E_2 = \frac{E_y}{1 - \mu_x \mu_y}$$

$$E_{12} = \mu_y E_1 = \mu_x E_2$$

$$G = \frac{\tau_{xy}}{\gamma_{xy}}$$

where

E_x = modulus of elasticity in the x-direction;
E_y = modulus of elasticity in the y-direction;
μ_x = contraction in the y-direction due to stress in the x-direction;
μ_y = contraction in the x-direction due to stress in the y-direction;
τ_{xy} = shearing stress in the $x-y$ direction;
γ_{xy} = shearing strain in the $x-y$ direction.

The strain deflection expressions given by Eq. (1-12) are also valid for orthotropic plates as they are a function of geometry only. Substituting Eq. (1-12) into Eq. (5-14) gives

$$\begin{bmatrix} \sigma_x \\ \sigma_y \\ \tau_{xy} \end{bmatrix} = -z \begin{bmatrix} E_1 & E_{12} & 0 \\ E_{12} & E_2 & 0 \\ 0 & 0 & 2G \end{bmatrix} \begin{bmatrix} \dfrac{\partial^2 w}{\partial x^2} \\ \dfrac{\partial^2 w}{\partial y^2} \\ \dfrac{\partial^2 w}{\partial x\,\partial y} \end{bmatrix}. \tag{5-15}$$

The moment expressions given by Eq. (1-17) become

$$\begin{bmatrix} M_x \\ M_y \\ M_{xy} \end{bmatrix} = - \begin{bmatrix} D_x & D_{xy} & 0 \\ D_{xy} & D_y & 0 \\ 0 & 0 & 2D_s \end{bmatrix} \begin{bmatrix} \dfrac{\partial^2 w}{\partial x^2} \\ \dfrac{\partial^2 w}{\partial y^2} \\ \dfrac{\partial^2 w}{\partial x\,\partial y} \end{bmatrix} \tag{5-16}$$

where D_x, D_y, D_{xy}, and D_s are bending stiffness constants defined as

$$D_x = \frac{E_1 t^3}{12} \qquad D_y = \frac{E_2 t^3}{12}$$
$$D_{xy} = \frac{E_{12} t^3}{12} \qquad D_s = \frac{G t^3}{12}. \tag{5-17}$$

The shear expressions Eqs. (1-27) and (1-28) become

$$Q_x = -\frac{\partial}{\partial x}\left(D_x \frac{\partial^2 w}{\partial x^2} + H \frac{\partial^2 w}{\partial y^2}\right) \tag{5-18}$$

$$Q_y = -\frac{\partial}{\partial y}\left(H \frac{\partial^2 w}{\partial x^2} + D_y \frac{\partial^2 w}{\partial y^2}\right) \tag{5-19}$$

where

$$H = D_{xy} + 2D_s. \tag{5-20}$$

The plate differential Eq. (1-25) becomes

$$D_x \frac{\partial^4 w}{\partial x^4} + 2H \frac{\partial^4 w}{\partial x^2\, \partial y^2} + D_y \frac{\partial^4 w}{\partial y^4} = p(x, y). \tag{5-21}$$

Equation (5-21) is normally referred to as the Huber's differential equation of orthotropic plates (Troitsky 1987).

For simply supported plates, the expressions for deflection and load are given by Eqs. (1-41) and (1-39). For a uniformly loaded plate, Eq. (1-40) gives

$$p_{mn} = \frac{16 p_o}{\pi^2 mn}$$

where m and n are odd.

Substituting this expression and Eqs. (1-39) and (1-41) into Eq. (5-21) gives

$$w_{mn} = \frac{16 p_o}{\pi^6} \frac{1}{mn\left(\dfrac{m^4}{a^4} D_x + \dfrac{2m^2n^2}{a^2b^2} H + \dfrac{n^4}{b^4} D_y\right)}. \tag{5-22}$$

The deflection is obtained by substituting this expression into Eq. (1-41).

For isotropic plates, $D_x = D_y = H = D$. Thus, Eq. (5-22) reduces to the value of w_{mn} given in Example 1-2 for isotropic plates.

For a plate simply supported along two sides and infinitely long along the other two sides (Fig. 5-6), we define the deflection as

$$w_h = \sum_{m=1}^{\infty} f_m(y) \sin \frac{m\pi x}{a}.$$

Then the homogeneous solution of Eq. (5-21) becomes

$$D_x\left(\frac{m\pi}{a}\right)^4 f_m - 2H\left(\frac{m\pi}{a}\right)^2 \frac{d^2 f_m}{dy^2} + D_y \frac{d^4 f_m}{dy^4} = 0.$$

Let $f_m = Ce^{gy}$.

Then,

$$D_x\left(\frac{m\pi}{a}\right)^4 - 2Hg^2\left(\frac{m\pi}{a}\right)^2 + D_y g^4 = 0$$

and

$$\left.\begin{aligned} g_{\substack{1\\2}} &= \pm \frac{m\pi}{a}\sqrt{\frac{1}{D_y}\left(H + \sqrt{H^2 - D_xD_y}\right)} \\ g_{\substack{3\\4}} &= \pm \frac{m\pi}{a}\sqrt{\frac{1}{D_y}\left(H - \sqrt{H^2 - D_xD_y}\right)}, \end{aligned}\right] \tag{5-23}$$

and the homogeneous solution becomes

$$w_h = \sum_{m=1}^{\infty} \left(A_m e^{g_1 y} + B_m e^{g_2 y} + C_m e^{g_3 y} + D_m e^{g_4 y}\right) \sin \frac{m\pi x}{a}. \tag{5-24}$$

The particular solution depends on the applied loads. If we assume a uniformly loaded plate, then

$$p(x) = \sum_{m=1}^{\infty} p_m \sin \frac{m\pi x}{a}$$

where

$$p_m = \frac{2}{a}\int_0^a p(x) \sin \frac{m\pi x}{a}\, dx$$

$$p_m = \frac{4p_o}{m\pi} \qquad m = 1, 3, \ldots$$

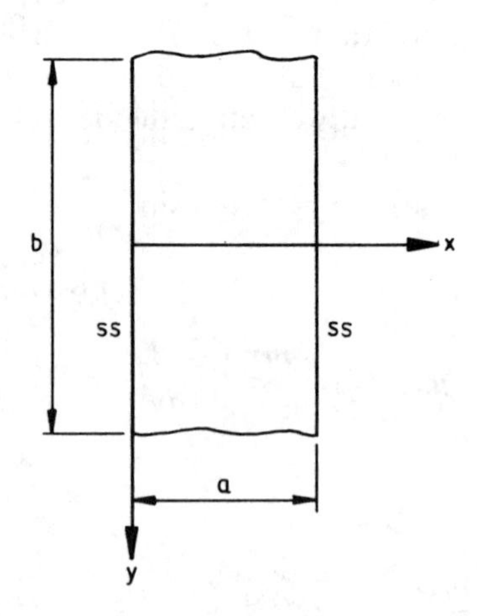

Figure 5-6.

Substituting this expression into Eq. (5-21) results in

$$D_x\left(\frac{m\pi}{a}\right)^4 a_m - 2H\left(\frac{m\pi}{a}\right)^2 \frac{d^2a_m}{dy^2} + D_y \frac{d^4a_m}{dy^4} = p_m.$$

The particular solution is

$$a_m = \frac{p_m}{D_x}\left(\frac{a}{m\pi}\right)^4$$

and

$$w_p = \sum_{m=1}^{\infty} \frac{p_m}{D_x}\left(\frac{a}{m\pi}\right)^4 \sin\frac{m\pi x}{a}. \tag{5-25}$$

The total solution becomes

$$w = \sum_{m=1}^{\infty} \left[A_m e^{g_1 y} + B_m e^{g_2 y} + C_m e^{g_3 y} + D_m e^{g_4 y} + \frac{p_m}{D_x}\left(\frac{a}{a\pi}\right)^4\right] \sin\frac{m\pi x}{a}. \tag{5-26}$$

As y gets larger, the quantities $e^{g_1 y}$ and $e^{g_3 y}$ tend to approach infinity. Thus, we must set A_m and C_m to zero and Eq. (5-26) reduces to

$$w = \sum_{m=1}^{\infty} \left[B_m e^{g_2 y} + D_m e^{g_4 y} + \frac{p_m}{D_x}\left(\frac{a}{m\pi}\right)^4\right] \sin\frac{m\pi x}{a}. \tag{5-27}$$

Problems

5-5 Find the maximum stress in a simply supported rectangular plate due to a uniform pressure of 8 psi. The plate is constructed of a boron/epoxy laminate with $E_x = 30{,}000$ ksi, $E_y = 3000$ ksi, $G = 1000$ ksi, $\mu_x = 0.30$, and $\mu_y = 0.03$. Let $a = 30$ inches, $b = 25$ inches, and $t = 1/4$ inch. Compare the result with that of an equivalent isotropic plate of $E = 16{,}500$ ksi and $\mu = 0.165$.

5-6 Use Eq. (5-27) to solve Problem 5-5. Let $a = 30$ inches and $b = 100$ inches in Fig. 5-6.

5-5 Orthotropic Materials and Structural Components

Solution of Eq. (5-25) requires specific values of the bending stiffness constants D_x, D_y, D_{xy}, and D_s. These constants must be determined experimentally or empirically for various materials and structural components. Many references are available for orthotropic plate design such as McFarland (McFarland et al. 1972), Timoshenko (Timoshenko and Woinowsky-Krieger 1959), Troitsky (Troitsky 1987), and Ugural (Ugural 1981). Some commonly encountered cases are given in this section.

Reinforced Concrete Slabs

In a two-way reinforced concrete slab (Fig. 5-7), the bending stiffness constants are usually taken as

$$\left.\begin{aligned} D_x &= \frac{E_c I_x}{1-\mu_c^2} \qquad & D_y &= \frac{E_c I_y}{1-\mu_c^2} \\ D_{xy} &= \mu_c \sqrt{D_x D_y} \qquad & D_s &= \frac{1-\mu_c}{2}\sqrt{D_x D_y} \\ H &= \sqrt{D_x D_y} \end{aligned}\right] \tag{5-28}$$

where

E_c = modulus of elasticity of concrete;
E_s = modulus of elasticity of steel;
I_x = moment of inertia about the neutral axis in an x = constant direction
= $(I_{cx} + (n-1)I_{sx})|_{x=\text{constant}}$;
I_y = moment of inertia about the neutral axis in a y = constant direction
= $(I_{cy} + (n-1)I_{sy})|_{y=\text{constant}}$;
I_{cx}, I_{cy} = moment of inertia of concrete about the neutral axis of composite section in the x- and y-axes, respectively;

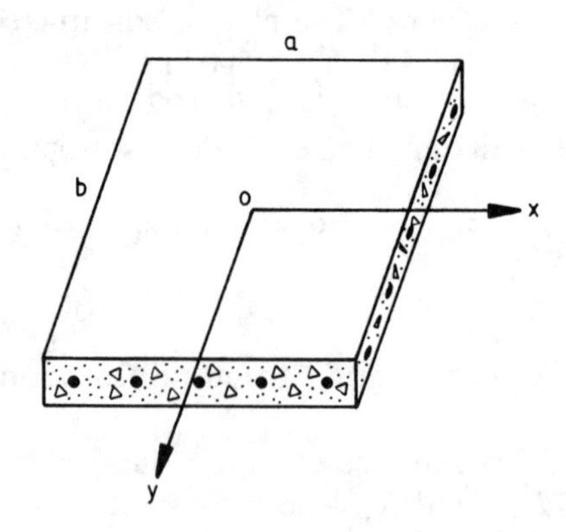

Figure 5-7.

I_{sx}, I_{sy} = moment of inertia of steel about the neutral axis of composite section in the x- and y-axes, respectively,
n = E_s/E_c;
μ_c = Poisson's ratio of concrete.

From Eqs. (5-20) and (5-26) it is seen that $H^2 = D_x D_y$. Hence, Eq. (5-23) reduces to two double roots of magnitude

$$g_1 = g_3 = g$$

$$g_2 = g_4 = -g$$

where

$$g = \frac{m\pi}{a}\sqrt[4]{\frac{D_x}{D_y}}$$

and Eq. (5-26) can be rewritten as

$$w = \sum_{m=1}^{\infty}\left[(C_m + D_m y)e^{-gy} + \frac{p_m}{D_x}\left(\frac{a}{m\pi}\right)\right]\sin\frac{m\pi x}{a}.$$

Corrugated Plate

For this type of construction (Fig. 5-8), the bending stiffnesses are given by

$$\left.\begin{aligned} D_x &= \frac{s}{L}\frac{Et^3}{12(1-\mu^2)} \\ D_y &= EI \qquad D_{xy} = H/2 \\ H &= \frac{L}{s}\frac{Et^3}{12(1+\mu)} \end{aligned}\right] \tag{5-29}$$

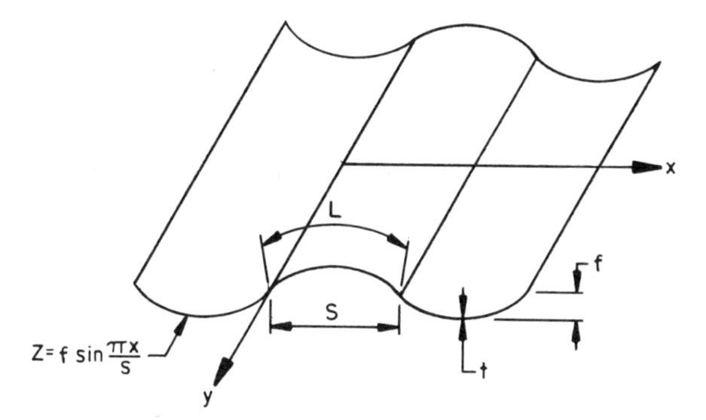

Figure 5-8.

where

$$L = s\left(1 + \frac{\pi^2 f^2}{4s^2}\right)$$

$$I = \frac{f^2 t}{2}\left[1 - \frac{0.81}{1 + 2.5\left(\frac{f}{2s}\right)^2}\right].$$

Stiffened Plates

For this type of construction (Fig. 5-9), the bending stiffnesses are given by

$$\left.\begin{aligned} D_x &= \frac{EI_x}{h} \qquad D_y = \frac{EI_y}{g} \\ D_{xy} &= \frac{G}{3}\left(\frac{d_1 t_1^3}{g} + \frac{d_2 t_2^3}{h}\right) \\ H &= \frac{Et^3}{12(1 - \mu^2)} + D_{xy} \end{aligned}\right] \tag{5-30}$$

where I_x and I_y are the composite moment of inertia of stiffeners and plate in the x- and y-directions, respectively.

Box-Type Bridge Decks

The stiffness values for this type of construction are discussed in detail by Troitsky (Troitsky 1987). The stiffness values of the configuration shown

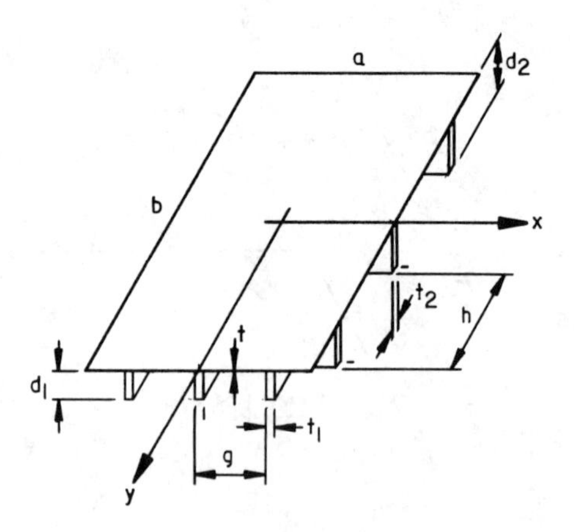

Figure 5-9.

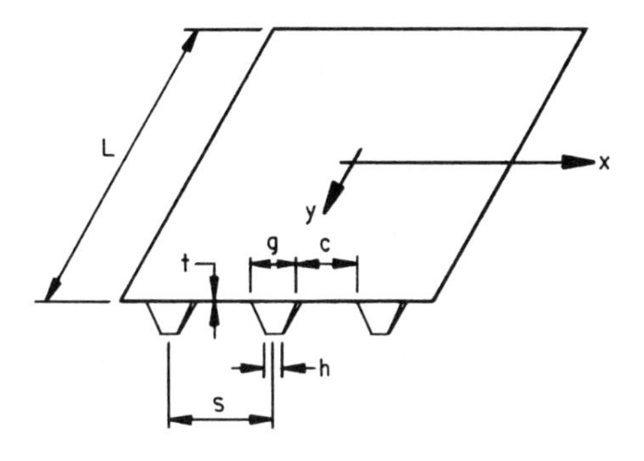

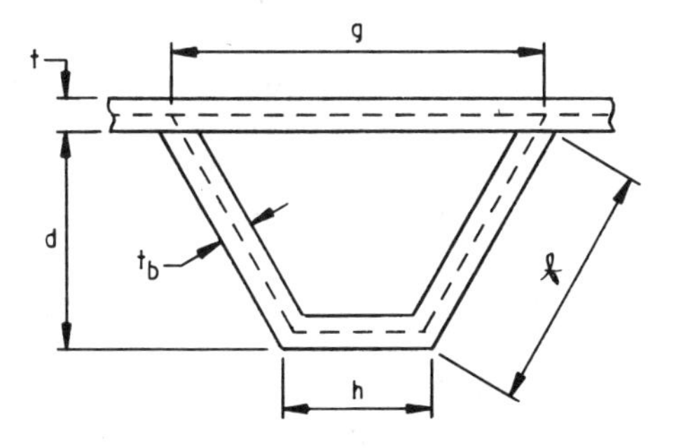

Figure 5-10.

in Fig. 5-10 are given by

$$\left.\begin{aligned} D_x &= \frac{Et^3}{12(1-\mu^2)} \\ D_y &= \frac{EI_B}{g+c} \\ H &= \frac{2\left[\dfrac{1}{2}(g+h)d\right]^2}{\dfrac{g}{t}+\dfrac{h+2f}{t_b}} \end{aligned}\right] \tag{5-31}$$

where

E = modulus of elasticity;
I_B = composite stiffness of rib and plate about the neutral axis in the y-direction;
t = thickness of the plate.

The results obtained from Eq. (5-31) tend to be on the unconservative side due to the distortion of the deflected cross section shown in Fig. 5-11. Thus, modifications are needed for the classical solution of Eq. (5-31). This modification, which is beyond the scope of this book, is referred to as the Pelikan–Esslinger method and is detailed by Troitsky.

Problems

5-7 Determine the maximum bending moment in the simply supported concrete slab shown in Fig. P5-7. The slab is 30 feet long and 18 feet wide. How does this moment compare with ACI's method of calculating the moment as a simply supported unit strip in the short direction only? Let p = 200 psf, E_s = 30,000 ksi, n = 10, and μ_c = 0.15.
5-8 A corrugated siding for a building (Fig. P5-8) is subjected to 40 psf wind load. Determine the maximum bending moment of the panel is assumed simply supported. Let E = 10,000 ksi and μ = 0.33.
5-9 Part of a ship deck is shown in Fig. P5-9. Determine the maximum bending moment assuming the deck to be simply supported. Let E = 29,000 ksi, p = 75 psf, and μ = 0.30.
5-10 Refer to Problem 2-8. The side plates have intermediate vertical stiffeners between the bulkheads. The stiffeners are 7C14.75 and spaced on 2 ft centers. The plate thickness is 11/32 inch. Calculate the maximum stress due to uniform internal pressure of 0.10 psi. Use Eq. (5-26) and let the top and bottom edges be simply supported and the sides fixed.
5-11 In Problem 5-10, let the pressure vary as described in Problem 2-8 with a maximum value of 1.25 psi. Calculate the maximum stress.
5-12 Oil barges are normally double-hulled due to environmental concerns. Thus, the barge discussed in Problems 2-8, 5-10, and 5-11 consists, in actuality, of an inner and outer skins as shown in Fig. P5-12. How should this sandwich construction be analyzed?

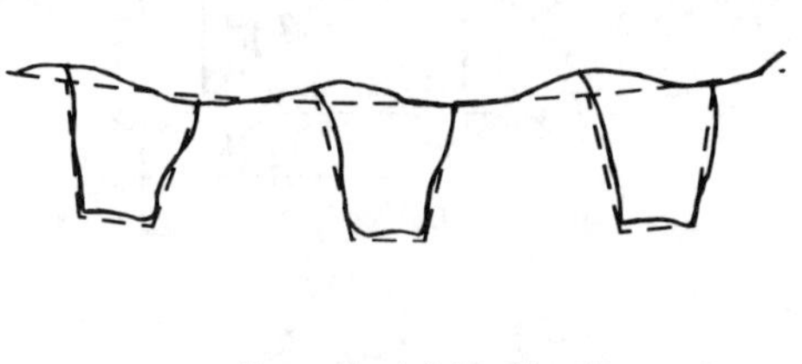

— — — theoretical deflection line

——— actual deflection line

Figure 5-11.

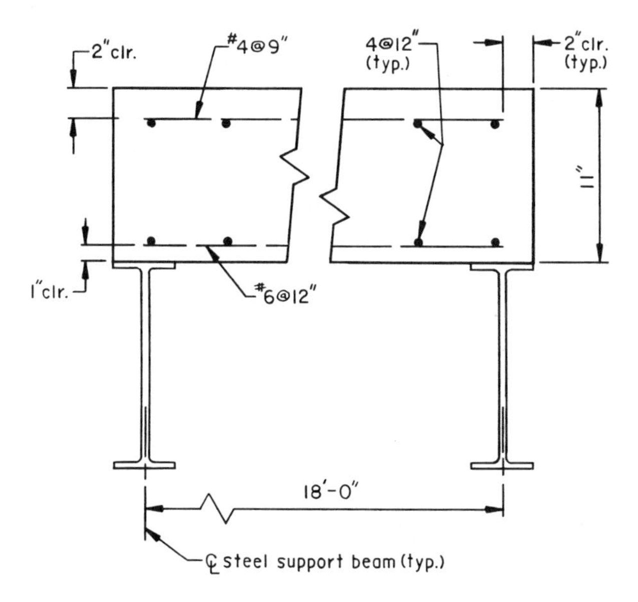

Figure P5-7.

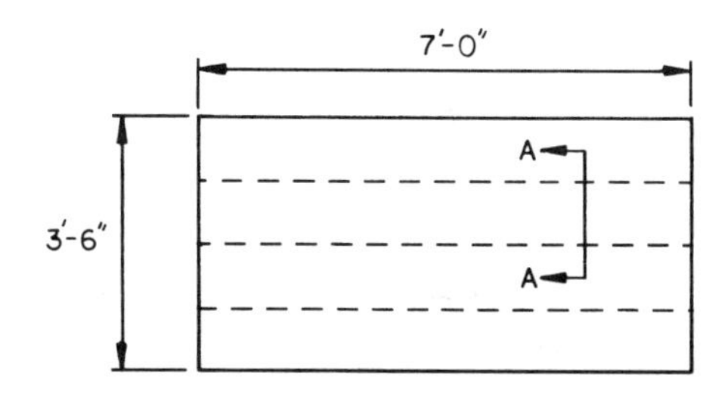

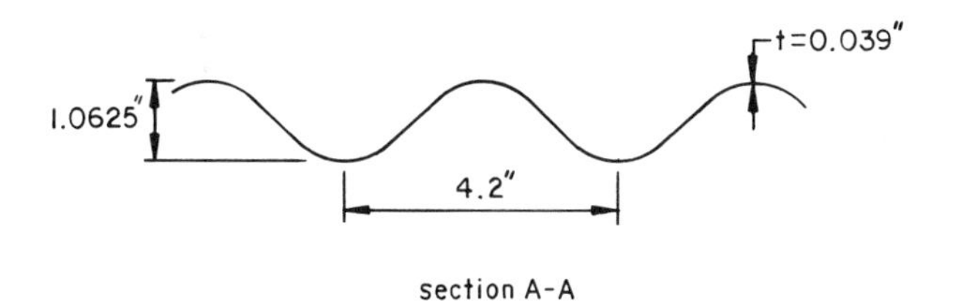

section A-A

Figure P5-8.

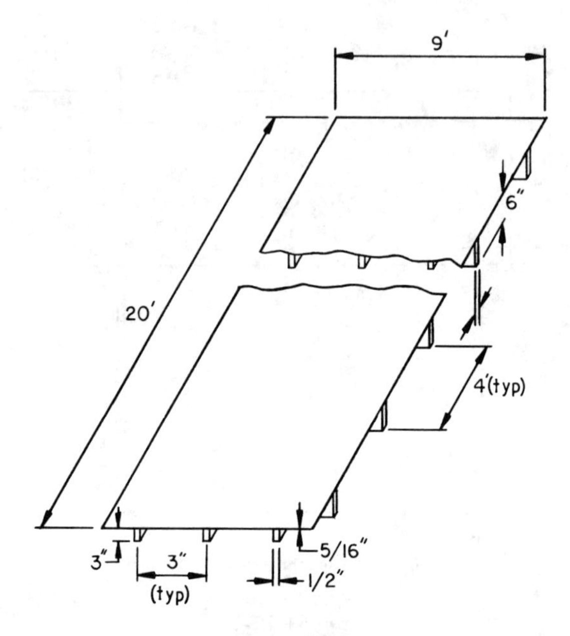

Figure P5-9.

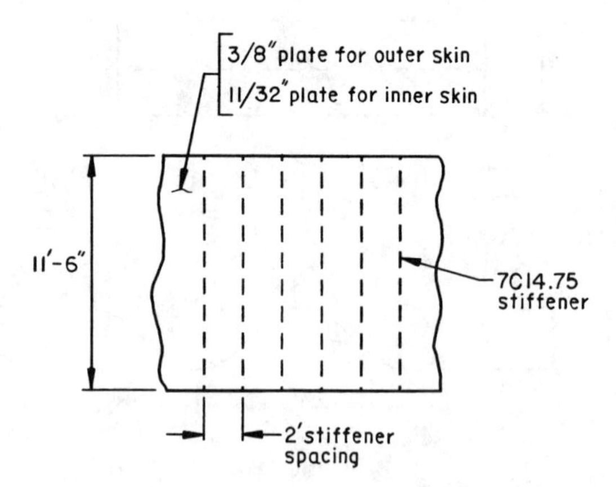

Figure P5-12.

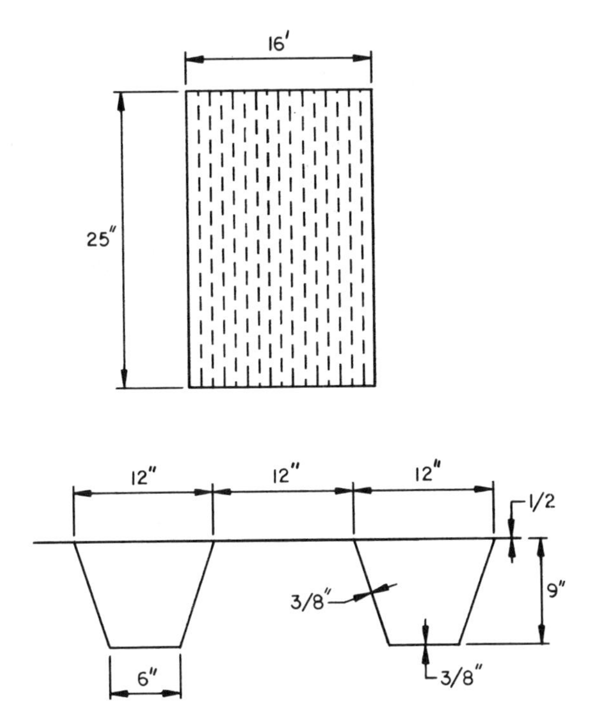

Figure P5-13.

5-13 A box-type bridge deck is shown in Fig. P5-13. Find the maximum bending moment due to a 200 psf uniform load. Let E = 30,000 ksi and μ = 0.29. Assume the deck to be simply supported.

5-6 Design of Plates with Various Shapes and Properties

With the exception of a few cases, the maximum bending of plates with other than rectangular or circular shapes cannot be obtained in a closed form solution based on existing theories. Accordingly, the engineer usually relies on approximate solutions such as those obtained from plastic theory or finite element analysis. Numerous finite element programs are available for personal computer applications. Such programs are used to solve simple plate problems with complicated geometries and boundary conditions.

Factors needed in Eq. (5-21) for solving orthotropic plate theory are presented in this chapter for commonly encountered types and configurations. Factors for plates with configurations and properties other than those discussed in Sections 5-4 and 5-5 or in other references must be obtained experimentally. Calculating these factors as well as the solution of the pertinent differential equation can be made expeditiously by writing a small computer program.

The methods given in this chapter for determining the maximum bending moment in reinforced concrete slabs have been simplified by ACI for design purposes. ACI-318 uses various factors to approximate the load distribution in the slabs. It then calculates the maximum moment in the slabs by treating them as unit strips in bending and then modifying the moments by various factors to approximate orthotropic plate solution.

6

Membrane Theory of Shells of Revolution

6-1 Basic Equations of Equilibrium

The membrane shell theory is used extensively in designing such structures as flat-bottom tanks, pressure vessel components (Fig. 6-1) and dome roofs. The membrane theory assumes that equilibrium in the shell is achieved by having the in-plane membrane forces resist all applied loads without any bending moments. The theory gives accurate results as long as the applied loads are distributed over a large area of the shell such as pressure and wind loads. The membrane forces by themselves cannot resist local concentrated loads. Bending moments are needed to resist such loads as discussed in Chapter 8. The basic assumptions made in deriving the membrane theory (Gibson 1965) are

1. The shell is homogeneous and isotropic.
2. The thickness of the shell is small compared to its radius of curvature.
3. The bending strains are negligible and only strains in the middle surface are considered.
4. The deflection of the shell due to applied loads is small.

In order to derive the governing equations for the membrane theory of shells, we need to define the shell geometry. The middle surface of a shell of constant thickness may be considered a surface of revolution. A surface of revolution is obtained by rotating a plane curve about an axis lying in the plane of the curve. This curve is called a meridian (Fig. 6-2). Any point in the middle surface can be described by first specifying the meridian on which it is located and second by specifying a quantity, called a parallel circle, that varies along the meridian and is constant on a circle around

Figure 6-1. A pressure vessel. (Courtesy of the Nooter Corp., St. Louis, MO.)

the axis of the shell. The meridian is defined by the angle θ and the parallel circle by ϕ as shown in Fig. 6-2.

Define r (Fig. 6-3) as the radius from the axis of rotation to any given point o on the surface; r_1 as the radius from point o to the center of curvature of the meridian; and r_2 as the radius from the axis of revolution to point o, and it is perpendicular to the meridian. Then from Fig. 6-3,

$$r = r_2 \sin \phi, \quad ds = r_1 \, d\phi, \quad \text{and} \quad dr = ds \cos \phi. \tag{6-1}$$

The interaction between the applied loads and resultant membrane forces is obtained from statics and is shown in Fig. 6-4. Shell forces N_ϕ and N_θ are membrane forces in the meridional and circumferential directions, respectively. Shearing forces $N_{\phi\theta}$ and $N_{\theta\phi}$ are as shown in Fig. 6-4. Applied load p_r is perpendicular to the surface of the shell; load p_ϕ is in the meridional direction; and load p_θ is in the circumferential direction. All forces are positive as shown in Fig. 6-4.

The first equation of equilibrium is obtained by summing forces parallel

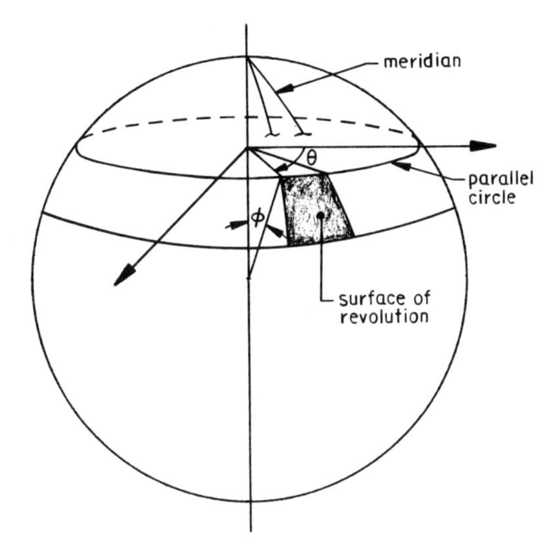

Figure 6-2.

to the tangent at the meridian. This yields

$$\begin{aligned} N_{\theta\phi} r_1 \, d\phi &- \left(N_{\theta\phi} + \frac{\partial N_{\theta\phi}}{\partial \phi} \, d\phi \right) r_1 \, d\phi - N_\phi r \, d\theta \\ &+ \left(N_\phi + \frac{\partial N_\phi}{\partial \phi} \, d\phi \right) \left(r + \frac{\partial r}{\partial \phi} \, d\phi \right) d\theta \\ &+ p_\phi r \, d\theta \, r_1 \, d\phi - N_\theta r_1 \, d\phi \, d\theta \cos \phi = 0. \qquad (6\text{-}2) \end{aligned}$$

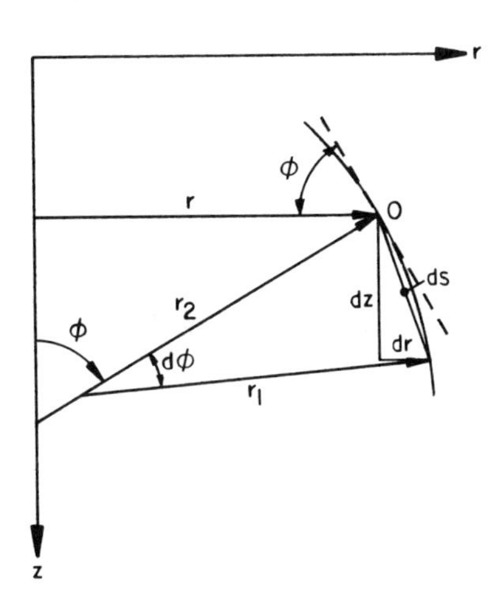

Figure 6-3.

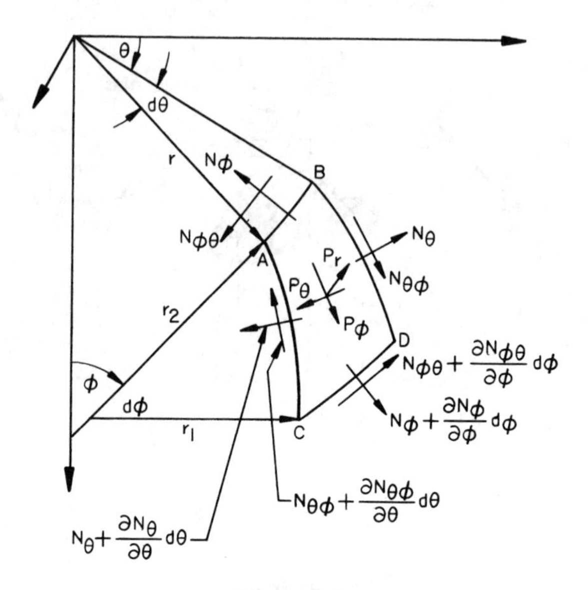

Figure 6-4.

The last term in Eq. (6-2) is the component of N_θ that is parallel to the tangent at the meridian. It is obtained from Fig. 6-5a by finding the components F_1 and F_2. These are expressed as

$$F_1 + F_2 = (N_\theta r_1\, d\phi)\,\frac{d\theta}{2} + \left(N_\theta + \frac{\partial N_\theta}{\partial \theta}\, d\theta\right) r_1\, d\phi\, \frac{d\theta}{2}.$$

Neglecting terms of higher order results in

$$F_1 + F_2 = N_\theta r_1\, d\phi\, d\theta.$$

The component of F_1 and F_2 that is parallel to the tangent at the meridian is shown in Fig. 6-5b and is given by

$$N_\theta r_1\, d\phi\, d\theta \cos\phi.$$

This value is shown as the last expression in Eq. (6-2). Equation (6-2) can be simplified as

$$\frac{\partial}{\partial \phi}(rN_\phi) - r_1\,\frac{\partial N_{\theta\phi}}{\partial \theta} - r_1 N_\theta \cos\phi + p_\phi r r_1 = 0. \qquad (6\text{-}3)$$

The second equation of equilibrium is obtained from summation of forces

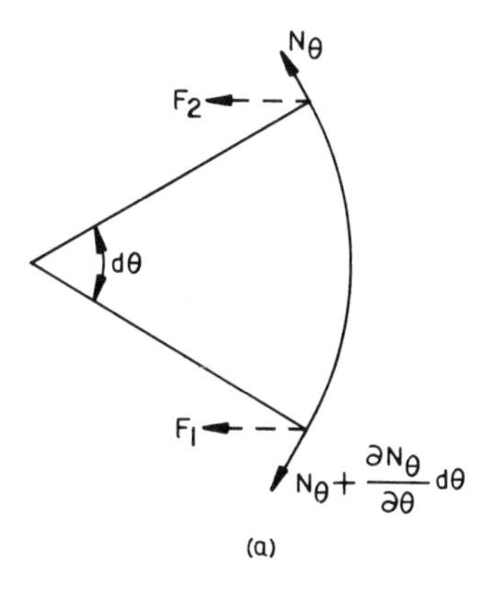

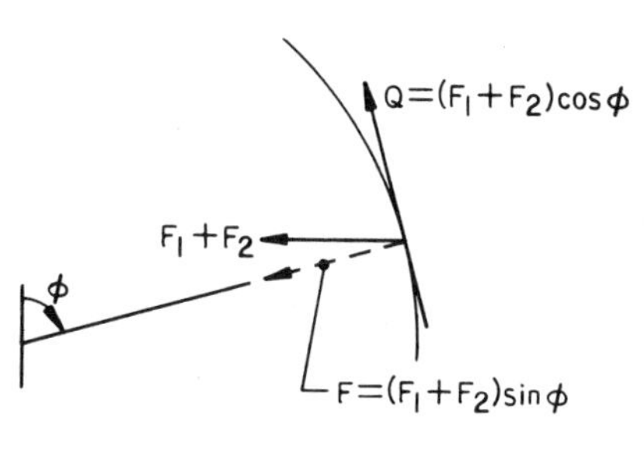

Figure 6-5.

in the direction of parallel circles. Referring to Fig. 6-4,

$$\begin{aligned} N_{\phi\theta} r \, d\theta &- \left(N_{\phi\theta} + \frac{\partial N_{\phi\theta}}{\partial \phi} d\phi\right)\left(r + \frac{\partial r}{\partial \phi} d\phi\right) d\theta \\ &- N_\theta r_1 \, d\phi + \left(N_\theta + \frac{\partial N_\theta}{\partial \theta} d\theta\right)(r_1 \, d\phi) \\ &+ p_\theta r \, d\theta \, r_1 \, d\phi - N_{\theta\phi} r_1 \, d\phi \, \frac{\cos \phi \, d\theta}{2} \\ &- \left(N_{\theta\phi} + \frac{\partial N_{\theta\phi}}{\partial \theta} d\theta\right)(r_1 \, d\phi) \, \frac{\cos \phi \, d\theta}{2} = 0. \end{aligned} \tag{6-4}$$

The last two expressions in Eq. (6-4) are obtained from Fig. 6-6 and are the component of $N_{\theta\phi}$ in the direction of parallel circles. Hence, from Fig.

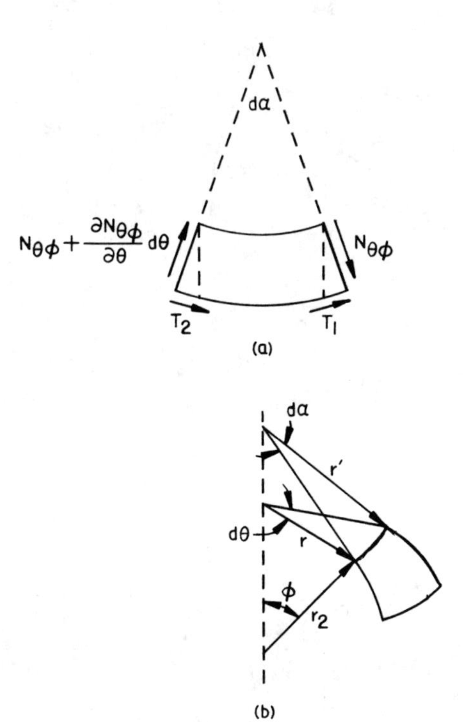

Figure 6-6.

6-6a,

$$T = T_1 + T_2 = N_{\theta\phi} r_1 \, d\phi \, \frac{d\alpha}{2} + \left(N_{\theta\phi} + \frac{\partial N_{\theta\phi}}{\partial \theta} \, d\theta \right) r_1 \, d\phi \, \frac{d\alpha}{2}. \tag{6-5}$$

The value of $d\alpha$ can be expressed in terms of θ and ϕ as shown in Fig. 6-6b.

$$r' \, d\alpha = r \, d\theta$$

or

$$d\alpha = \frac{r \, d\theta}{r_2 \tan \phi}$$

$$= \frac{r_2 \sin \phi \, d\theta}{r_2 \tan \phi}$$

$$d\alpha = \cos \phi \, d\theta. \tag{6-6}$$

Substituting Eq. (6-6) into Eq. (6-5) results in the expression that is shown as the last term in Eq. (6-4). Equation (6-4) can now be simplified to read

$$\frac{\partial}{\partial\phi}(rN_{\phi\theta}) - r_1\frac{\partial N_\theta}{\partial\theta} + r_1 N_{\theta\phi}\cos\phi - p_\theta r r_1 = 0. \tag{6-7}$$

This is the second equation of equilibrium of the infinitesimal element shown in Fig. 6-4. The last equation of equilibrium is obtained by summing forces perpendicular to the middle surface. Referring to Figs. 6-4, 6-5, and 6-7,

$$(N_\theta r_1\, d\phi\, d\theta)\sin\phi - p_r r\, d\theta\, r_1\, d\phi + N_\phi r\, d\theta\, d\phi = 0$$

or

$$N_\theta r_1 \sin\phi + N_\phi r = p_r r r_1. \tag{6-8}$$

Equations (6-3), (6-7), and (6-8) are the three equations of equilibrium of a shell of revolution subjected to axisymmetric loads.

6-2 Ellipsoidal and Spherical Shells Subjected to Axisymmetric Loads

In many structural applications, loads such as dead weight, snow, and pressure are symmetric around the axis of the shell. Hence, all forces and deformations must also be symmetric around the axis. Accordingly, all loads and forces are independent of θ and all derivatives with respect to

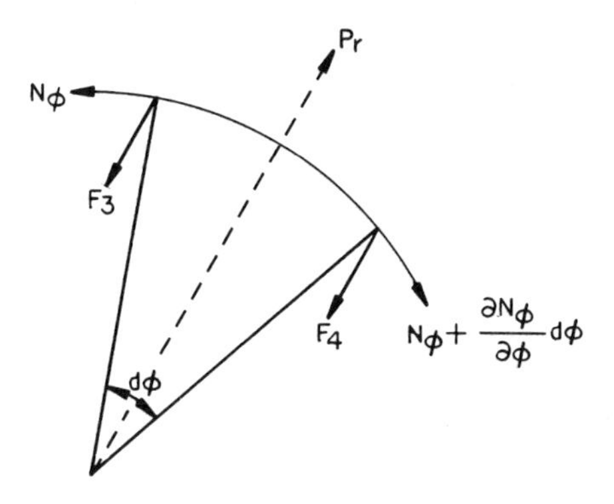

Figure 6-7.

θ are zero. Equation (6-3) reduces to

$$\frac{\partial}{\partial \phi}(rN_\phi) - r_1 N_\theta \cos \phi = -p_\phi r r_1. \tag{6-9}$$

Equation (6-7) becomes

$$\frac{\partial}{\partial \phi}(rN_{\theta\phi}) - r_1 N_{\theta\phi} \cos \phi = p_\theta r r_1. \tag{6-10}$$

In this equation, we let the cross shears $N_{\phi\theta} = N_{\theta\phi}$ in order to maintain equilibrium.

Equation (6-8) can be expressed as

$$\frac{N_\theta}{r_2} + \frac{N_\phi}{r_1} = p_r. \tag{6-11}$$

Equation (6-10) describes a torsion condition in the shell. This condition produces deformations around the axis of the shell. However, the deformation around the axis is zero due to axisymmetric loads. Hence, we must set $N_{\theta\phi} = p_\theta = 0$ and we disregard Eq. (6-10) from further consideration.

Substituting Eq. (6-11) into Eq. (6-9) gives

$$N_\phi = \frac{1}{r_2 \sin^2 \phi}\left[\int r_1 r_2 (p_r \cos \phi - p_\phi \sin \phi) \sin \phi \, d\phi + C\right]. \tag{6-12}$$

The constant of integration C in Eq. (6-12) is additionally used to take into consideration the effect of any additional applied loads that cannot be defined by p_r and p_ϕ such as weight of contents.

Equations (6-11) and (6-12) are the two governing equations for designing double-curvature shells under membrane action.

Example 6-1

Determine the expressions for N_ϕ and N_θ due to internal pressure p in an ellipsoidal shell (Fig. 6-8) of radii a and b.

Solution

For internal pressure we define $p_r = p$ and $p_\phi = 0$. Then from Eqs. (6-1) and (6-11)

$$N_\phi = \frac{1}{r_2 \sin^2 \phi}\left(p \int r \, dr + C\right)$$

$$N_\phi = \frac{1}{r_2 \sin^2 \phi}\left(\frac{pr^2}{2} + C\right). \tag{1}$$

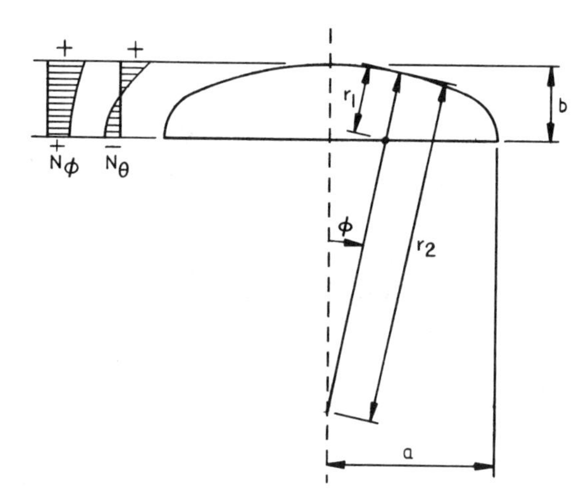

Figure 6-8.

The constant C is obtained from the following boundary condition:

$$\text{At } \phi = \pi/2, \quad r_2 = r \quad \text{and} \quad N_\phi = pr/2.$$

Hence, from Eq. (1) we get $C = 0$ and N_ϕ can be expressed as

$$N_\phi = \frac{pr^2}{2r_2 \sin^2 \phi}$$

or

$$N_\phi = pr_2/2. \tag{2}$$

From Eq. (6-11),

$$N_\theta = pr_2\left(1 - \frac{r_2}{2r_1}\right). \tag{3}$$

From analytical geometry, the relationship between the major and minor axes of an ellipse and r_1 and r_2 is given by

$$r_1 = \frac{a^2b^2}{(a^2 \sin^2 \phi + b^2 \cos^2 \phi)^{3/2}}$$

$$r_2 = \frac{a^2}{(a^2 \sin^2 \phi + b^2 \cos^2 \phi)^{1/2}}.$$

Substituting these expressions into Eqs. (2) and (3) gives the following expressions for membrane forces in ellipsoidal shells due to internal pressure:

$$N_\phi = \frac{pa^2}{2}\frac{1}{(a^2 \sin^2\phi + b^2 \cos^2\phi)^{1/2}} \tag{4}$$

$$N_\theta = \frac{pa^2}{2b^2}\frac{b^2 - (a^2 - b^2)\sin^2\phi}{(a^2 \sin^2\phi + b^2 \cos^2\phi)^{1/2}}. \tag{5}$$

A plot of Eqs. (4) and (5) is shown in Fig. 6-8. Equation (4) for the longitudinal force, N_ϕ, is always in tension regardless of the a/b ratio. Equation (5) for N_θ on the other hand gives compressive circumferential, hoop, forces near the equator when the value $a/b \geq \sqrt{2}$. For large a/b ratios under internal pressure, the compressive circumferential force tends to increase in magnitude and instability may occur for large a/t ratios. Thus extreme case must be exercised by the engineer to avoid buckling failure. The ASME VIII-2 code contains design rules that take into account the instability of shallow ellipsoidal shells due to internal pressure as discussed in Section 6-5.

For spherical shells under axisymmetric loads, the differential equations can be simplified by letting $r_1 = r_2 = R$. Equations (6-11) and (6-12) become

$$N_\phi + N_\theta = p_r R \tag{6-13}$$

and

$$N_\phi = \frac{R}{\sin^2\phi}\left[\int (p_r \cos\phi - p_\phi \sin\phi)\sin\phi \, d\phi + C\right]. \tag{6-14}$$

These two expressions form the basis for developing solutions to various loading conditions in spherical shells. For any loading condition, expressions for p_r and p_ϕ are first determined and then the above equations are solved for N_ϕ and N_θ.

Example 6-2

A concrete dome with thickness t has a dead load of γ psf. Find the expressions for N_ϕ and N_θ.

Solution

From Fig. 6-9a and Eq. (6-14),

$$p_r = -\gamma \cos\phi \quad \text{and} \quad p_\phi = \gamma \sin\phi$$

$$N_\phi = \frac{R}{\sin^2\phi}\left[\int(-\gamma \cos^2\phi - \gamma \sin^2\phi)\sin\phi \, d\phi + C\right]$$

$$N_\phi = \frac{R}{\sin^2\phi}(\gamma \cos\phi + C). \tag{1}$$

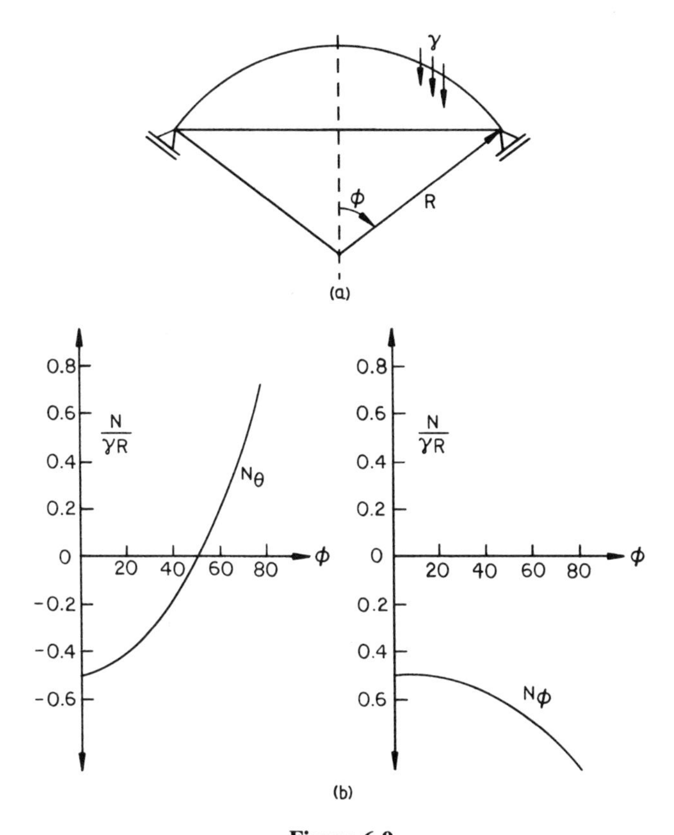

Figure 6-9.

As ϕ approaches zero, the denominator in Eq. (1) approaches zero. Accordingly, we must let the bracketed term in the numerator equal zero. This yields $C = -\gamma$. Equation (1) becomes

$$N_\phi = \frac{-R\gamma(1 - \cos\phi)}{\sin^2\phi}. \tag{2}$$

The convergence of Eq. (2) as ϕ approaches zero can be checked by l'Hopital's rule. Thus,

$$N_\phi|_{\phi=0} = \left.\frac{-R\gamma\sin\phi}{2\sin\phi\cos\phi}\right|_{\phi=0} = \frac{-\gamma R}{2}$$

Equation (2) can be written as

$$N_\phi = \frac{-\gamma R}{1 + \cos\phi}. \tag{3}$$

From Eq. (6-13), N_θ is given by

$$N_\theta = \gamma R\left(\frac{1}{1 + \cos\phi} - \cos\phi\right).$$

A plot of N_ϕ and N_θ for various values of ϕ is shown in Fig. 6-9b. The plot shows that for angles ϕ greater than 52°, the hoop force, N_θ, is in tension and special attention is needed for concrete reinforcing details.

Example 6-3

Find the forces in a spherical dome due to a lantern load P_o applied at an angle $\phi = \phi_o$ as shown in Fig. 6-10a.

Solution

Since $p_r = p_\phi = 0$, Eq. (6-14) becomes

$$N_\phi = \frac{RC}{\sin^2\phi}. \tag{1}$$

From statics at $\phi = \phi_o$, we get from Fig. 6-10b

$$N_\phi = \frac{P_o}{\sin\phi_o}.$$

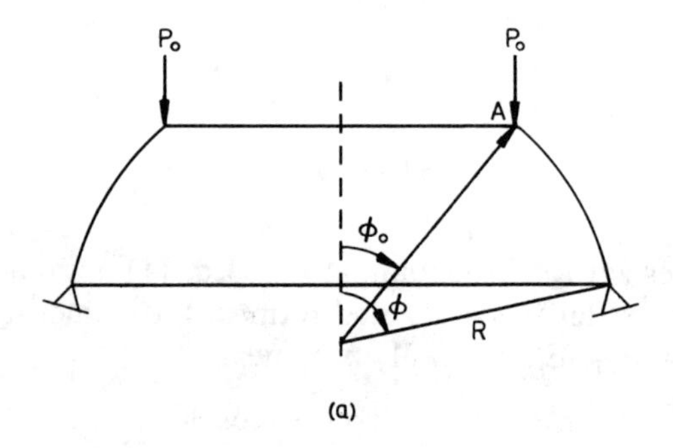

(a)

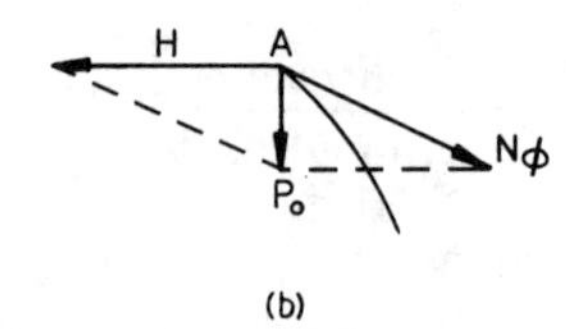

(b)

Figure 6-10.

Substituting this expression into Eq. (1), and keeping in mind that it is a compressive membrane force, gives

$$C = \frac{-P_o}{R} \sin \phi_o$$

and Eq. (1) yields

$$N_\phi = -P_o \frac{\sin \phi_o}{\sin^2 \phi}.$$

From Eq. (6-13),

$$N_\theta = P_o \frac{\sin \phi_o}{\sin \phi}.$$

In this example there is another force that requires consideration. Referring to Fig. 6-10b, it is seen that in order for P_o and N_ϕ to be in equilibrium, another horizontal force, H, must be considered. The direction of H is inwards in order for the force system to have a net resultant force P_o downwards. This horizontal force is calculated as

$$H = \frac{-P_o \cos \phi_o}{\sin \phi_o}.$$

Example 6-4

The sphere shown in Fig. 6-11a is filled with a liquid of density γ. Hence, p_r and p_ϕ can be expressed as

$$p_r = \gamma R(1 - \cos \phi)$$
$$p_\phi = 0.$$

(a) Determine the expressions for N_ϕ and N_θ throughout the sphere.
(b) Plot N_ϕ and N_θ for various values of ϕ when $\phi_o = 110°$.
(c) Plot N_ϕ and N_θ for various values of ϕ when $\phi_o = 130°$.
(d) If $\gamma = 62.4$ pcf, $R = 30$ ft, and $\phi_o = 110°$, determine the magnitude of the unbalanced force H at the cylindrical shell junction. Design the sphere, support cylinder, and the junction ring. Let the allowable stress in tension be 20 ksi and that in compression be 10 ksi.

Solution

(a) From Eq. (6-14), we obtain

$$N_\phi = \frac{\gamma R^2}{\sin^2 \phi}\left(\frac{1}{2}\sin^2 \phi + \frac{1}{3}\cos^3 \phi + C\right). \tag{1}$$

As ϕ approaches zero, the denominator approaches zero. Hence, the bracketed term in the numerator must be set to zero. This gives $C = -1/3$ and Eq. (1) becomes

$$N_\phi = \frac{\gamma R^2}{6 \sin^2 \phi} (3 \sin^2 \phi + 2 \cos^3 \phi - 2). \tag{2}$$

The corresponding N_θ from Eq. (6-13) is

$$N_\theta = \gamma R^2 \left[\frac{1}{2} - \cos \phi - \frac{1}{3 \sin^2 \phi} (\cos^3 \phi - 1) \right]. \tag{3}$$

As ϕ approaches π, we need to evaluate Eq. (1) at that point to ensure a finite solution. Again the denominator approaches zero and the bracketed term in the numerator must be set to zero. This gives $C = 1/3$ and Eq. (1) becomes

$$N_\phi = \frac{\gamma R^2}{6 \sin^2 \phi} (3 \sin^2 \phi + 2 \cos^3 \phi + 2). \tag{4}$$

The corresponding N_θ from Eq. (6-13) is

$$N_\theta = \gamma R^2 \left[\frac{1}{2} - \cos \phi - \frac{1}{3 \sin^2 \phi} (\cos^3 \phi + 1) \right]. \tag{5}$$

Equations (2) and (3) are applicable between $0 < \phi < \phi_o$, and Eqs. (4) and (5) are applicable between $\phi_o < \phi < \pi$.

(b) A plot of Eqs. (2) through (5) for $\phi_o = 110°$ is shown in Fig. 6-11b. N_ϕ below circle $\phi_o = 110°$ is substantially larger than that above circle

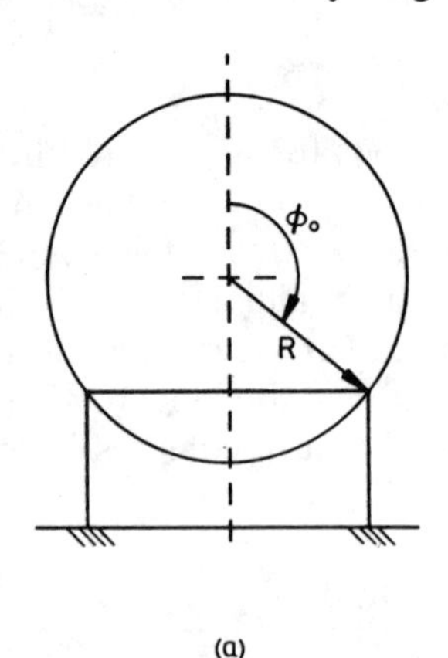

Figure 6-11.

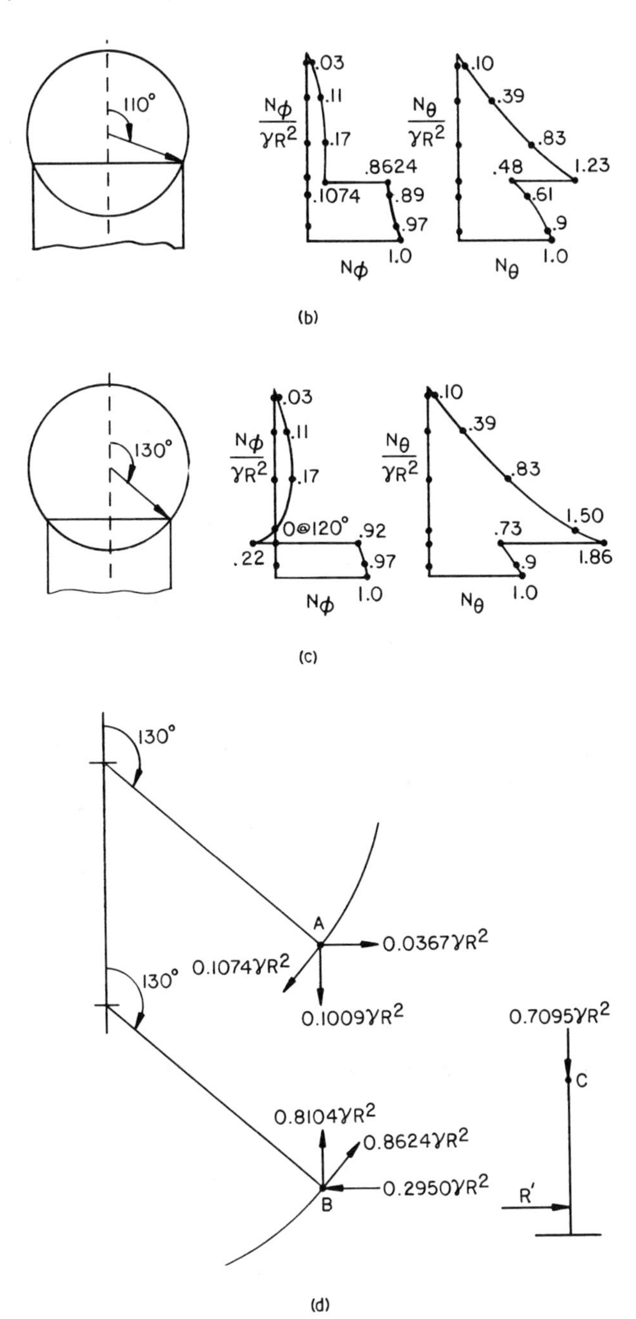

Figure 6-11 (continued)

110°. This is due to the fact that most of the weight of the contents is supported by the spherical portion that is below the circle $\phi_o = 110°$. Also, because N_ϕ does not increase in proportion to the increase in pressure as ϕ increases, Eq. (6-13) necessitates a rapid increase in N_θ in order to maintain the relationship between the left- and righthand sides. This is illustrated in Fig. 6-11b.

A plot of N_ϕ and N_θ for $\phi_o = 130°$ is shown in Fig. 6-11c. In this case, N_ϕ is in compression just above the circle $\phi_o = 130°$. This indicates that as the diameter of the supporting cylinder gets smaller, the weight of the water above circle $\phi_o = 130°$ must be supported by the sphere in compression. This results in a much larger N_θ value just above $\phi_o = 130°$. Buckling of the sphere becomes a consideration in this case.

(c) From Fig. 6-11b for $\phi_o = 110°$, the maximum force in the sphere is $N_\theta = 1.23\ \gamma R^2$. The required thickness of the sphere is

$$\begin{aligned} t &= \frac{1.23(62.4)(30)^2/12}{20{,}000} \\ &= 0.29 \text{ inch.} \end{aligned}$$

A free body diagram of the spherical and cylindrical junction at $\phi_o = 110°$ is shown in Fig. 6-11d. The values of N_ϕ at points A and B are obtained from Eqs. (2) and (4), respectively. The vertical and horizontal components of these forces are shown at points A and B in Fig. 6-11d. The unbalanced vertical forces result in a downward force at point C of magnitude 0.7095 γR^2. The total force on the cylinder is $(0.7095\ \gamma R^2)(2\pi)(R)(\sin(180 - 110))$. This total force is equal to the total weight of the contents in the sphere given by $(4/3)(\pi R^3)\gamma$. The required thickness of the cylinder is

$$\begin{aligned} t &= \frac{0.7095(62.4)30^2/12}{10{,}000} \\ &= 0.33 \text{ inch.} \end{aligned}$$

Summation of horizontal forces at points A and B results in a compressive force of magnitude 0.2583 γR^2. The needed area of compression ring at the cylinder to sphere junction is

$$\begin{aligned} A &= \frac{Hr}{\sigma} = \frac{0.2583 \times 62.4 \times 30^2(30 \sin 70)}{10{,}000} \\ &= 40.89 \text{ in.}^2 \end{aligned}$$

This area is furnished by a large ring added to the sphere or an increase in the thickness of the sphere at the junction.

Problems

6-1 Derive Eq. (6-3).
6-2 Derive Eq. (6-12).
6-3 Plot the values of N_ϕ and N_θ as a function of ϕ in an ellipsoidal shell with a ratio of 3:1.
6-4 The nose of a submersible titanium vehicle is made of a 2:1 ellipsoidal shell. Calculate the required thickness due to an external pressure of 300 psi. Let $a = 30$ inches, $b = 15$ inches, and the allowable compressive stress = 10 ksi.
6-5 Determine the forces in the spherical shelter (Fig. P6-5) due to snow load.
6-6 Determine the values of N_ϕ and N_θ of the roof of the underwater habitat (Fig. P6-6). For hydrostatic pressure, let

$$p_\phi = 0$$
$$p_r = \gamma[H + R(1 - \cos\phi)]$$

6-7 Determine the magnitude of the reaction R in the dome shelter shown in Fig. P6-7 due to dead load γ.
6-8 Determine the forces in the missile head (Fig. P6-8) due to load, γ, induced by acceleration. The equivalent pressure is expressed as

$$p_\phi = \gamma \sin\phi$$
$$p_r = -\gamma \cos\phi.$$

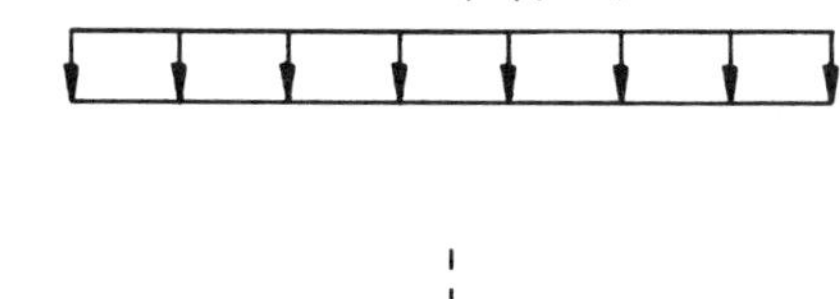

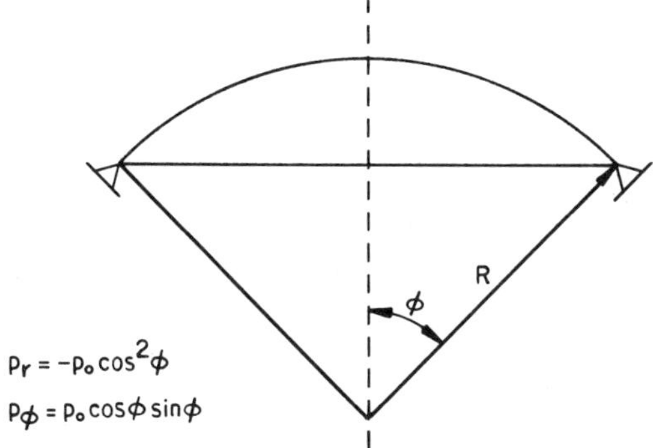

Figure P6-5.

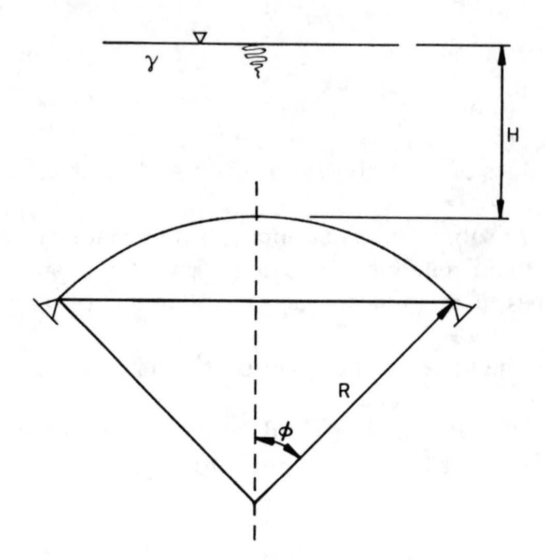

Figure P6-6.

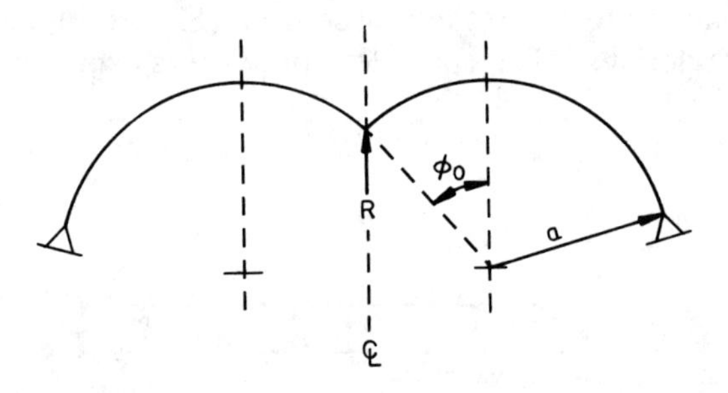

Figure P6-7.

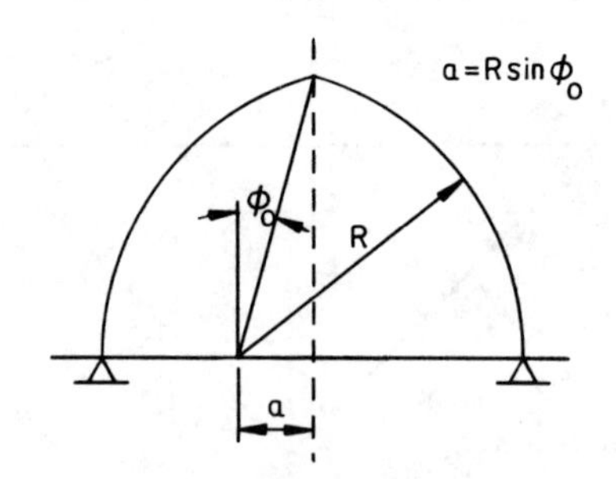

Figure P6-8.

6-3 Conical Shells

Equations (6-11) and (6-12) cannot readily be used for analyzing conical shells because the angle ϕ in a conical shell is constant. Hence, the two equations have to be modified accordingly. Referring to Fig. 6-12, it can be shown that

$$\left.\begin{aligned} \phi &= \beta = \text{constant} \\ r_1 &= \infty \qquad r_2 = s \tan \alpha \qquad r = s \sin \alpha \\ N_\phi &= N_s. \end{aligned}\right] \tag{6-15}$$

Equation (6-11) can be written as

$$\frac{N_s}{r_1} + \frac{N_\theta}{s \tan \alpha} = p_r$$

or since $r_1 = \infty$,

$$\left.\begin{aligned} N_\theta &= p_r s \tan \alpha \\ &= p_r r_2 \\ N_\theta &= \frac{p_r r}{\cos \alpha} \end{aligned}\right] . \tag{6-16}$$

Similarly from Eqs. (6-1) and (6-9),

$$\frac{d}{ds} r_1(s \sin \alpha N_s) - r_1 N_\theta \sin \alpha = -p_s s \sin \alpha r_1.$$

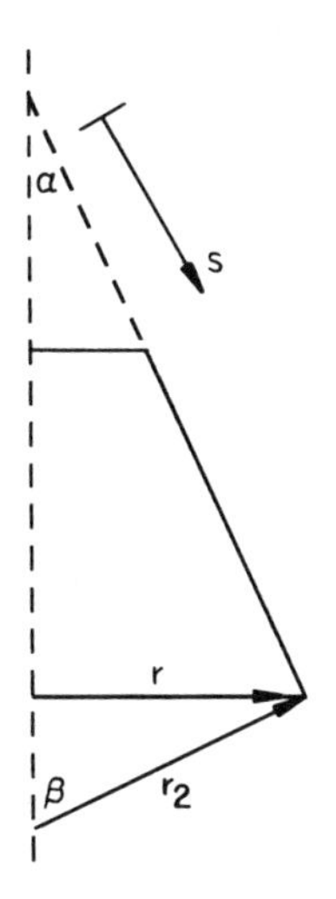

Figure 6-12.

Substituting Eq. (6-16) into this equation results in

$$N_s = \frac{-1}{s}\left[\int (p_s - p_r \tan \alpha) s\, ds + C\right]. \tag{6-17}$$

It is of interest to note that while N_θ is a function of N_ϕ for shells with double curvature, it is independent of N_ϕ for conical shells as shown in Eqs. (6-16) and (6-17). Also, as α approaches 0°, Eq. (6-17) becomes

$$N_\theta = p_r r_2,$$

which is the expression for the circumferential hoop force in a cylindrical shell.

The analysis of conical shells consists of solving the forces in Eqs. (6-16) and (6-17) for any given loading condition. The thickness is then determined from the maximum forces and a given allowable stress.

Example 6-5

Determine the longitudinal and circumferential forces, N_s and N_θ, of the mushroom-like concrete shelter shown in Fig. 6-13a due to a dead load, γ.

Solution

From Fig. 6-13, $p_s = \gamma \cos \alpha$ and $p_r = -\gamma \sin \alpha$

From Eq. (6-16),

$$N_\theta = -\gamma s \sin \alpha \tan \alpha.$$

From Eq. (6-17),

$$N_s = \frac{-1}{s}\left[\int\left(\gamma \cos \alpha + \gamma \sin \alpha \frac{\sin \alpha}{\cos \alpha}\right) s\, ds + C\right]$$

$$= -\frac{\gamma s}{2 \cos \alpha} - \frac{C}{s}.$$

At $s = L$, $N_s = 0$.

Hence,

$$C = \frac{-\gamma L^2}{2 \cos \alpha}$$

and

$$N_s = \frac{\gamma (L^2 - s^2)}{2s \cos \alpha}.$$

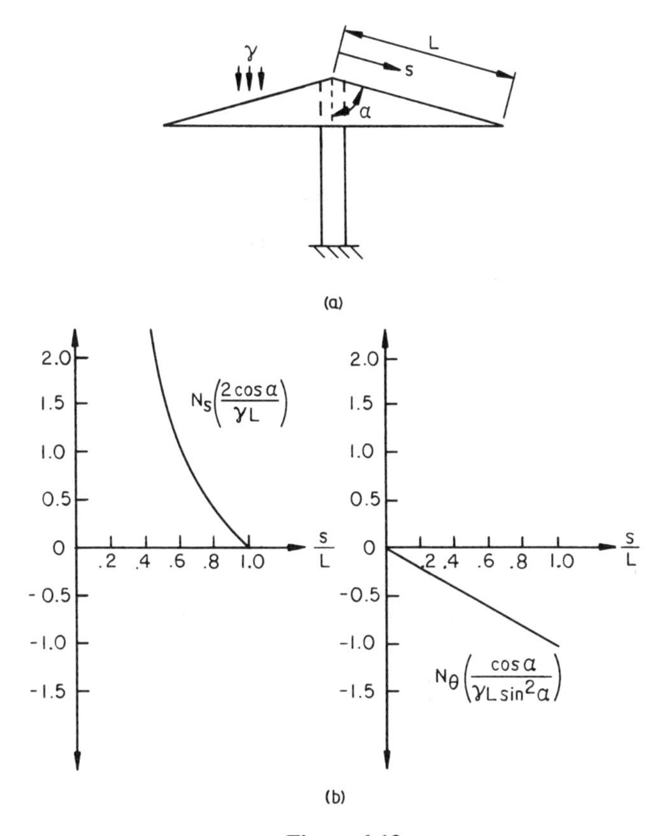

Figure 6-13.

A plot of N_θ and N_s is shown in Fig. 6-13b. The figure illustrates the need for reinforcing bars along the entire length of the shelter and especially near the support. Theoretically, reinforcing bars are not needed in the hoop direction. They are needed, however, as temperature reinforcement.

Example 6-6

Determine the maximum longitudinal and circumferential forces in the conical hopper shown in Fig. 6-14 due to internal pressure p.

Solution

From Eq. (6-16), the maximum N_θ occurs at the large end of the cone and is given by

$$N_\theta = p\left(\frac{r_o}{\sin\alpha}\right)\tan\alpha = \frac{pr_o}{\cos\alpha}.$$

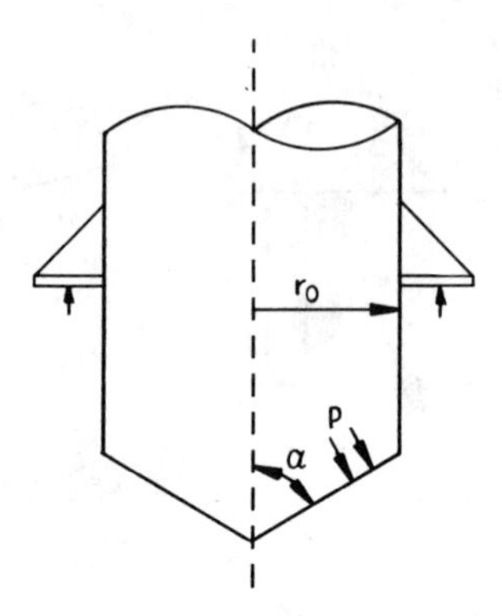

Figure 6-14.

From Eq. (6-17),

$$N_s = \frac{-1}{s}\left(\int -p \tan\alpha\, s\, ds + C\right) = \frac{-1}{s}\left(-p \tan\alpha \frac{s^2}{2} + C\right). \tag{1}$$

At $s = L$, $N_s = \dfrac{pr_o}{2}\dfrac{1}{\cos\alpha}$

Substituting this expression into Eq. (1), and using the relationships of Eq. (6-15), gives $C = 0$. Equation (1) becomes

$$N_s = \frac{pr}{2 \cos \alpha}$$

$$\text{and max } N_s = \frac{pr_o}{2 \cos \alpha}.$$

It is of interest to note that the longitudinal and hoop forces are identical to those of a cylinder with equivalent radius of $r_o/\cos \alpha$.

Problems

6-9 Determine the maximum values and location of N_s and N_θ in the wine glass shown in Fig. P6-9. Let $L = 3.00$ inches, $L_o = 0.25$ inches, $\gamma = 0.0289$ lb/in^3, and $\alpha = 30°$.

6-10 The lower portion of a reactor is subjected to a radial nozzle load as shown in Fig. P6-10. Determine the required thickness of the conical

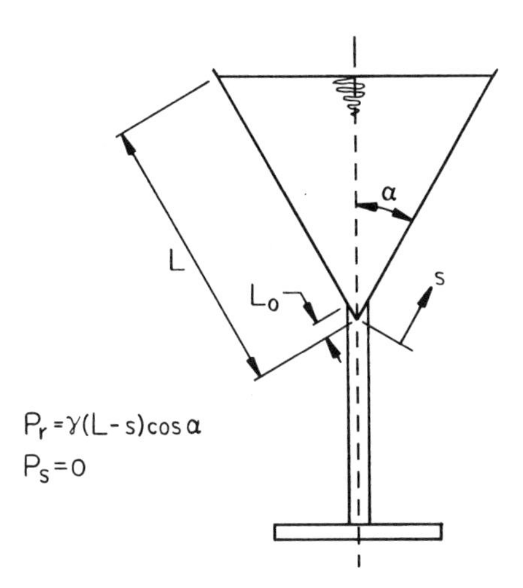

Figure P6-9.

section. Use an allowable stress of 10 ksi. What is the required area of the ring at the point of application of the load?

6-11 Solve Example 6-4 for a snow load, q.

6-4 Wind Loads

The distribution of wind pressure on shells of revolution is assumed (Flugge 1967) perpendicular to the surface (Fig. 6-15), and is usually expressed as

$$\left.\begin{aligned} p_r &= -p \sin\phi \cos\theta \\ p_\theta &= p_\theta = 0. \end{aligned}\right] \tag{6-18}$$

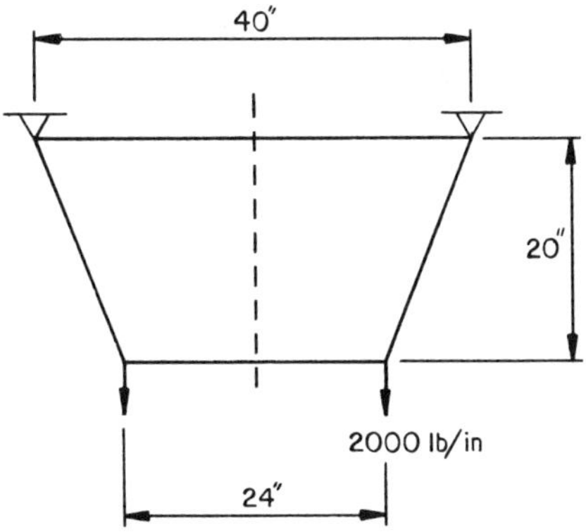

Figure P6-10.

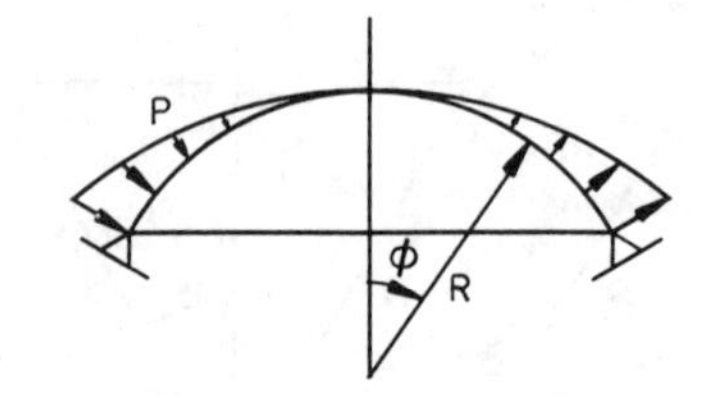

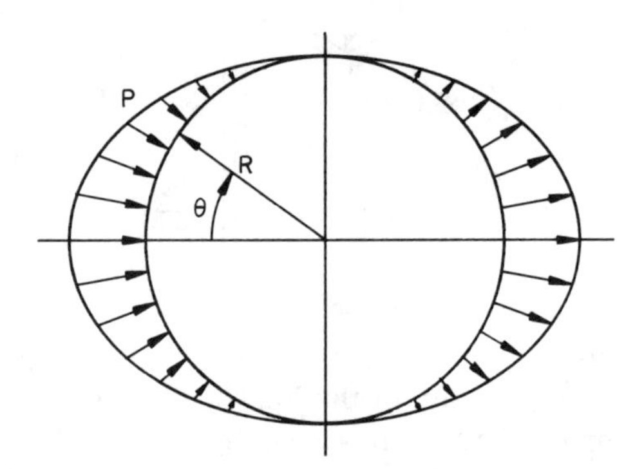

Figure 6-15.

This distribution is an approximation of the actual wind pressure on a structure. More accurate approximations can be made. However, they are too complicated to solve mathematically in a closed-form equation. The result obtained from an approximate distribution is well within most acceptable engineering tolerances and is discussed in this section. The wind load (Fig. 6-15) causes a compressive pressure on the windward side of the shell and suction on the leeward side of the shell. The total sum of the pressure, i.e., windward plus leeward, is equal to the numerical value published by various standards for wind loads on structures. The load distribution is a function of both ϕ and θ and the equilibrium Eqs. (6-3), (6-7), and (6-8) are valid for obtaining membrane forces in the shell due to wind loads. Substituting Eq. (6-18) into Eq. (6-8) gives

$$N_\theta = -pr_2 \sin\phi \cos\theta - \frac{r_2}{r_1} N_\phi. \tag{6-19}$$

Using this equation to eliminate N_θ from Eqs. (6-3) and (6-7) and using $N_{\theta\phi} = N_{\phi\theta}$ results in

$$\frac{\partial N_\phi}{\partial_\phi} + \left(1 + \frac{r_1}{r_2}\right) N_\phi \cot\phi - \frac{r_1}{r} \frac{\partial N_{\phi\theta}}{\partial\theta} = -pr_1 \cos\phi \cos\theta \tag{6-20}$$

and

$$\frac{\partial N_{\phi\theta}}{\partial \phi} + 2\frac{r_1}{r_2} N_{\phi\theta} \cot \phi + \frac{1}{\sin \phi} \frac{\partial N_{\phi}}{\partial \theta} = pr_1 \sin \theta. \qquad (6\text{-}21)$$

Thus, membrane forces in shells of revolution due to wind loads are obtained by solving first Eqs. (6-20) and (6-21) for the two forces N_{ϕ} and $N_{\phi\theta}$. Then Eq. (6-19) is solved for N_{θ}.

For spherical shells, $r_1 = r_2 = R$. Also, the membrane forces are assumed distributed so that N_{θ} and N_{ϕ} have their maximum value at $\theta = 0°$ and 180°, as these are the locations for maximum wind load (Fig. 6-15). The shell acts similar to a beam under bending. The shearing force $N_{\phi\theta}$ is a maximum at $\theta = 90°$ where N_{ϕ} and N_{θ} are zero. The three membrane forces can be expressed as

$$\left.\begin{aligned} N_{\phi} &= C_{\phi} \cos \theta \\ N_{\phi\theta} &= C_{\phi\theta} \sin \theta \\ N_{\theta} &= C_{\theta} \cos \theta \end{aligned}\right] \qquad (6\text{-}22)$$

where C_{ϕ}, $C_{\phi\theta}$, and C_{θ} are functions of ϕ only. Substituting Eq. (6-22) into Eq. (6-20) results in

$$\frac{dC_{\phi}}{d\phi} + 2C_{\phi} \cot \phi - \frac{1}{\sin \phi} C_{\phi\theta} = -pR \cos \phi. \qquad (6\text{-}23)$$

Similarly, substituting Eq. (6-22) into Eq. (6-21) gives

$$\frac{dC_{\phi\theta}}{d\phi} + 2C_{\phi\theta} \cot \phi - \frac{1}{\sin \phi} C_{\phi} = pR. \qquad (6\text{-}24)$$

The solution of Eqs. (6-23) and (6-24) can best be obtained by defining functions U_{ϕ} and V_{ϕ} as

$$\left.\begin{aligned} U_{\phi} &= C_{\phi} + C_{\phi\theta} \\ V_{\phi} &= C_{\phi} - C_{\phi\theta}. \end{aligned}\right] \qquad (6\text{-}25)$$

Adding Eqs. (6-23) and (6-24) gives

$$\frac{dU_{\phi}}{d\phi} + \left(2 \cot \phi - \frac{1}{\sin \phi}\right) U_{\phi} = pR(1 - \cos \phi). \qquad (6\text{-}26)$$

This is a partial differential equation of the form

$$\frac{dy}{dx} + A(x)y = B(x),$$

the solution of which is given by

$$y = \frac{1}{e^{\int A(x)dx}} \int B(x) e^{\int A(x)dx}\, dx.$$

$$\begin{aligned} \text{The quantity } e^{\int A(x)\,dx} &= e^{\int (2\cot\phi - (1/\sin\phi))\,dx} \\ &= e^{2\ln\sin\phi - \ln\tan(\phi/2)} \\ &= \sin^2\phi \cot(\phi/2). \end{aligned}$$

$$\begin{aligned} \text{The quantity } \int \mathrm{B}(x) e^{\int A(x)\,dx} &= \int_0^{\phi} pR(1 - \cos\phi)\sin^2\phi \cot(\phi/2) \\ &= \int_0^{\phi} pR \sin^3\phi \\ &= \frac{-pR}{3}[\cos\phi(\sin^2\phi + 2) - 2]. \end{aligned}$$

Thus, U_ϕ becomes

$$U_\phi = \frac{-pR}{3\sin^2\phi\cot(\phi/2)}[\cos\phi(\sin^2\phi + 2) - 2].$$

Similarly, V_ϕ in Eq. (6-25) is obtained by subtracting Eqs. (6-23) and (6-24) and following the above procedure. This gives

$$V_\phi = \frac{pR}{3\sin^2\phi\tan(\phi/2)}[\cos\phi(\sin^2\phi + 2) - 2]$$

and

$$\begin{aligned} C_\phi &= \frac{U_\phi + V_\phi}{2} \\ &= \frac{pR[\cos\phi(\sin^2\phi + 2) - 2]\cos\phi}{3\sin^3\phi}. \end{aligned}$$

Similarly

$$\begin{aligned} C_{\phi\theta} &= \frac{U_\phi - V_\phi}{2} \\ &= \frac{pR[\cos\phi(\sin^2\phi + 2) - 2]}{3\sin^3\phi}. \end{aligned}$$

With C_ϕ and $C_{\phi\theta}$ known, Eq. (6-22) is solved for N_ϕ and $N_{\phi\theta}$. The value of N_θ is obtained from Eq. (6-19).

For conical shells, Fig. 6-12 and Eq. (6-15), define the geometry. Replacing $N_{\phi\theta}$ with $N_{s\theta}$ and defining $ds = r\,d\,\phi$, we get from Eq. (6-21),

$$\frac{dN_{s\theta}}{ds} + \frac{2}{s} N_{s\theta} = p \sin \theta$$

which upon integrating gives

$$N_{s\theta} = \frac{p \sin \theta}{s^2}\left(\frac{s^3}{3} + C\right) \tag{6-27}$$

where C is a constant obtained from the boundary conditions. Similarly, Eq. (6-20) reduces to

$$N_s = p\left(\frac{s}{6 \sin \alpha} - \frac{C}{s^2 \sin \alpha} - \frac{s \sin \alpha}{2}\right) \cos \theta \tag{6-28}$$

and Eq. (6-19) for the hoop force becomes

$$N_\theta = -ps \sin \alpha \cos \theta. \tag{6-29}$$

Problems

6-12 Derive Eqs. (6-20) and (6-21).
6-13 Derive Eqs. (6-27) and (6-28).
6-14 The auditorium dome (Fig. P6-14) is subjected to a wind pressure, p, of 12 psf. Determine the maximum forces and their location.

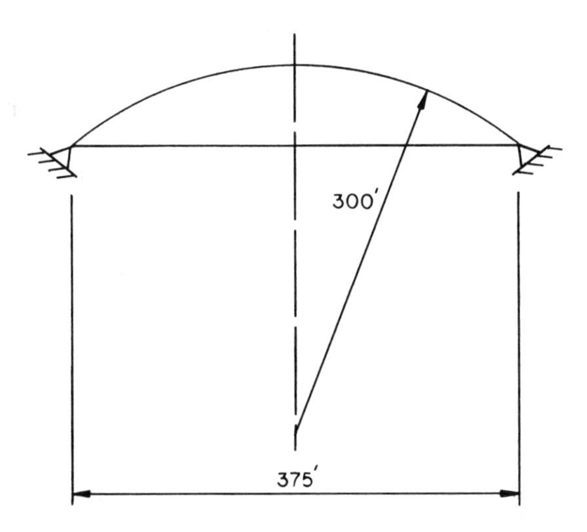

Figure P6-14.

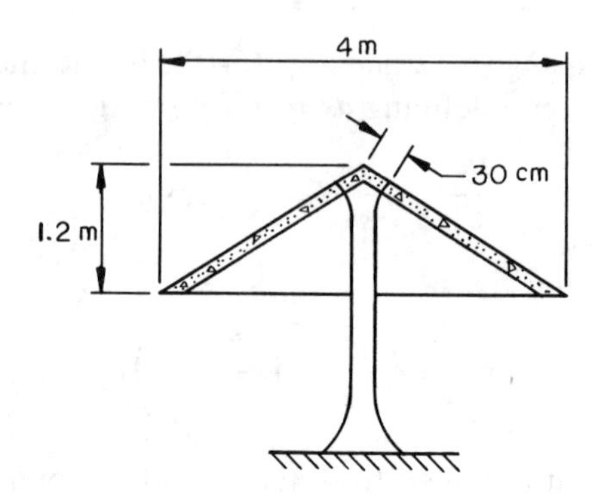

Figure P6-15.

6-15 The picnic shed shown in Fig. P6-15 is subjected to a wind pressure, p, of 50 kgf/m^2. Determine the maximum forces and their location.

6-5 Design of Shells of Revolution

The maximum forces for various shell geometries subjected to commonly encountered loading conditions are listed in numerous references. One such reference is by NASA (Baker et al. 1968), where extensive tables and design charts are listed. Flugge (1967) contains a thorough coverage of a wide range of applications to the membrane theory, as does Roark and Young (1975).

Extra care should be taken in the design of shallow ellipsoidal heads subjected to internal pressure. Example 6-1 illustrated the possibility of inelastic instability in the circumferential direction of an ellipsoidal head due to internal pressure. Many codes have provisions and design aids for avoiding such instability, which tends to occur in heads subjected to low pressures and having large diameter-to-thickness ratios of over about 230. The ASME VIII-2 code provides a chart (Fig. 6-16) for designing ellipsoidal heads with various a/b ratios. The design is based on approximating the geometry of a head with a spherical radius, L, and a knuckle radius, r, as defined in Fig. 6-16. The required thickness of a specific head is determined from Fig. 6-16 by knowing the values of L, r, base diameter of the head, applied pressure, and allowable membrane stress.

Example 6-7

Some of the commonly used boiler heads in the United States have an a/b ratio of about 2.95. This corresponds to $L = D$ and $r = 0.06D$. Determine

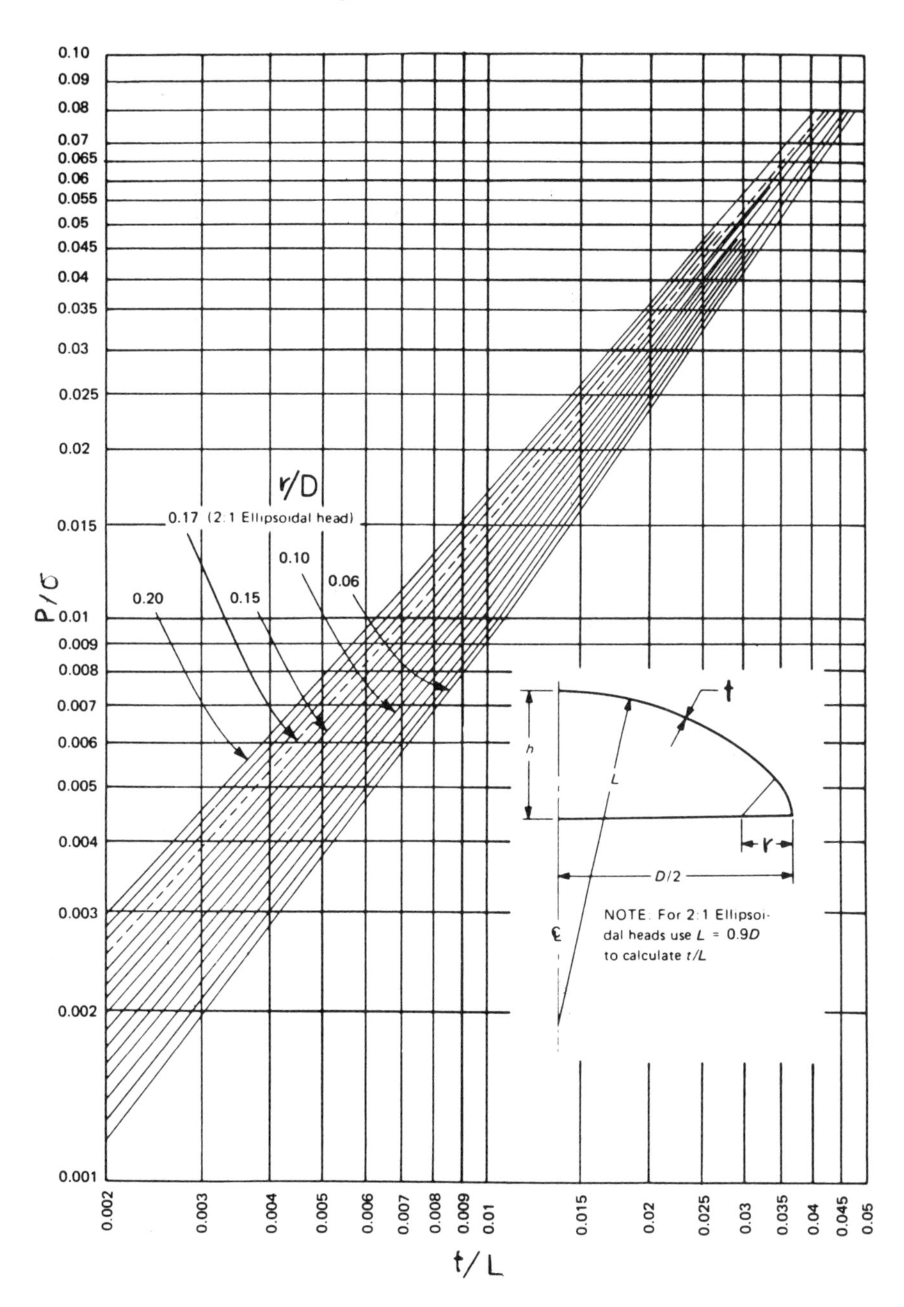

Figure 6-16. (Courtesy of ASME.)

the required thickness of this head if $D = 14$ ft, $p = 75$ psi, and $\sigma =$ 20,000 psi. Calculate the required thickness from Eq. (4) of Example 6-1 and also from Fig. 6-16.

Solution

From Eq. (4) of Example 6-1, the maximum force is at $\phi = 0°$. Thus,

$$t = \frac{pa^2}{2b\sigma}$$

$$= \frac{75 \times (168/2)^2}{2 \times (84/2.95) \times 20{,}000}$$

$$= 0.46 \text{ inch.}$$

From Fig. 6-16 with $p/\sigma = 0.0038$, and $r/D = 0.06$ we get $t/L = 0.005$,

$$\text{or } t = 0.005 \times 168 = 0.84 \text{ inch (controls).}$$

Notice that the thickness obtained from Fig. 6-16 is almost double that determined from theoretical membrane equations that do not take into consideration any instability due to internal pressure.

7

Various Applications of the Membrane Theory

7-1 Analysis of Multi-Component Structures

The quantity $N_\phi r_2 \sin^2 \phi$ in Eq. (6-12), when multiplied by 2π, represents the total applied force acting on a structure at a given parallel circle of angle ϕ. Hence, for complicated geometries, the value of N_ϕ in Eq. (6-12) at any given location can be obtained by taking a free-body diagram of the structure. The value of N_θ at the same location can then be determined from Eq. (6-11). This method is widely used (Jawad and Farr 1989) in designing pressure vessels, flat-bottom tanks, elevated water towers (Fig. 7-1), and other similar structures. Example 7-1 illustrates the application of this method to the design of a water tower. The American Petroleum Institute (API 620 1991) Standard has various equations and procedures for designing components by the free-body method. This method is also useful in obtaining an approximate design at the junction of two shells of different geometries. A more accurate analysis utilizing bending moments may then be performed to establish the discontinuity stresses of the selected members at a junction if a more exact analysis is needed.

Example 7-1

The tank shown in Fig. 7-2 is filled with a liquid up to point *a*. The specific gravity is 1.0. Above point *a* the tank is subjected to a gas pressure of 0.5 psi. Determine the forces and thicknesses of the various components of the steel tank disregarding the dead weight of the tank. Use an allowable tensile stress of 12,000 psi and an allowable compressive stress of 8000 psi.

Figure 7-1. Elevated water tank. (Courtesy of Chicago Bridge and Iron Company.)

Solution

Tank Roof

The maximum force in the roof is obtained from Fig. 7-3a. Below section a–a, a 0.5 psi pressure is needed to balance the pressure above section a–a. Force N_ϕ in the roof has a vertical component V around the perimeter

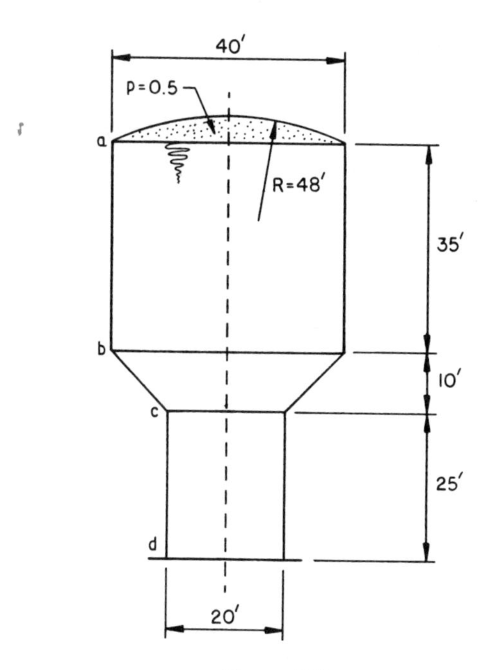

Figure 7-2.

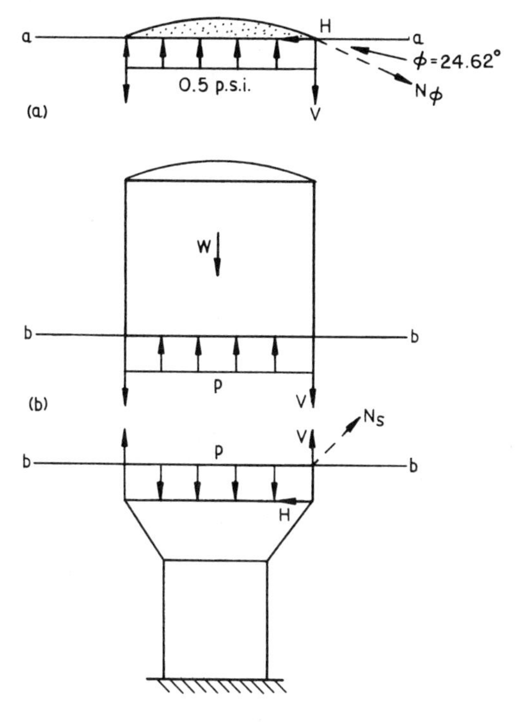

Figure 7-3.

of the roof. Summation of forces in the vertical direction gives

$$2\pi RV - \pi R^2 P = 0$$
$$V = 60 \text{ lbs/inch.}$$

Hence,

$$N_\phi = \frac{V}{\sin \phi} = 60/0.42$$
$$= 144 \text{ lbs/inch.}$$

From Eq. (6-11) with $r_1 = r_2 = R$

$$N_\theta = 144 \text{ lbs/inch.}$$

Since N_θ and N_ϕ are the same, use either one to calculate the thickness. The required thickness is

$$t = N_\phi/\sigma = 144/12{,}000$$
$$= 0.012 \text{ inch.}$$

Because this thickness is impractical to handle during fabrication of a tank with such a diameter, use $t = 1/4$ inch.

40-Ft Shell

The maximum force in the shell is at section b–b as shown in Fig. 7-3b. The total weight of liquid at section b–b is

$$W = 62.4(\pi)(20)^2(35)$$
$$= 2{,}744{,}500 \text{ lbs.}$$

Total pressure at b–b is

$$P = 0.5 + (62.4/144)(35)$$
$$P = 15.67 \text{ psi.}$$

The total sum of the vertical forces at b–b is equal to zero. Hence,

$$2{,}744{,}500 - (15.67)(\pi)(240)^2 + V(\pi)(480) = 0$$

or

$$V = 60 \text{ lbs/inch}$$

and

$$N_\phi = 60 \text{ lbs/inch.}$$

In a cylindrical shell, $r_1 = \infty$ and $r_2 = R$. Hence, Eq. (6-11) becomes

$$N_\theta = pR = (15.67)(240) = 3761 \text{ lb/inch.}$$

The required thickness $t = N_\theta/\sigma = 3761/12{,}000$.

$$t = .031 \text{ inch.}$$

Use $t = 3/8$ inch in order to match the conical transition section discussed later.

The unbalanced force at the roof-to-shell junction is

$$H = 131 \text{ lb/inch (inwards).}$$

The area required to contain the unbalanced force, H, is given by

$$\begin{aligned} A &= (H)\text{ (shell radius)/allowable compressive stress} \\ &= 131 \times 240/8000 = 3.93 \text{ inch}^2. \end{aligned}$$

Use 1 inch thick $\times$ 4 inch wide ring as shown in Fig. 7-4a.

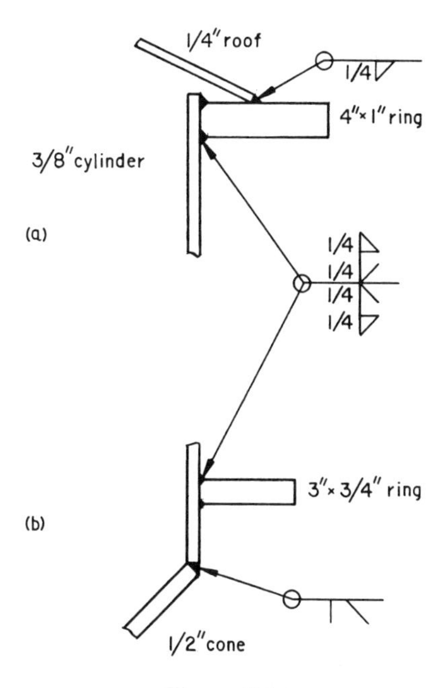

Figure 7-4.

Conical Transition

At section $b-b$, force V in the 40-ft shell must equal force V in the cone in order to maintain equilibrium as shown in Fig. 7-3b. Thus,

$$V = 60 \text{ lb/inch.}$$

and

$$N_s = 60/0.707 = 85 \text{ lb/inch.}$$

In a conical shell $r_1 = \infty$ and $r_2 = R/\cos\alpha$. Hence from Eq. (6-15)

$$N_\theta = pR/\cos\alpha = 240(15.67)/0.707 = 5319 \text{ lb/inch.}$$

The required thickness at the large end of the cone is

$$t = 5319/12{,}000 = 0.44 \text{ inch.}$$

The horizontal force at point b is N_s cos 45.

$$H = 60 \text{ lbs/inch (inwards)}$$

The required area is

$$A = 60 \times 240/8000 = 1.8 \text{ in}^2.$$

Use 3/4 inch thick by 3 inch wide ring as shown in Fig. 7-4b.

The forces at the small end of the cone are shown in Fig. 7-5a. The weight of the liquid in the conical section at point c is

$$\begin{aligned} W &= \pi\gamma H(r_1^2 + r_1 r_2 + r_2^2)/3 \\ &= \pi \times 62.4 \times 10(10^2 + 10 \times 20 + 20^2)/3 \\ &= 457{,}400 \text{ lbs.} \end{aligned}$$

Total liquid weight is

$$W = 2{,}744{,}500 + 457{,}400 = 3{,}201{,}900 \text{ lbs.}$$

Pressure at section $c-c$ is

$$\begin{aligned} P &= 0.5 + (62.4/144)(45) \\ &= 20.0 \text{ psi.} \end{aligned}$$

Summing forces at section $c-c$ gives

$$20.0 \times \pi \times 120^2 - 3{,}201{,}900 - (V \times \pi \times 240) = 0$$
$$V = -3047 \text{ lb/inch.}$$

The negative sign indicates that the vertical component of N_s is opposite to that assumed in Fig. 7-5a and is in compression rather than tension. This is caused by the column of liquid above the cone whose weight is greater than the net pressure force at section $c-c$.

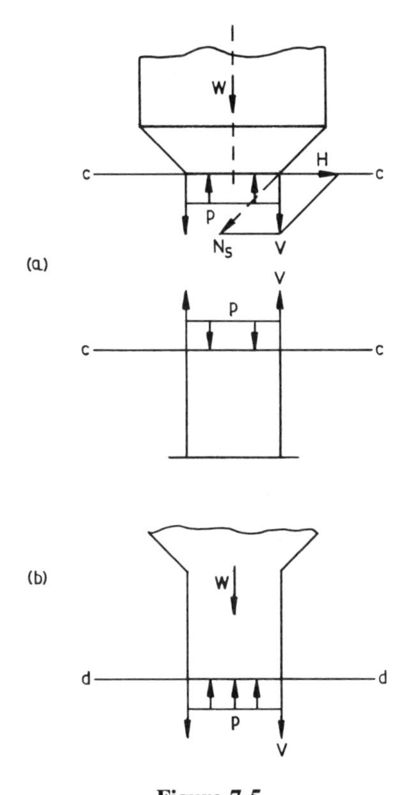

Figure 7-5.

$$
\begin{aligned}
N_s &= -3047/0.707 \\
&= -4309 \text{ lb/inch (compressive)} \\
N_\theta &= pR/\cos\alpha = 20.0 \times 120/0.707 \\
&= 3395 \text{ lb/inch.}
\end{aligned}
$$

N_θ at the small end is smaller than N_θ at the large end. Hence, the thickness at the small end need not be calculated for N_θ. Since N_s at the small end is in compression, the thickness due to this force needs to be calculated because the allowable stress in compression is smaller than that in tension. Hence,

$$t = 4309/8000 = 0.54 \text{ inch.}$$

Use $t = 5/8$ inch for the cone.

The horizontal force at section c–c (Fig. 7-5) is given by

$$H = 3047 \text{ lb/inch inwards.}$$

The required area of the ring is

$$A = 3047 \times 120/8000 = 45.71 \text{ in.}^2$$

This large required area is normally distributed around the junction as shown in Fig. 7-6.

20-Ft Shell

At section c–c (Fig. 7-5), the value of V in the 20-ft shell is the same as V in the cone due to continuity. Thus,

$$N_\phi = V = -3047 \text{ lb/inch}$$

$$N_\theta = pR = 20.0 \times 120 = 2400 \text{ lb/inch.}$$

At section d–d (Fig. 7-5), the liquid weight is given by

$$W = 3{,}201{,}900 + (62.4)(\pi)(10)^2(25)$$
$$= 3{,}692{,}000 \text{ lb}$$

and the pressure is calculated as

$$p = 0.5 + (62.4/144)(70)$$
$$= 30.83 \text{ psi.}$$

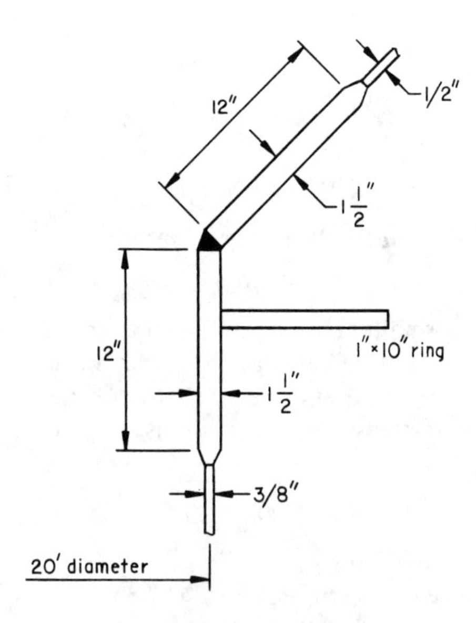

Figure 7-6.

From Fig. 7-5b, the summation of forces about d–d gives

$$3{,}692{,}000 - 30.83 \times \pi \times 120^2 + V \times \pi \times 240 = 0$$

or

$$N_\phi = V = -3047 \text{ lb/inch},$$

which is the same as that at point c.

$$N_\theta = pR = 30.83 \times 120$$
$$= 3700 \text{ lb/inch}.$$

The required thickness of the shell is governed by N_ϕ at section d–d.

$$t = 3047/8000 = 0.38 \text{ inch}.$$

Use $t = 3/8$ inch for bottom cylindrical shell.

Problems

7-1 Determine the thickness of all components including stiffener rings at points A and B of the steel tower shown in Fig. P7-1. The tower is full

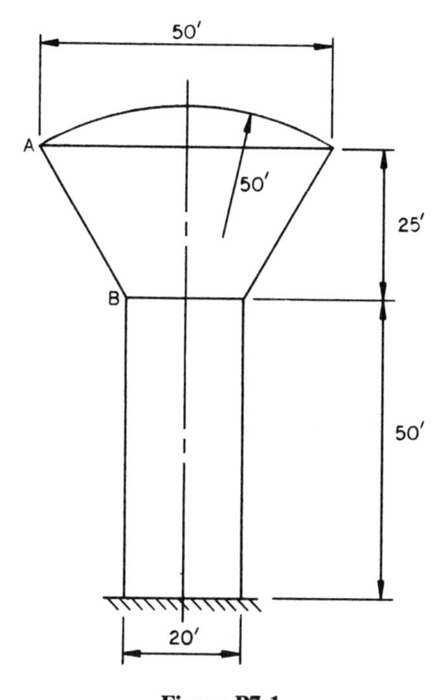

Figure P7-1.

of liquid to point A. The roof is subjected to a snow load of 25 psf. The allowable stress in tension is 15 ksi and that in compression is 10 ksi.

7-2 Determine the thickness of all components including stiffener rings of the elevated water tower shown in Fig. P7-2. The tower is full of liquid between points A and D. The allowable stress in tension is 15 ksi and that in compression is 10 ksi.

7-2 Pressure-Area Method of Analysis

The membrane theory is very convenient in determining thicknesses of major components such as cylindrical, conical, hemispherical, and ellipsoidal shells. The theory, however, is inadequate for analyzing complicated geometries such as nozzle attachments, transition sections, and other details similar to those shown in Fig. 7-7. An approximate analysis of these components can be obtained by using the pressure-area method. A more accurate analysis can then be performed based on the bending theories of Chapters 8 and 9 or the finite element theory.

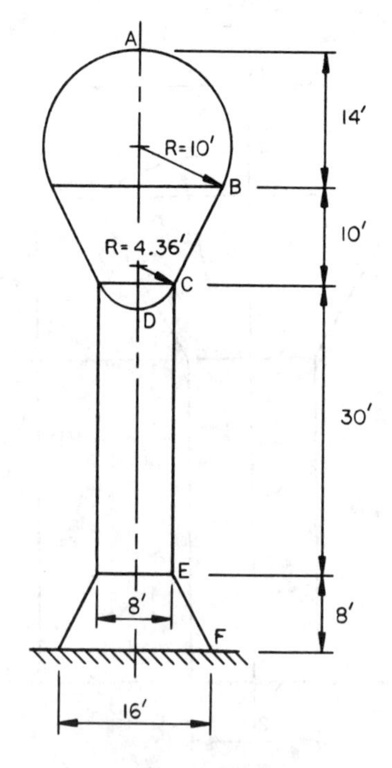

Figure P7-2.

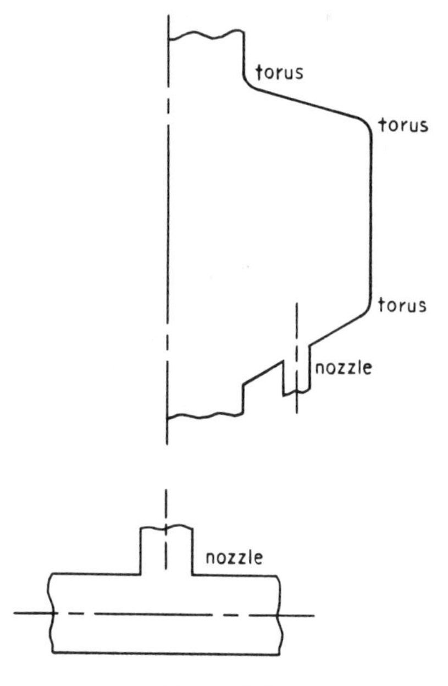

Figure 7-7.

The pressure-area analysis is based on the concept (Zick 1963) that the pressure contained in a given area within a shell must be resisted by the metal close to that area. Referring to Fig. 7-8a, the total force in the shaded area of the cylinder is $(r)(P)(L)$ while the force supported by the available metal is $(L)(t)(\sigma)$. Equating these two expressions results in $t = Pr/\sigma$ which is the equation for the required thickness of a cylindrical shell. Similarly for spherical shells, Fig. 7-8b gives

$$(R\phi)(R)(P)(1/2) = (R\phi)(t)(\sigma)$$

$$t = PR/2\sigma.$$

Referring to Fig. 7-9a, it is seen that pressure area A is supported by the cylinder wall and pressure area B is supported by the nozzle wall. However, pressure area C is not supported by any material. Thus it must be supported by adding material, M, at the junction. The area of material M is given by

$$(P)(R)(r) = (\sigma)(M)$$

$$M = (P(R)(r)/\sigma.$$

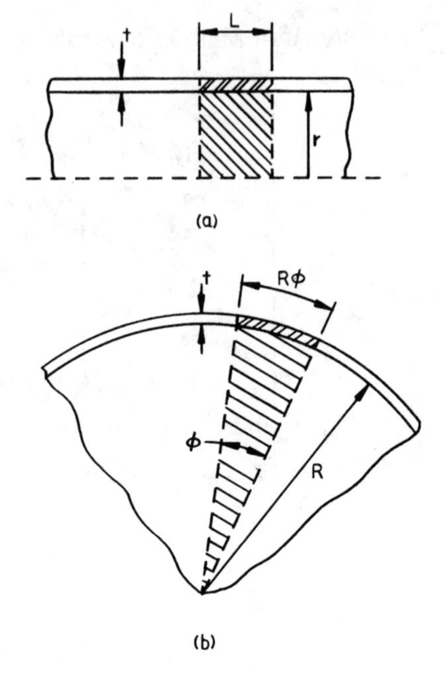

Figure 7-8.

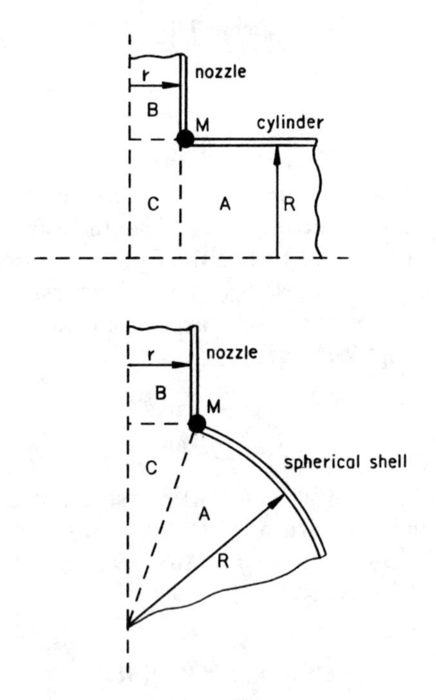

Figure 7-9.

For a spherical shell, the required area, M, from Fig. 7-9b is

$$(P)(R)(r)(1/2) = (\sigma)(M)$$

$$M = (1/2)(P)(R)(r)/\sigma.$$

The required area is added either to the shell, nozzle, or as a reinforcing pad as shown in Fig. 7-10.

The pressure-area method can also be applied for junctions between components as shown in Fig. 7-11. Referring to Fig. 7-11a, the spherical shell must support the pressure within area ABC. The cylindrical shell supports the pressure within area $AOCD$. At point A where the spherical and cylindrical shells intersect, the pressure area to be supported at point A is given by AOC. However, because area AOC is used both in the ABC area for the sphere and $AOCD$ for the cylinder, and because it can be used only once, this area must be subtracted from the total calculated pressure in order to maintain equilibrium. In other words, this area causes compressive stress at point A. The area required is given by

$$A = (r)(\sqrt{R^2 - r^2})(1/2)(P)/\sigma$$

where σ is compressive stress.

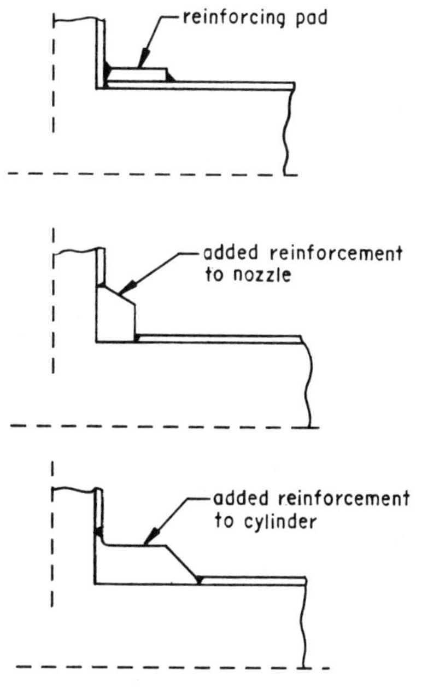

Figure 7-10.

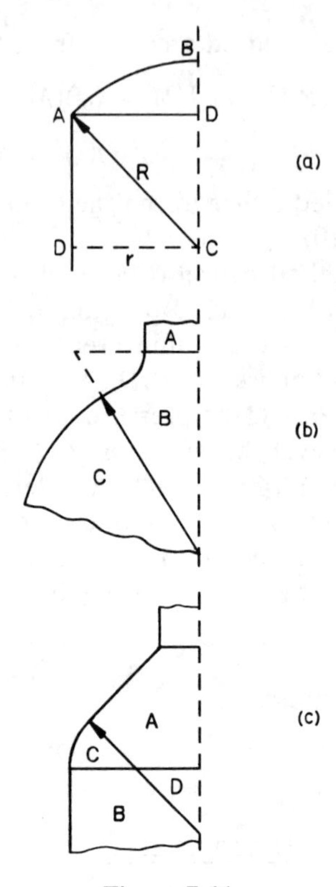

Figure 7-11.

In Fig. 7-11b, pressure area *A* is supported by the cylindrical shell and area *C* by the spherical shell. Area *B* is supported by the transition shell. The transition shell is in tension because area *B* is used neither in the area *A* nor area *B* calculations.

In Fig. 7-11c, pressure area *A* is supported by the cone and area *B* is supported by the cylinder. The transition shell between the cone and the cylinder supports pressure area *C* which is in tension and area *D* which is in compression. Summation of areas *C* and *D* will determine the state of stress in the transition shell.

The pressure-area method is also commonly used to design fittings and other piping components. Figure 7-12 shows one design method for some components.

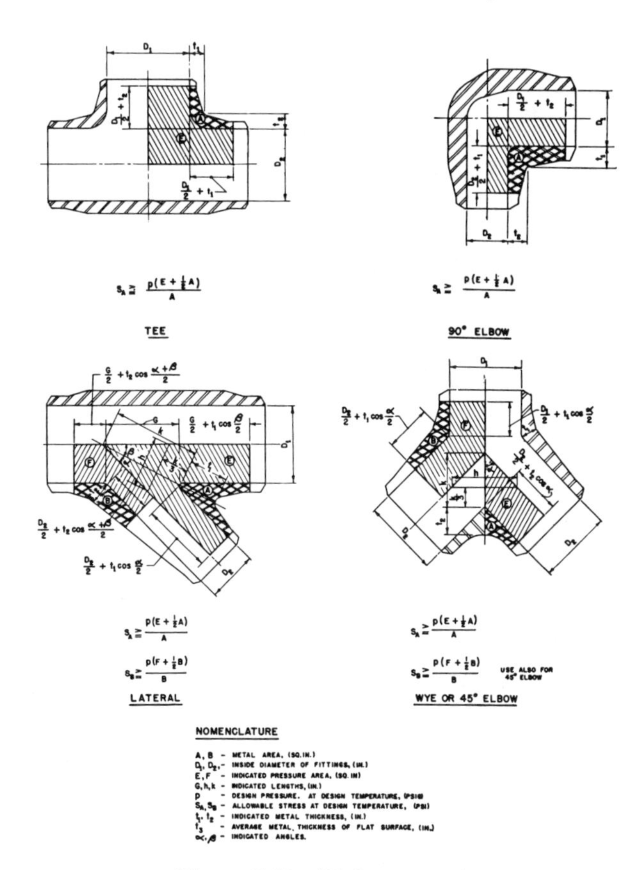

Figure 7-12. (Kellogg 1961.)

Example 7-2

Find the required thickness of the cylindrical, spherical, and transition shells shown in Fig. 7-13. Let $p = 150$ psi and $\sigma = 15{,}000$ psi.

Solution

For the cylindrical shell,

$$t = pr/\sigma = 150 \times 36/15{,}000 = 0.36 \text{ inch.}$$

For the spherical shell,

$$t = pR/2\sigma = 150 \times 76/2 \times 15{,}000 = 0.38 \text{ inch.}$$

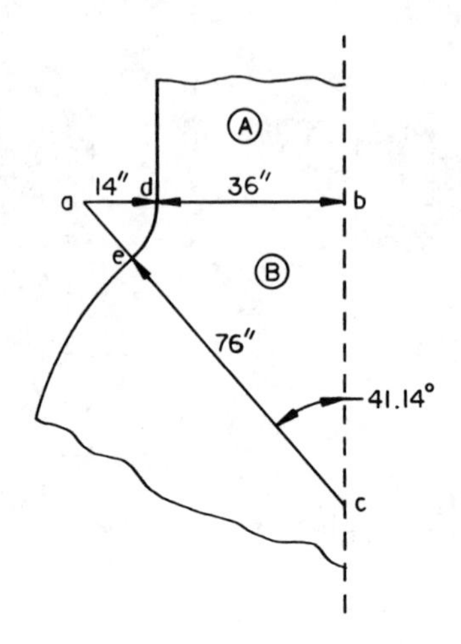

Figure 7-13.

For the transition shell, pressure area B = area of triangle abc − area of segment ade

$$= 150[(76 \cos 41.14 \times 50/2 - 14^2(48.86 \times \pi/180)/2]$$
$$= 202{,}110 \text{ lbs-in.}^2$$

$$t \times \sigma \times 14(48.86\pi/180) = 202{,}110$$

or

$$t = 1.13 \text{ inch.}$$

Another application of the pressure-area method is in analyzing conduit bifurcations (Fig. 7-14). The analysis of such structures is discussed by Swanson (Swanson, et al. 1955) and AISI (AISI 1981).

Problems

7-3 Using the pressure-area method, show that the required thickness of the bellows expansion joint (Fig. P7-3) due to internal pressure is given by

$$t = \frac{p(d + w)}{s(1.14 + 4w/q)}$$

where

$d >> w$;

s = allowable stress;

p = internal pressure.

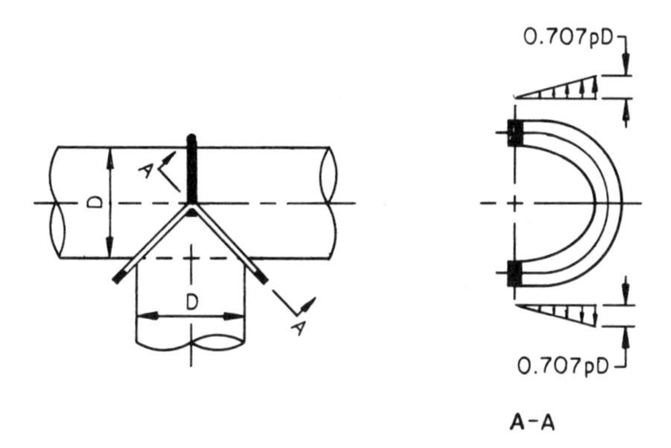

Figure 7-14. (Courtesy of Steel Plate Fabricators Association.)

7-4 Calculate the area needed to reinforce the nozzle shown in Fig. P7-4. Let $p = 800$ psi and $\sigma = 20$ ksi.
7-5 Calculate the thicknesses t_1, t_2, t_3, t_4, and area A of the section shown in Fig. P7-5 based on an average allowable stress of 18 ksi. Let $p = 250$ psi.
7-6 The pipe elbow shown in Fig. P7-6 is subjected to a pressure of 4000 psi. Show that the average stress in the outer surface is equal to 10,170 psi and the stress in the inner surface is equal to 28,230 psi.

7-3 One-Sheet Hyperboloids

Many structures (Fig. 7-15) are shaped in the form of one-sheet hyperboloids. These include cooling towers in power plants as well as water

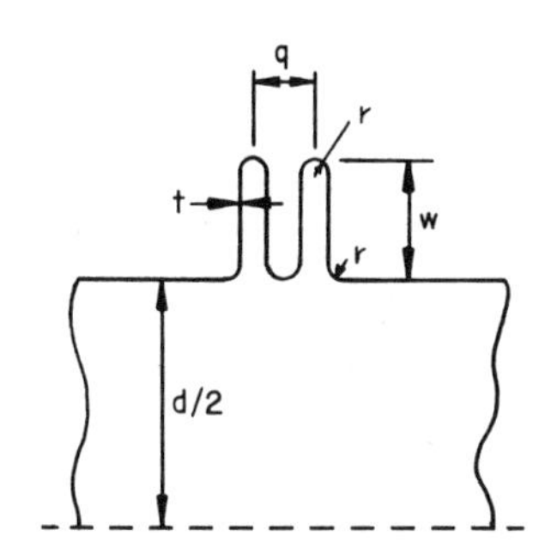

Figure P7-3.

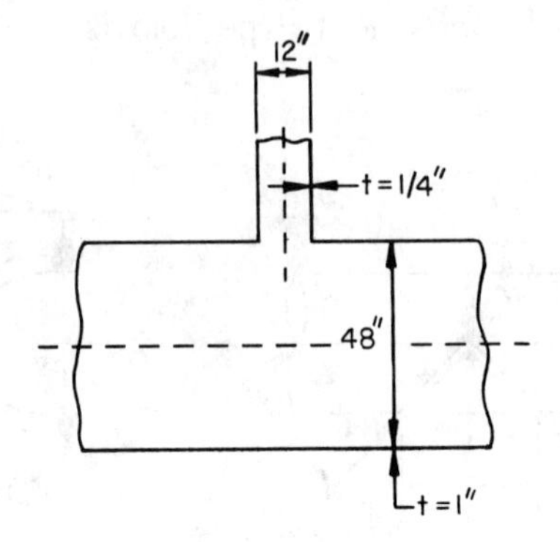

Figure P7-4.

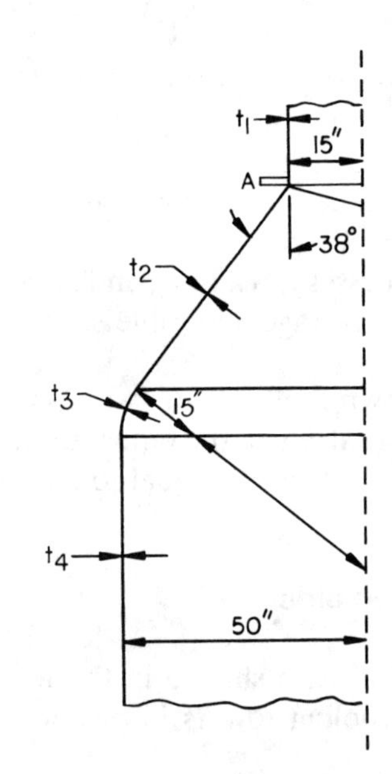

Figure P7-5.

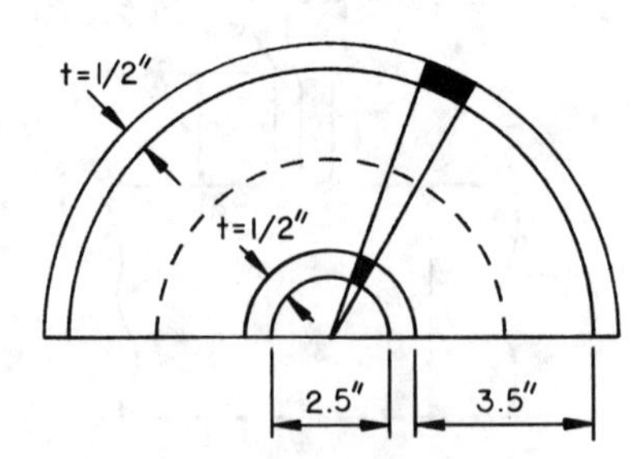

Figure P7-6.

Figure 7-15. St. Louis Planetarium. (Courtesy of Jennifer Jawad.)

storage tanks. The governing equation for such structures is

$$\frac{x^2 + y^2}{a^2} - \frac{z^2}{b^2} = 1. \tag{7-1}$$

This equation is plotted in Fig. 7-16. At any constant elevation $z = c$, the cross section is a circle. At the surface of this circle, such as $x = a$, Eq. (7-1) reduces to

$$\frac{y^2}{a^2} = \frac{z^2}{b^2} \tag{7-2}$$

or

$$z = \pm(b/a)y \tag{7-3}$$

which is the equation of a pair of straight lines that lie along the surface at that point.

Equation (7-1) can also be written in polar coordinates as

$$\frac{r^2}{a^2} - \frac{z^2}{b^2} = 1$$

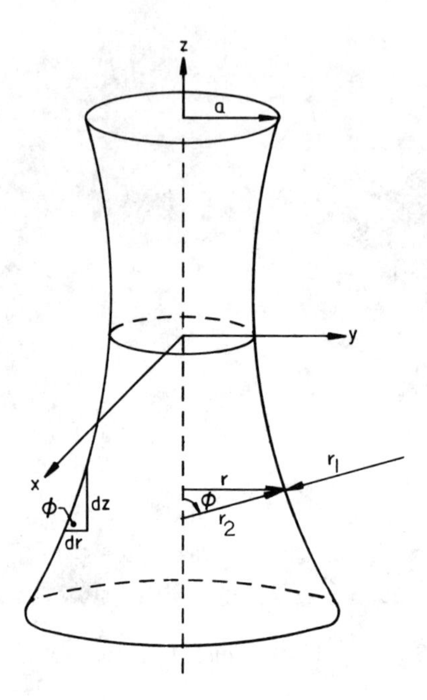

Figure 7-16.

or

$$r = \pm\frac{a}{b}(b^2 + z^2)^{1/2}. \tag{7-4}$$

This equation can also be written as

$$z = \pm\frac{b}{a}(r^2 - a^2)^{1/2}. \tag{7-5}$$

From Fig. 7-16 we get the relationship

$$\tan\phi = dz/dr.$$

Substituting Eq. (7-5) into this expression gives

$$\tan\phi = \pm\frac{b}{a}\left(\frac{r^2}{r^2 - a^2}\right)^{1/2}. \tag{7-6}$$

The equations for r_1, r_2, and r are expressed as (Flugge 1967)

$$r_1 = \frac{-a^2b^2}{(a^2\sin^2\phi - b^2\cos^2\phi)^{3/2}} \tag{7-7}$$

$$r_2 = \frac{a^2}{(a^2 \sin^2 \phi - b^2 \cos^2 \phi)^{1/2}} \tag{7-8}$$

and

$$r = \frac{a^2 \sin \phi}{(a^2 \sin^2 \phi - b^2 \cos^2 \phi)^{1/2}}. \tag{7-9}$$

A hyperboloid of a given shape can be expressed by Eq. (7-1) once a and b are established. The value of a is obtained at the equator where $z = 0$. The value of b is determined at a given elevation z. At any given elevation, the value of r is obtained from Eq. (7-4) and the value of ϕ is calculated from Eq. (7-6). At the same elevation, the values of r_1 and r_2 are obtained from Eqs. (7-7) and (7-8).

The dead load of the hyperboloid at any given elevation can be obtained from Fig. 7-17 as

$$W = \int 2\pi r \, dz = \int 2\pi r |r_1| \, d\phi \tag{7-10}$$

Substituting the values of r_1 and r from Eqs. (7-7) and (7-9) into this expression and defining Q and dQ (Kelkar and Sewell 1987) as

$$Q = \frac{\sqrt{a^2 + b^2}}{a} \cos \phi \tag{7-11}$$

$$dQ = -\frac{\sqrt{a^2 + b^2}}{a} \sin \phi \, d\phi \tag{7-12}$$

gives

$$W = -\frac{\pi \gamma a b^2}{2\sqrt{a^2 + b^2}} \left[\frac{2Q}{1 - Q^2} + \ln \frac{1 + Q}{1 - Q} \right]_{Q_o}^{Q} \tag{7-13}$$

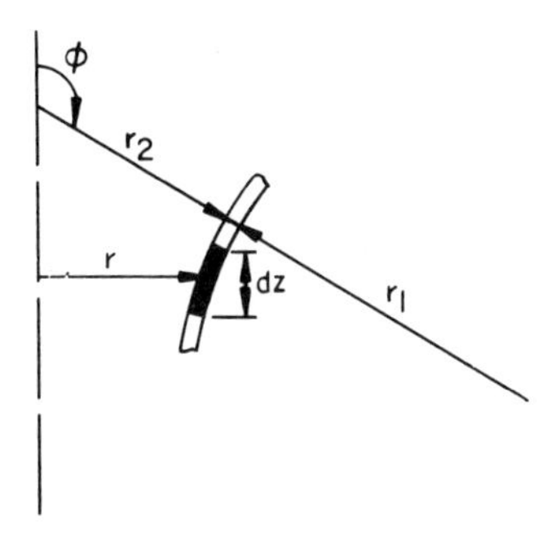

Figure 7-17.

where Q_o is evaluated at the top of the tower and γ is the dead weight per unit surface area.

The direction of r_1 in Fig. 7-16 and Eq. (7-7) is opposite that shown in Fig. 6-4 due to the negative curvature of the hyperboloid. Accordingly, r_1 must be entered as a negative value in Eqs. (6-10) through (6-12) for the equilibrium of shells of revolution. Thus, Eq. (6-11) must be written as

$$\frac{N_\theta}{r_2} - \frac{N_\phi}{|r_1|} = p_r. \tag{7-14}$$

For axisymmetric loads, Eq. (6-10) can be discarded and Eq. (6-12) defines the equilibrium of the shell at a given elevation as described in Section 7-1. Accordingly, the analysis of a one-sheet hyperboloid due to axisymmetric loads consists of establishing first the values of N_ϕ at various elevations in the structure. Then N_θ is calculated from Eq. (7-14) at these elevations.

For nonsymmetric loads such as wind and earthquake forces, the solution becomes more complicated. References such as Flugge (Flugge 1967) and Gould (Gould 1988) discuss such loading conditions and their solutions.

Problems

7-7 Determine the forces N_ϕ and N_θ in the natural draft cooling tower shown in Fig. P7-7a due to dead weight. The thickness profile is shown in Fig. P7-7b. Calculate these forces at 20-ft increments in height. Assume the weight of the concrete as 150 pcf.

7-8 Determine the forces in the water tower shown in Fig. P7-8. Disregard the dead weight of the tower and consider weight of the water only. Calculate the forces at 3-meter increments.

7-4 Deflection Due to Axisymmetric Loads

The deflection of a shell due to membrane forces caused by axisymmetric loads can be derived from Fig. 7-18. The change of length AB due to deformation is given by

$$\frac{dv}{d\phi}\,d\phi - w\,d\phi.$$

The strain is obtained by dividing this expression by the original length $r_1\,d\phi$

$$\varepsilon_\phi = \frac{1}{r_1}\frac{dv}{d\phi} - \frac{w}{r_1}. \tag{7-15}$$

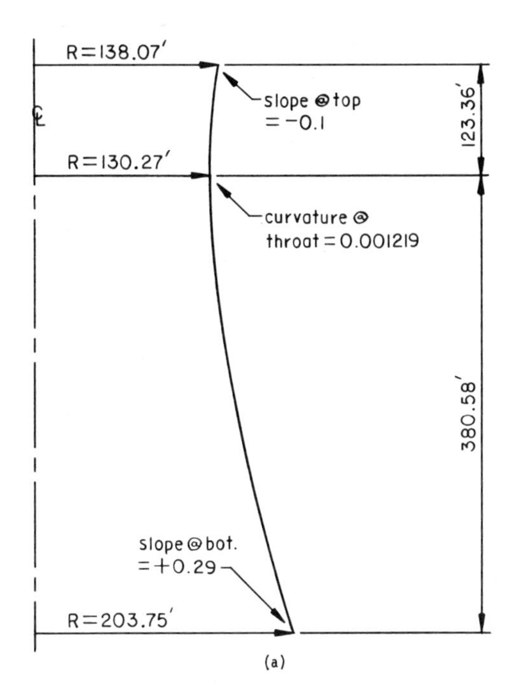

(a)

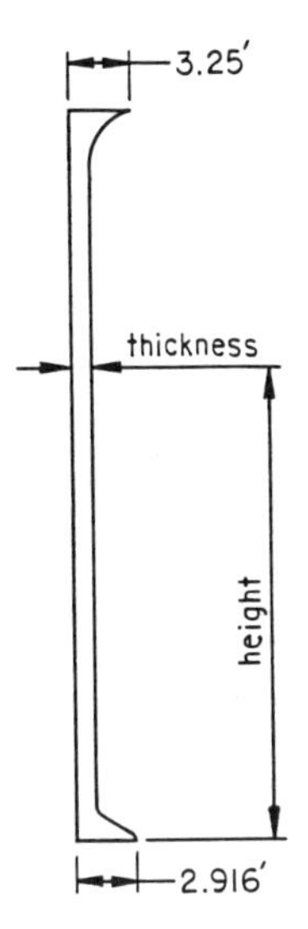

height feet	thickness feet
503.94	3.25
498.16	2.22
491.90	1.45
479.37	0.7083
340.00	0.7083
320.00	0.765
270.00	0.84
230.00	0.854
110.00	0.854
20.00	0.902
15.00	0.975
9.843	1.52
2.881	2.666
0.00	2.916

(b)

Figure P7-7. (Courtesy of Zurn Balcke-Durr, Tampa, FL.)

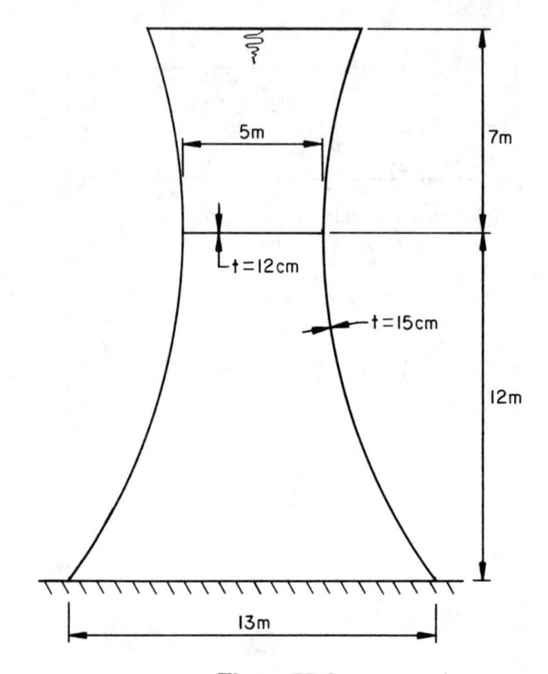

Figure P7-8.

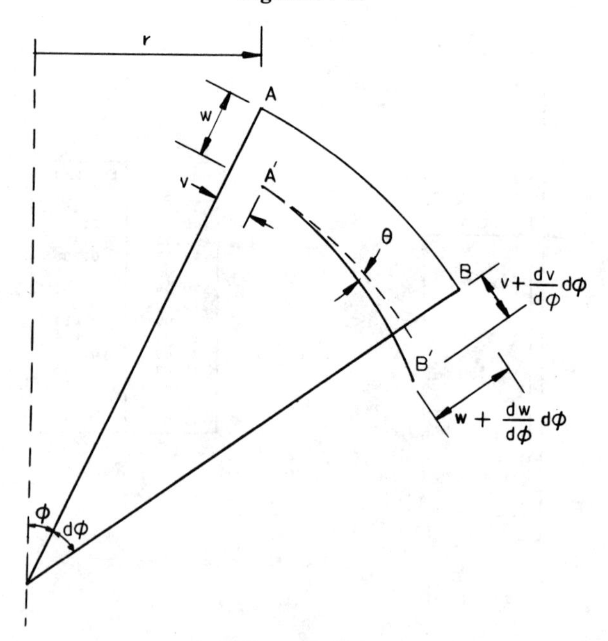

Figure 7-18.

The increase in radius r due to deformation (Fig. 7-19) is given by

$$v \cos \phi - w \sin \phi$$

or

$$\varepsilon_\theta = \frac{1}{r}(v \cos \phi - w \sin \phi).$$

Substituting in this equation the value

$$r = r_2 \sin \phi$$

gives

$$\varepsilon_\theta = \frac{v}{r_2} \cot \phi - \frac{w}{r_2} \tag{7-16}$$

or

$$w = v \cot \phi - r_2 \varepsilon_\theta. \tag{7-17}$$

From Eqs. (7-15) and (7-16) we get

$$\frac{dv}{d\phi} - v \cot \phi = r_1 \varepsilon_\phi - r_1 \varepsilon_\theta. \tag{7-18}$$

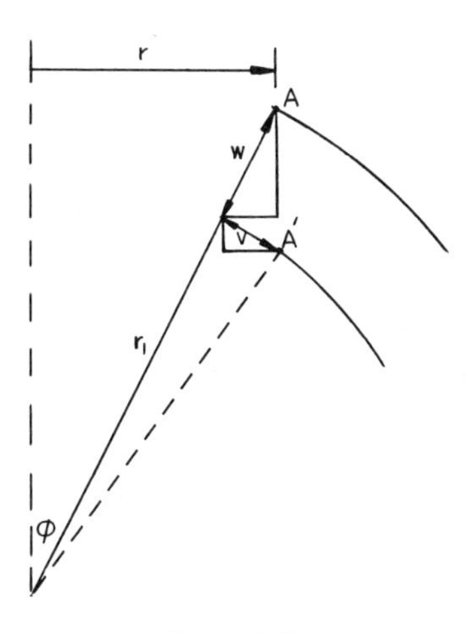

Figure 7-19.

From Eq. (1-13) for two-dimensional problems,

$$\left.\begin{aligned}\varepsilon_\phi &= \frac{1}{Et}(N_\phi - \mu N_\theta)\\ \varepsilon_\theta &= \frac{1}{Et}(N_\theta - \mu N_\phi).\end{aligned}\right] \tag{7-19}$$

Hence, Eq. (7-18) becomes

$$\frac{dv}{d\phi} - v \cot \phi = \frac{1}{Et}[N_\phi(r_1 + \mu r_2) - N_\theta(r_1 + \mu r_2)] \tag{7-20}$$

Let the righthand side of this equation be expressed as $g(\phi)$,

$$g(\phi) = \frac{1}{Et}[N_\phi(r_1 + \mu r_2) - N_\theta(r_1 + \mu r_2)] \tag{7-21}$$

and Eq. (7-6) is reduced to

$$v = \sin \phi\left(\int \frac{g(\phi)}{\sin \phi} + C\right). \tag{7-22}$$

In order to solve for the deflections in a structure due to a given loading condition, we first obtain N_ϕ and N_θ from Eqs. (6-12) and (6-11). We then calculate v from Eqs. (7-21) and (7-22). The normal deflection, w, is then calculated from Eq. (7-17).

The rotation at any point is obtained from Figs. 7-18 and 7-19 as

$$\psi = \frac{1}{r_1}\left(\frac{dw}{d\phi} + v\right). \tag{7-23}$$

Example 7-3

Find the deflection at point A of the dome roof shown in Fig. 7-20 due to an internal pressure p. Let $\mu = 0.3$.

Solution

For a spherical roof with internal pressure,

$$r_1 = r_2 = R$$

$$p_r = p \quad \text{and} \quad p_\phi = 0.$$

The membrane forces are obtained from Eqs. (6-12) and (6-11) as

$$N_\phi = N_\theta = pR/2.$$

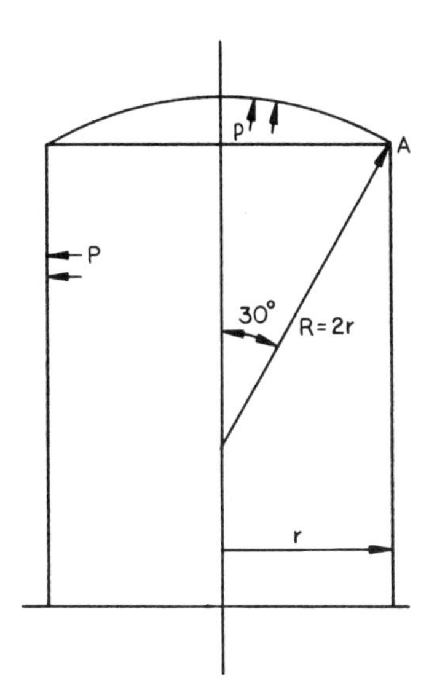

Figure 7-20.

From Eq. (7-21),

$$g(\phi) = \frac{1}{Et}[(pR/2)(R)(1 + \mu) - (pR/2)(R)(1 + \mu)]$$
$$= 0$$

and Eq. (7-22) gives

$$v = C \sin \phi$$

at the point of support, $\phi = 30°$, and the deflection $v = 0$. Hence, $C = 0$.

From Eq. (7-17)

$$w = -r_2\varepsilon_\theta$$

and from Eq. (7-19),

$$w = \frac{-pR^2}{2Et}(1 - \mu). \qquad (1)$$

At point A, the horizontal component of the deflection is given by

$$w_h = \frac{pR^2}{2Et}(1 - \mu) \sin \phi$$

Hence, with $\mu = 0.3$ and $\phi = 30°$

$$w_h = \frac{0.175pR^2}{Et}.$$

If we substitute into this expression the quantity $R = 2r$ which is obtained from Fig. 7-20, we get

$$w_h = \frac{0.7pr^2}{Et}. \tag{2}$$

It will be seen in the next chapter that the deflection of a cylinder due to internal pressure is given by

$$w = \frac{pr^2}{Et}(1 - \mu/2)$$

or, for $\mu = 0.3$,

$$w = \frac{0.85pr^2}{Et} \tag{3}$$

A comparison of Eqs. (2) and (3) shows that at the roof-to-cylinder junction there is an offset in the calculated horizontal deflection. However, this offset does not exist in a real structure because of discontinuity forces that normally develop at the junction. These forces consist of local bending moments and shear forces. Although these forces eliminate the deflection offset, they do create high localized bending stresses. It turns out that these localized stresses are secondary in nature and can be discarded in the design of most structures as described in the next chapter.

8

Bending of Thin Cylindrical Shells due to Axisymmetric Loads

8-1 Basic Equations

The membrane forces discussed in the last two chapters are sufficient to resist many commonly encountered loading conditions. At locations where the deflection is restricted or there is a change in geometry such as cylindrical-to-spherical shell junction, the membrane theory is inadequate to maintain deflection and rotation compatibility between the shells as illustrated in Example 7-3. At these locations discontinuity forces are developed which result in bending and shear stresses in the shell. These discontinuity forces are localized over a small area of the shell and dissipate rapidly along the shell. Many structures such as missiles (Fig. 8-1), pressure vessels, and storage tanks are designed per the membrane theory and the total stress at discontinuities is determined from the membrane and bending theories. In this chapter the bending theory of cylindrical shells is developed and in Chapter 9 the bending theory of spherical and conical shells is discussed.

We begin the derivation of the bending of thin cylindrical shells by assuming the applied loads to be symmetric with respect to angle θ. A free-body diagram of an infinitesimal section of a cylindrical shell is shown in Fig. 8-2. The applied loads p can vary in the x-direction only. At edges $x = 0$ and $x = dx$ the axial membrane force N_x, bending moments M_x, and shearing forces Q_x are axisymmetric. In the circumferential direction, only the hoop membrane force N_θ and bending moments M_θ are needed for equilibrium. There are no shearing forces, Q_θ, because the applied loads are symmetric in the circumferential direction. Summation of forces in the x-direction gives the first equation of equilibrium:

$$(N_x r \, d\theta) - \left(N_x + \frac{dN_x}{dx}\, dx\right) r \, d\theta = 0$$

Figure 8-1. The space shuttle Endeavour. (Courtesy of NASA.)

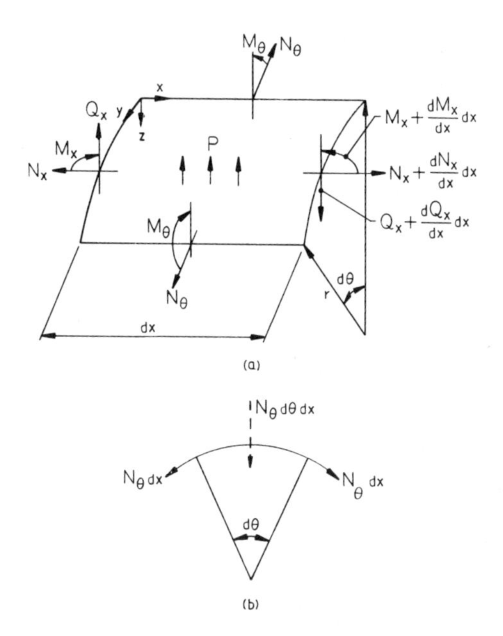

Figure 8-2.

or

$$\frac{dN_x}{dx} r\, dx\, d\theta = 0. \tag{8-1}$$

This equation indicates that N_x must be a constant. We assume a cylinder with open ends and set $N_x = 0$. In Section 8-3, we will discuss the case where N_x is not zero.

Summation of forces in the z-direction gives the second equation of equilibrium:

$$Q_x r\, d\theta - \left(Q_x + \frac{dQ_x}{dx} dx\right) r\, d\theta - N_\theta\, dx\, d\theta + pr\, dx\, d\theta = 0$$

or

$$\frac{dQ_x}{dx} + \frac{N_\theta}{r} = p. \tag{8-2}$$

Summation of moments around the y-axis gives the third equation of equilibrium:

$$M_x r\, d\theta - \left(M_x + \frac{dM_x}{dx} dx\right) r\, d\theta + \left(Q_x + \frac{dQ_x}{dx} dx\right) (r\, d\theta)\, dx$$
$$- pr\, d\theta \frac{dx}{2} dx + 2N_\theta \frac{d\theta}{2} dx \frac{dx}{2} = 0.$$

After simplifying and deleting terms of higher order we get

$$\frac{dM_x}{dx} - Q_x = 0. \tag{8-3}$$

Eliminating Q_x from Eqs. (8-2) and (8-3) gives

$$\frac{d^2M_x}{dx^2} + \frac{N_\theta}{r} = p. \tag{8-4}$$

Equation (8-4) contains two unknowns, N_θ and M_x. Both of these unknowns can be expressed in terms of deflection, w. Define axial strain as

$$\varepsilon_x = \frac{du}{dx}. \tag{8-5}$$

The circumferential strain is obtained from Fig. 8-3 as

$$\varepsilon_\theta = \frac{2\pi(r + \Delta r) - 2\pi r}{2\pi r}$$

$$\varepsilon_\theta = \frac{\Delta r}{r}$$

or

$$\varepsilon_\theta = \frac{-w}{r} \tag{8-6}$$

where w is the deflection and is taken as positive inwards. the stress–strain relationship given by Eq. (1-14) can be written in terms of force–strain

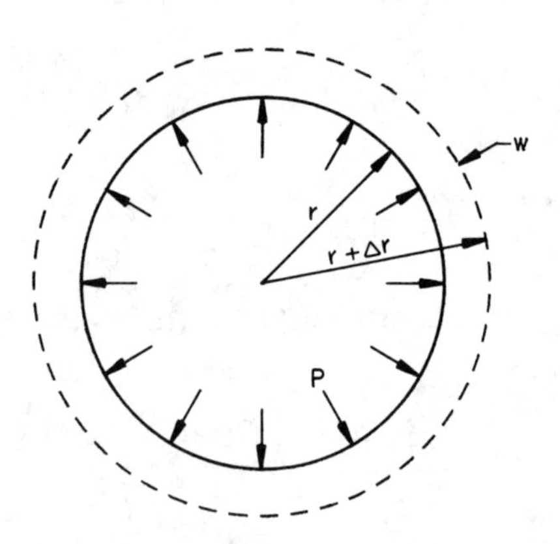

Figure 8-3.

relationship as

$$\begin{bmatrix} N_x \\ N_\theta \end{bmatrix} = \frac{Et}{1 - \mu^2} \begin{bmatrix} 1 & \mu \\ \mu & 1 \end{bmatrix} \begin{bmatrix} \varepsilon_x \\ \varepsilon_\theta \end{bmatrix}. \tag{8-7}$$

Notice that the shearing strain, $\gamma_{x\theta}$, is zero in this case due to load symmetry in the θ-direction. Substituting Eqs. (8-5) and (8-6) into the first expression of Eq. (8-7) results in

$$N_x = \frac{Et}{1 - \mu^2} \left(\frac{du}{dx} - \mu \frac{w}{r} \right).$$

Substituting into this expression the value $N_x = 0$ from Eq. (8-1), we get

$$\frac{du}{dx} = \mu \frac{w}{r}. \tag{8-8}$$

Similarly, the second term of Eq. (8-7) can be written as

$$N_\theta = \frac{Et}{1 - \mu^2} \left(- \frac{w}{r} + \mu \frac{du}{dx} \right),$$

or upon inserting Eq. (8-8), it becomes

$$N_\theta = - \frac{Etw}{r}. \tag{8-9}$$

The basic moment-deflection relationships of Eq. (1-17) are also applicable to thin cylindrical shells. Referring to the two axes as x and θ rather than x and y, the first two expressions in Eq. (1-17) become

$$\begin{bmatrix} M_x \\ M_\theta \end{bmatrix} = -D \begin{bmatrix} 1 & \mu \\ \mu & 1 \end{bmatrix} \begin{bmatrix} \dfrac{\partial^2 w}{\partial x^2} \\ \dfrac{\partial^2 w}{\partial \theta^2} \end{bmatrix}. \tag{8-10}$$

It should be noted that x and θ in Eq. (8-10) are not in polar coordinates but redefined x- and y-axes. Polar transformation of Eq. (8-10) is given in Chapter 11.

The third expression, M_{xy}, in Eq. (1-17) vanishes because the rate of change of deflection with respect to θ is zero due to symmetry of applied loads. Also, due to symmetry with respect to θ, all derivatives with respect to θ vanish and the first expression in Eq. (8-10) reduces to

$$M_x = -D\frac{d^2w}{dx^2} \tag{8-11}$$

and the second expression in Eq. (8-10) becomes

$$M_\theta = -\mu D\frac{d^2w}{dx^2}. \tag{8-12}$$

From Eqs. (8-11) and (8-12) it can be concluded that

$$M_\theta = \mu M_x. \tag{8-13}$$

Substituting Eqs. (8-9) and (8-11) into Eq. (8-4) gives

$$\frac{d^4w}{dx^4} + \frac{Et}{Dr^2}w = -p(x)/D$$

which is the differential equation for the bending of cylindrical shells due to loads that are variable in the x-direction and uniformly distributed in the θ-direction. Defining

$$\beta^4 = \frac{Et}{4Dr^2} = \frac{3(1-\mu^2)}{r^2t^2} \tag{8-14}$$

the differential equation becomes

$$\frac{d^4w}{dx^4} + 4\beta^4 w = -p(x)/D \tag{8-15}$$

where p is a function of x.

Solution of Eq. (8-15) results in an expression for the deflection, w. The longitudinal and circumferential moments are then obtained from Eqs. (8-11) and (8-13), respectively. The circumferential membrane force, N_θ, is determined from Eq. (8-9).

One solution of Eq. (8-15) that is commonly used for long cylindrical shells is expressed as

$$\begin{aligned} w = {} & e^{\beta x}(C_1 \cos\beta x + C_2 \sin\beta x) \\ & + e^{-\beta x}(C_3\cos\beta x + C_4 \sin\beta x) + f(x). \end{aligned} \tag{8-16}$$

Where $f(x)$ is the particular solution and C_1 to C_4 are constants that are evaluated from the boundary conditions.

A different solution of Eq. (8-15) that is commonly used for short cylindrical shells is expressed as

$$\begin{aligned} w = {} & C_1 \sin\beta x \sinh\beta x + C_2 \sin\beta x \cosh\beta x \\ & + C_3\cos\beta x \sinh\beta x + C_4 \cos\beta x\cosh\beta x. \end{aligned} \tag{8-17}$$

'he procedure for establishing moments and forces in cylindrical shells vell as defining long and short cylinders is discussed in the following sections.

For sign convention, M_x, Q_x, N_θ, and M_θ are all positive as shown in Fig. 8-2. The deflection, w, is positive inwards and the rotation, ψ, in the x-direction is positive in the direction of positive bending moments.

8-2 Long Cylindrical Shells

One application of Eq. (8-16) is of shear forces and bending moments applied at the edge of a cylindrical shell, (Fig. 8-4). Referring to Eq. (8-16) for the deflection of a shell, we can set the function $f(x)$ to zero as there are no applied loads along the cylinder. Also, the deflection due to the term $e^{\beta x}$ in Eq. (8-16) tends to approach infinity as x gets larger. However, the deflection due to moments and forces applied at one end of an infinitely long cylinder tend to dissipate as x gets larger. Thus, constants C_1 and C_2 must be set to zero and Eq. (8-16) becomes

$$w = e^{-\beta x}(C_3 \cos \beta x + C_4 \sin \beta x). \tag{8-18}$$

The boundary conditions for the infinitely long cylinder (Fig. 8-4) are obtained from Eq. (8-11) as

$$M_x|_{x=0} = M_o = -D \left.\frac{d^2w}{dx^2}\right|_{x=0}$$

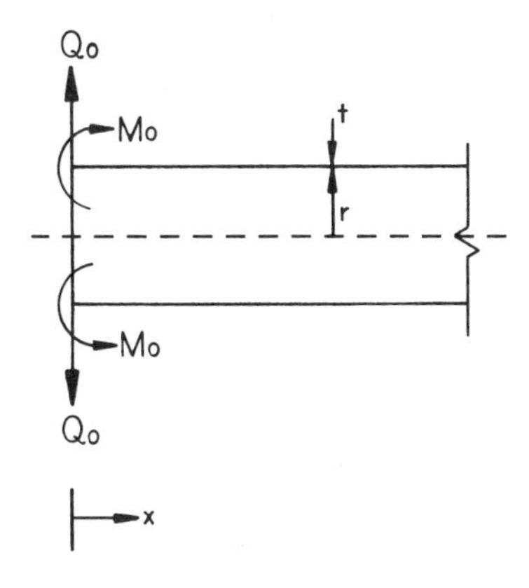

Figure 8-4.

Table 8-1. Values of functions $A_{\beta x}$, $B_{\beta x}$, $C_{\beta x}$, $D_{\beta x}$

βx	$A_{\beta x}$	$B_{\beta x}$	$C_{\beta x}$	$D_{\beta x}$
0	1.0000	1.0000	1.0000	0.0000
0.05	0.9976	0.9025	0.9500	0.0475
0.10	0.9907	0.8100	0.9003	0.0903
0.15	0.9797	0.7224	0.8510	0.1286
0.20	0.9651	0.6398	0.8024	0.1627
0.30	0.9267	0.4888	0.7077	0.2189
0.40	0.8784	0.3564	0.6174	0.2610
0.50	0.8231	0.2415	0.5323	0.2908
0.55	0.7934	0.1903	0.4919	0.3016
0.60	0.7628	0.1431	0.4530	0.3099
0.80	0.6354	−0.0093	0.3131	0.3223
1.00	0.5083	−0.1108	0.1988	0.3096
1.20	0.3899	−0.1716	0.1091	0.2807
1.40	0.2849	−0.2011	0.0419	0.2430
1.60	0.1959	−0.2077	−0.0059	0.2018
1.80	0.1234	−0.1985	−0.0376	0.1610
2.00	0.0667	−0.1794	−0.0563	0.1231
2.50	−0.0166	−0.1149	−0.0658	0.0491
3.00	−0.0423	−0.0563	−0.0493	0.0070
3.5	−0.0389	−0.0177	−0.0283	−0.0106
4.0	−0.0258	0.0019	−0.0120	−0.0139
5.0	−0.0045	0.0084	0.0019	−0.0065
6.0	0.0017	0.0031	0.0024	−0.0007
7.0	0.0013	0.0001	0.0007	0.0006

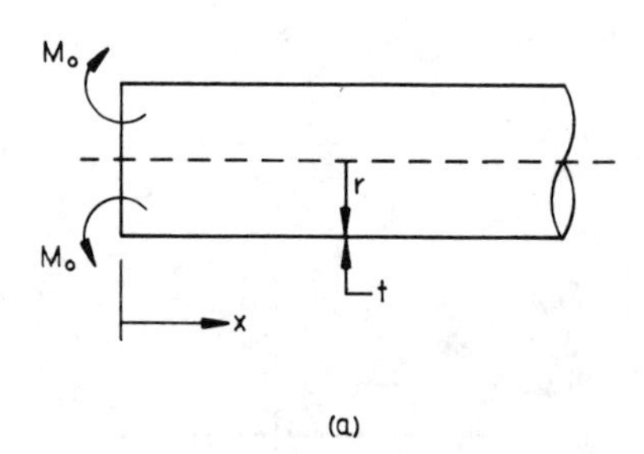

(a)

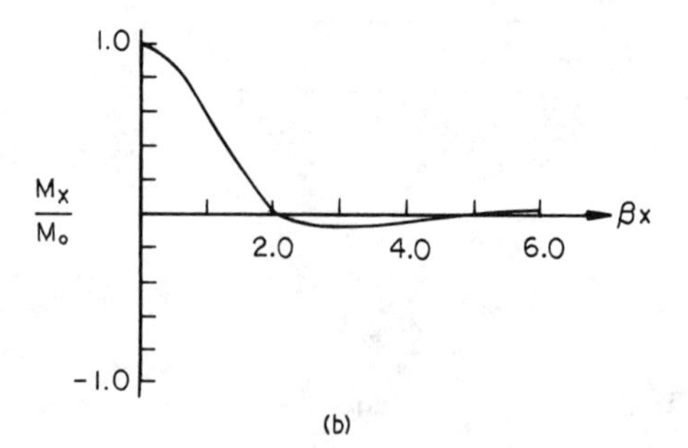

(b)

Figure 8-5.

and from Eq. (8-3) as

$$Q_x|_{x=0} = Q_o = \left.\frac{dM}{dx}\right|_{x=0}.$$

Substituting Eq. (8-18) into the first boundary condition gives

$$C_4 = \frac{M_o}{2\beta^2 D}$$

and from the second boundary condition

$$C_3 = -\frac{1}{2\beta^3 D}(Q_o + \beta M_o).$$

Hence, the deflection equation for the long cylinder shown in Fig. 8-4 is

$$w = \frac{e^{-\beta x}}{2\beta^3 D}[M_o(\sin\beta x - \cos\beta x) - Q_o\cos\beta x]. \tag{8-19}$$

By defining

$$A_{\beta x} = e^{-\beta x}(\cos\beta x + \sin\beta x) \tag{8-20}$$

$$B_{\beta x} = e^{-\beta x}(\cos\beta x - \sin\beta x) \tag{8-21}$$

$$C_{\beta x} = e^{-\beta x}\cos\beta x \tag{8-22}$$

$$D_{\beta x} = e^{-\beta x}\sin\beta x \tag{8-23}$$

the expression for the deflection and its derivative becomes

$$\text{deflection} = w_x = \frac{-1}{2\beta^3 D}(\beta M_o B_{\beta x} + Q_o C_{\beta x}) \tag{8-24}$$

$$\text{slope} = \psi_x = \frac{1}{2\beta^2 D}(2\beta M_o C_{\beta x} + Q_o A_{\beta x}) \tag{8-25}$$

$$\text{moment} = M_x = \frac{1}{2\beta}(2\beta M_o A_{\beta x} + 2Q_o D_{\beta x}) \tag{8-26}$$

$$\text{shear} = Q_x = -(2\beta M_o D_{\beta x} - Q_o B_{\beta x}). \tag{8-27}$$

The functions $A_{\beta x}$ through $D_{\beta x}$ are calculated in Table 8-1 for various values of βx.

Example 8-1

A long cylindrical shell is subjected to end moment M_o as shown in Fig. 8-5a. Plot the value of M_x from $\beta x = 0$ to $\beta x = 5.0$. Also determine the distance x at which the moment is about 1% of the original applied moment M_o.

Solution

From Eq. (8-26),

$$M_x = M_o A_{\beta x}.$$

The values of $A_{\beta x}$ are obtained from Eq. (8-20) and a plot of M_x is shown in Fig. 8-5b. From Eq. (8-26) and Table 8-1, the value of βx at which M_x is equal to about 1% of M_o is about 2.285. Hence,

$$\beta x = 2.285$$

or

$$x = 1.78\sqrt{rt} \quad \text{for } \mu = 0.3.$$

The significance of the quantity $1.78\sqrt{rt}$ is apparent from Fig. 8-5b. It shows that a moment applied at the end of a long cylinder dissipates very rapidly as x increases and it reduces to about 1% of the original value at a distance of $17.8\sqrt{rt}$ from the edge. Many design codes such as the ASME VIII-1 use a similar criterion for defining long cylinders where the forces applied at one end have a negligible effect at the other end.

Example 8-2

Determine the maximum stress in a long cylinder due to the radial load shown in Fig. 8-6a.

Solution

From the free-body diagram of Fig. 8-6b, the end load at point A is equal to $Q_o/2$. Also, from symmetry the slope is zero at point A, Thus,

$$\text{Slope due to } M_o + \text{slope due to } Q_o/2 = 0.$$

From Eq. (8-25)

$$\frac{M_o}{\beta D} - \frac{Q_o}{4\beta^2 D} = 0$$

or

$$M_o = \frac{Q_o}{4\beta}.$$

$$\text{The maximum longitudinal bending stress} = 6M_o/t^2 = \frac{3Q_o}{2\beta t^2}$$

$$\text{Maximum deflection at point } A = \frac{-M_o}{2\beta^2 D} + \frac{Q_o}{4\beta^3 D} = \frac{Q_o}{8\beta^3 D}$$

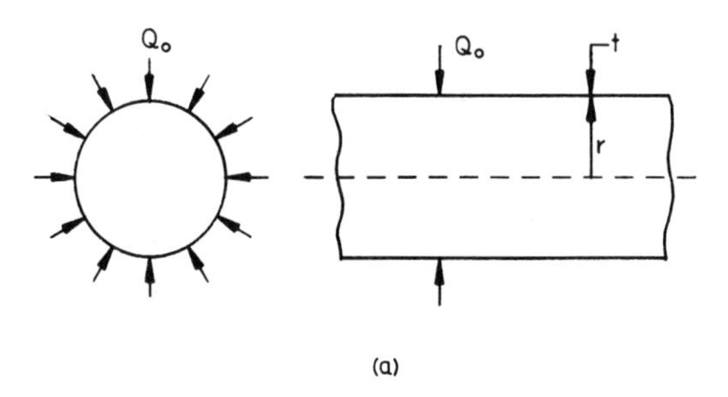

(a)

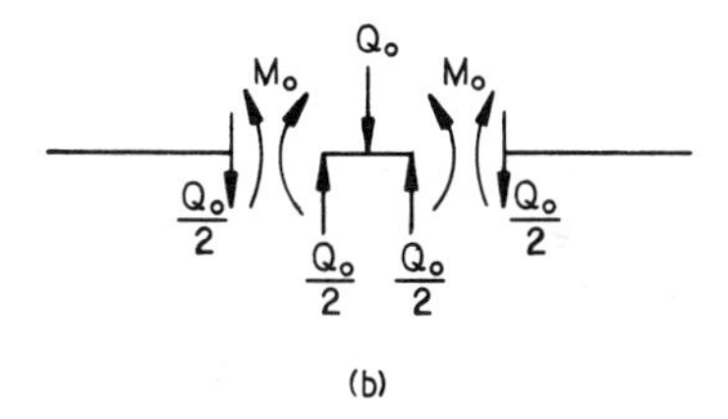

(b)

Figure 8-6.

The circumferential membrane force is obtained from Eq. (8-9) as

$$N_\theta = -\frac{Etw}{r} = -\frac{EtQ_o}{8r\beta^3 D}.$$

The circumferential bending moment is

$$M_\theta = \mu M_x = \frac{3\mu Q_o}{2\beta t^2}$$

and the total maximum circumferential stress is

$$\sigma_\theta = -\frac{EQ_o}{8r\beta^3 D} - \frac{3\mu Q_o}{2\beta t^2}.$$

Example 8-3

Determine the expression for stress in the long cylinder shown in Fig. 8-7a due to an internal pressure of 100 psi. The cylinder is supported by rigid bulkheads.

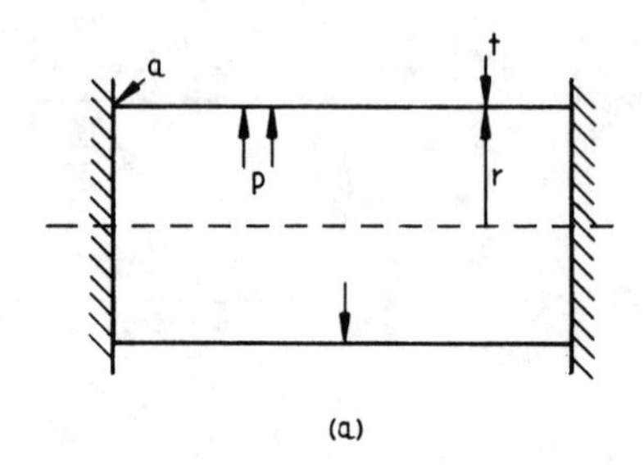

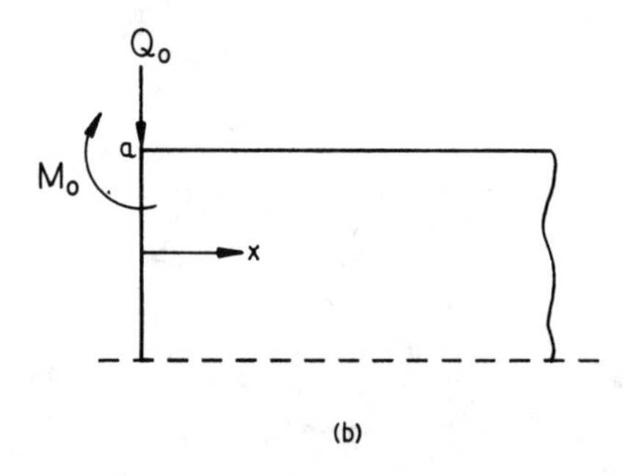

Figure 8-7.

Solution

Because the internal pressure is considered applied load, Eq. (8-18) can be written as

$$w = e^{-\beta x}\,(C_3 \cos \beta x + C_4 \sin \beta x) + f(x)$$

and because the pressure is constant along the length of the cylinder, the particular solution can be expressed as

$$w_p = K.$$

Substituting this expression in Eq. (8-15) results in

$$w_p = -\frac{p}{4\beta^4 D}$$

which can also be expressed as

$$w_p = -\frac{pr^2}{Et}.$$

A free-body diagram of the discontinuity forces at the end is shown in Fig. 8-7b. The unknown moment and shear forces are assumed in a given direction as shown. A negative answer will indicate that the true direction is opposite the assumed one. The compatibility condition requires that the deflection at the edge due to pressure plus moment plus shear is equal to zero. Referring to Eq. (8-24) and Fig. 8-7,

$$\frac{-p}{4\beta^4 D} - \frac{M_o}{2\beta^2 D} + \frac{Q_o}{2\beta^3 D} = 0$$

or

$$M_o - \frac{Q_o}{\beta} = \frac{-p}{2\beta^2}. \tag{1}$$

Similarly the slope due to pressure plus moment plus shear is equal to zero at the edge. Hence, from Eq. (8-24) and Fig. 8-7,

$$0 + \frac{M_o}{\beta D} - \frac{Q_o}{2\beta^2 D} = 0$$

or

$$M_o - \frac{Q_o}{2\beta} = 0. \tag{2}$$

Solving Eqs. (1) and (2) gives

$$M_o = \frac{p}{2\beta^2} \quad \text{and} \quad Q_o = \frac{p}{\beta}.$$

Thus, the deflection is given by Eq. (8-24) and Fig. 8-7 as

$$w = \frac{p}{2\beta^4 D}\left(-\frac{1}{2}B_{\beta x} + C_{\beta x}\right) - \frac{p}{4\beta^4 D}.$$

At the bulkhead attachment, the circumferential membrane force N_θ is zero because the deflection is zero in accordance with Eq. (8-9). The axial bending moment is given by Eq. (8-11) as

$$M_x = -D\frac{d^2w}{dx^2} = \frac{pe^{-\beta x}}{2\beta^2}(\sin\beta x - \cos\beta x).$$

The circumferential bending moment is given by Eq. (8-13) as

$$M_\theta = \mu M_x = \frac{\mu p e^{-\beta x}}{2\beta^2}(\sin\beta x - \cos\beta x).$$

The circumferential membrane force is given by Eq. (8-9) as

$$N_\theta = \frac{-Et}{r}\left(\frac{p}{2\beta^4 D}\right)(-B_{\beta x} + 2C_{\beta x} - 1).$$

longitudinal bending stress $\sigma_{Lb} = \pm 6M_x/t^2$
longitudinal membrane stress $\sigma_{Lm} = 0$
circumferential bending stress $\sigma_{\theta b} = \mu\sigma_{Lb}$
circumferential membrane stress $\sigma_{\theta m} = N_\theta/t$

Problems

8-1 In Example 8-1, assume a shearing force Q_o is applied at the end of the cylinder rather than M_o. Find the maximum longitudinal moment and its location from the edge. Let $\mu = 0.30$.
8-2 A cylindrical container is filled with a fluid to a level *a–a* (Fig. P8-2). The metal temperature at a given time period is 400°F above section *a–a* an 100°F below section *a–a*. Determine the discontinuity forces in the cylinder at section *a–a*. Let $\alpha = 6.5 \times 10^{-6}$ inch/inch/°F, $E = 30{,}000$ ksi, and $\mu = 0.30$.
8-3 Calculate the longitudinal bending stress at points *a* and *b* due to the applied loads shown in Fig. P8-3. Let $E = 27{,}000$ ksi and $\mu = 0.32$.
8-4 Show that for a uniform load over a small length, *a*, the deflection at point *A* in Fig. P8-4 is given by

$$w = \frac{pr^2}{2Et}(2 - e^{-\beta c}\cos\beta c - e^{-\beta b}\cos\beta b).$$

8-5 Find the expression for the bending moment in the water tank shown in Fig. P8-5. Let $E = 20{,}000$ kgf/mm^2 and $\mu = 0.29$. *Hint*: Calculate first

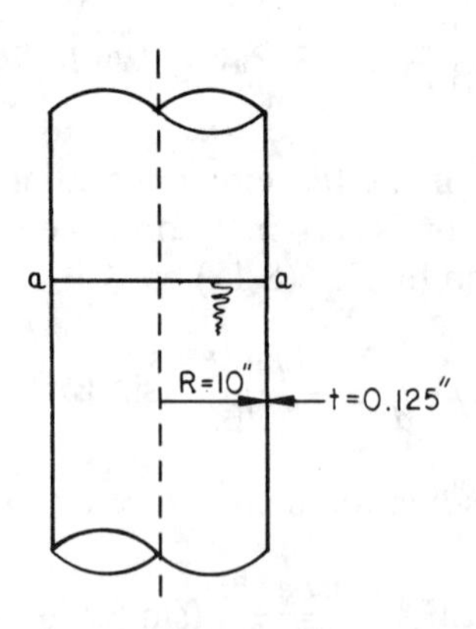

Figure P8-2.

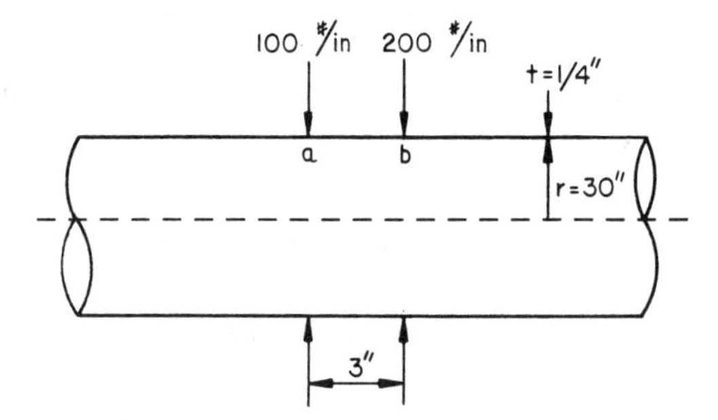

Figure P8-3.

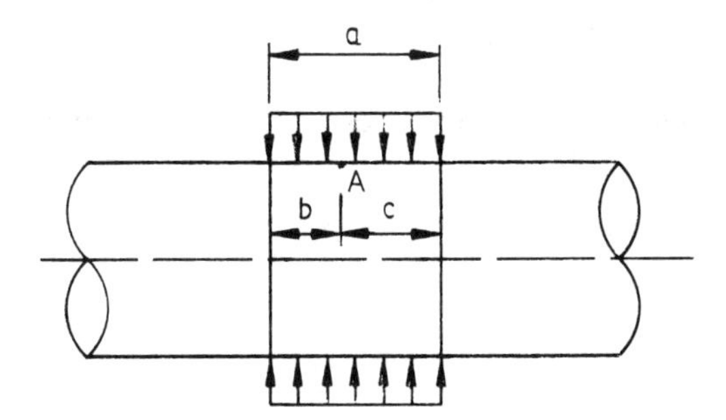

Figure P8-4.

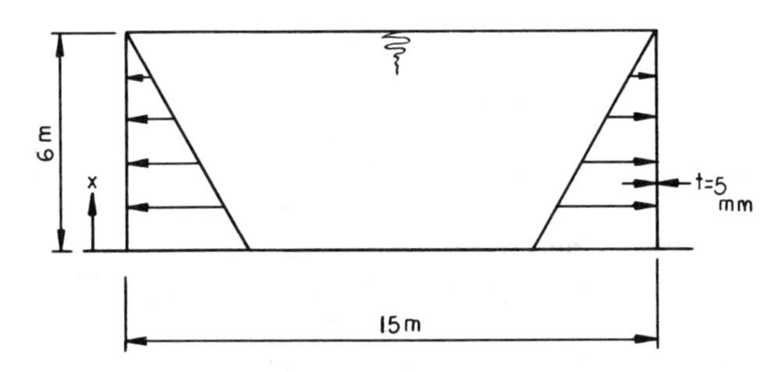

Figure P8-5.

N_θ in terms of $p(x)$. Next use Eq. (8-9) to calculate w and dw/dx. Then use the fixed boundary condition at the bottom to calculate M_o. M_x is then obtained from Eq. (8-26).

8-3 Long Cylindrical Shells with End Loads

The radial deflection obtained from Eq. (8-18) is based on the assumption that the axial membrane force, N_x, in Eq. (8-1) is negligible. However, many applications involve pressure and hydrostatic loads that result in axial forces. The deflections and slopes due to these axial forces must be determined first and then the deflections and slopes due to edge effects described in the previous section are superimposed for a final solution. This procedure is illustrated here for a cylindrical shell with end closures and subjected to internal pressure.

Let a cylindrical shell (Fig. 8-8a) be subjected to internal pressure p. Then the circumferential, hoop, stress at a point away from the ends is obtained from Fig. 8-8b as

$$2\sigma_\theta tL = p(2r)L$$

or

$$\sigma_\theta = \frac{pr}{t}. \tag{8-28}$$

Similarly the longitudinal stress is obtained from Fig. 8-8c as

$$\pi r^2 p = \sigma_x(2\pi r)$$

or

$$\sigma_x = \frac{pr}{2r}. \tag{8-29}$$

The maximum stresses given by Eqs. (8-28) and (8-29) for thin cylindrical shells subjected to internal pressure are valid as long as the shell is allowed to grow freely. Any restraints such as end closures and stiffening rings that prevent the shell from growing freely will result in bending moments and shear forces in the vicinity of the restraints. The magnitude of these moments and forces is determined subsequent to solving Eq. (8-18). The total stress will then be a summation of those obtained from Eqs. (8-28) and (8-29) plus those determined as a result of solving Eq. (8-18).

The circumferential and axial strains are obtained from Eqs. (1-14), with

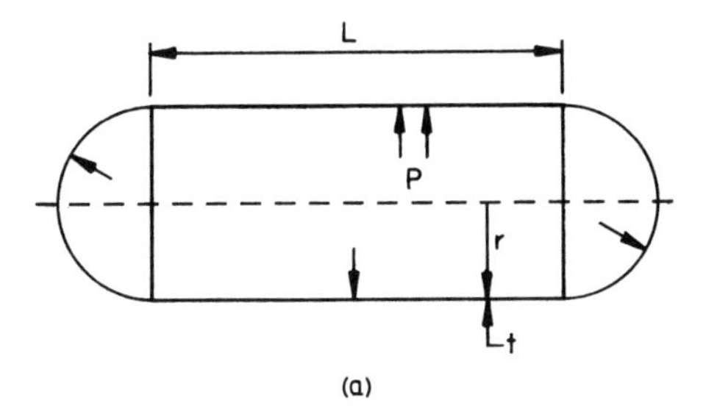

(a)

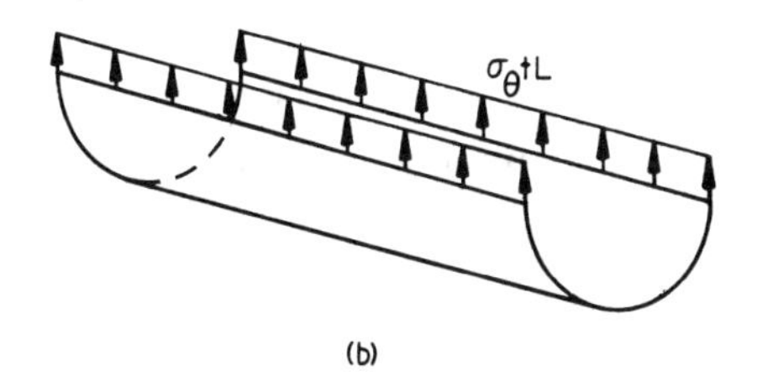

(b)

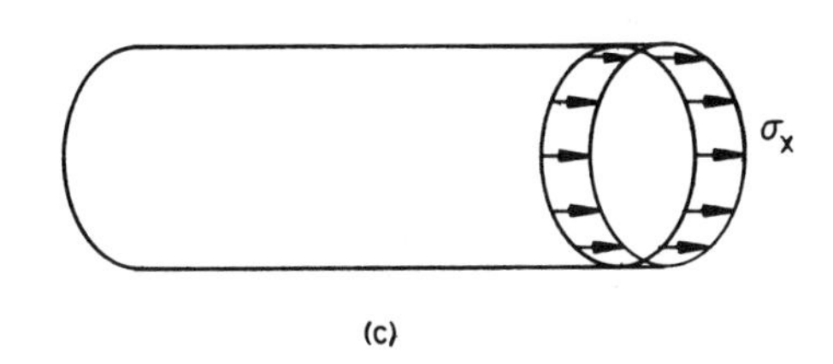

(c)

Figure 8-8.

$\tau_{xy} = 0$, (8-28), and (8-29) as

$$\varepsilon_\theta = \frac{pr}{Et}(1 - \mu/2) \tag{8-30}$$

$$\varepsilon_x = \frac{pr}{2Et}(1 - 2\mu). \tag{8-31}$$

The radial deflection is obtained from Fig. 8-3 as

$$\varepsilon_\theta = \frac{2\pi(r + \Delta r) - 2\pi r}{2\pi r}$$

$$= \frac{\Delta r}{r} = \frac{-w}{r}$$

or

$$w = -r\varepsilon_\theta.$$

From this expression and Eq. (8-30), we get

$$w = -\frac{pr^2}{Et}(1 - \mu/2). \tag{8-32}$$

Example 8-4

A stiffening ring is placed around a cylinder at a distance removed from the ends as shown in Fig. 8-9a. The radius of the cylinder is 50 inches and its thickness is 0.25 inch. Also, the internal pressure = 100 psi, E = 30,000 ksi, and μ = 0.3. Find the discontinuity stresses if (a) the ring is assumed to be infinitely rigid and (b) the ring is assumed to be 4 inches wide × 3/8 inch thick.

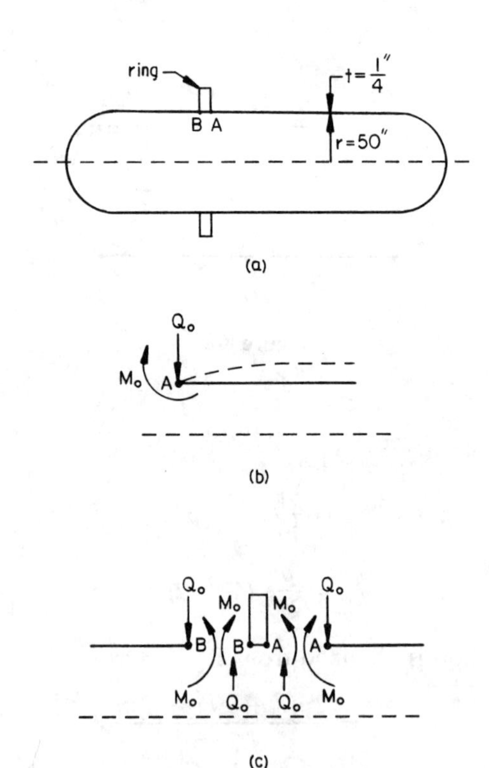

Figure 8-9.

Solution

(a)

$$\beta = \sqrt[4]{\frac{3(1 - 0.3^2)}{50^2 \times 0.25^2}} = 0.36357$$

$$D = \frac{30{,}000{,}000 \times 0.25^3}{12(1 - 0.3^2)} = 42{,}925.82.$$

From Eq. (8-32),

$$w_p = \frac{100 \times 50^2}{30{,}000{,}000 \times 0.25}(1 - 0.3/2) = 0.0283 \text{ inch.}$$

From Fig. 8-9b it is seen that for an infinitely rigid ring, the deflection is zero. Also from symmetry, the slope is zero and Eqs. (8-24) and (8-25) give

$$w_M = \frac{M_o}{2\beta^2 D} \qquad \theta_M = \frac{M_o}{\beta D}$$

$$w_Q = \frac{Q_o}{2\beta^3 D} \qquad \theta_Q = \frac{Q_o}{2\beta^2 D}.$$

Total deflection at the ring attachment is equal to zero

$$0.0283 + \frac{M_o}{2\beta^2 D} - \frac{Q_o}{2\beta^3 D}. \tag{1}$$

Similarly, the slope at the ring attachment is zero

$$\frac{M_o}{\beta D} - \frac{Q_o}{2\beta^2 D} = 0 \tag{2}$$

From Eq. (2),

$$Q_o = 2\beta M_o.$$

From Eq. (1),

$$M_o = 321.53 \text{ inch-lbs/inch.}$$

and

$$Q_o = 233.79 \text{ lbs/inch.}$$

Stress in the ring is zero because it is infinitely rigid.

longitudinal bending stress in cylinder $= 6M_o/t^2 = 30{,}870$ psi circumferential bending stress = 0.3 × 30,870 = 9260 psi

circumferential membrane stress at ring junction = 0 psi

$$\text{total longitudinal stress} = 30{,}870 + \frac{100 \times 50}{2 \times 0.25} = 40{,}870 \text{ psi}$$

total circumferential stress = 9260 psi.

(b) From Fig. 8-9c, and symmetry, we can conclude that the shear and moment in the shell to the left of the ring are the same as the shear and moment to the right of the ring. Accordingly, we can solve only one unknown shear and one unknown moment value by taking the discontinuity forces of a shell on one side of the ring only. The deflection of the ring due to pressure can be ignored because the ring width is 16 times that of the shell. The deflection of the ring due to $2Q_o$ is given by

$$w_R = \frac{2Q_o r^2}{AE}.$$

Compatibility of the shell and ring deflections require that

deflection of shell = deflection of ring

or, from Fig. 8-9c,

(deflection due to p − deflection due to Q + deflection due to M)$|_{\text{shell}}$

= deflection due to $2Q_o|_{\text{ring}}$

$$\frac{pr^2}{Et}(1 - \mu/2) - \frac{Q_o}{2\beta^3 D} + \frac{M_o}{2\beta^2 D} = \frac{2Q_o r^2}{AE}$$

$$M_o - 4.01Q_o = -321.4. \qquad (1)$$

From symmetry, the rotation of the shell due to pressure plus Q_o, plus M_o must be set to zero.

$$\frac{Q_o}{2\beta^2 D} = \frac{M_o}{\beta D}$$

or

$$Q_o = 2\beta M_o. \qquad (2)$$

Solving Eqs. (1) and (2) gives

$$Q_o = 122 \text{ lbs/inch}$$

$$M_o = 167.7 \text{ inch-lbs/inch}.$$

Notice that this moment is about half of the moment in the case of an infinitely rigid ring.

$$\text{stress in ring} = \frac{2Q_o r}{A} = 8130 \text{ psi.}$$

Maximum longitudinal stress in shell occurs at the ring attachment and is given by

$$\sigma_x = \frac{pr}{2t} + \frac{6M}{t^2} = 10{,}000 + 16{,}100 = 26{,}100 \text{ psi.}$$

Deflection of shell at ring junction is given by

$$w = w_p - w_Q + w_M$$

$$w = \frac{pr^2}{Et}(0.85) - \frac{Q_o}{2\beta^3 D} + \frac{M_o}{2\beta^2 D}$$

$$w = \frac{406{,}120}{E}.$$

The circumferential membrane force is

$$N_\theta = \frac{Etw}{(1 - \mu^2)r}.$$

The circumferential bending moment is

$$M_\theta = \mu M_x = 0.3 \times 167.7 = 50.3 \text{ inch-lbs/inch}$$

$$\sigma_\theta = \frac{2030}{0.25} + \frac{6 \times 50.3}{0.25^2} = 8120 + 4830 = 12{,}950 \text{ psi.}$$

Problems

8-6 The cross section of an aluminum beer can is shown in Fig. P8-6. Assume the top of the can to be a flat plate connected to an equivalent cylinder as shown by the shaded line. Determine the maximum permissible internal pressure. Let the shell allowable membrane stress = 30 ksi and the shell allowable bending plus membrane stress = 90 ksi. Let the allowable bending stress in the top plate = 45 ksi. Assume E = 11,000 ksi and μ = 0.33.

The pressure obtained from the assumptions made above is much lower than the actual pressure in the aluminum can. In the actual can there is an extension, or an expansion joint, between the top plate and the cylinder. Explain the effect of the expansion joint in increasing the permissible pressure. Also, determine the deflection of the top plate due to calculated pressure and calculate the membrane stress in the plate if it were treated as a spherical shell having a shape of the deflected plate. How does the membrane stress in the deflected plate compare with the bending stress of a flat plate?

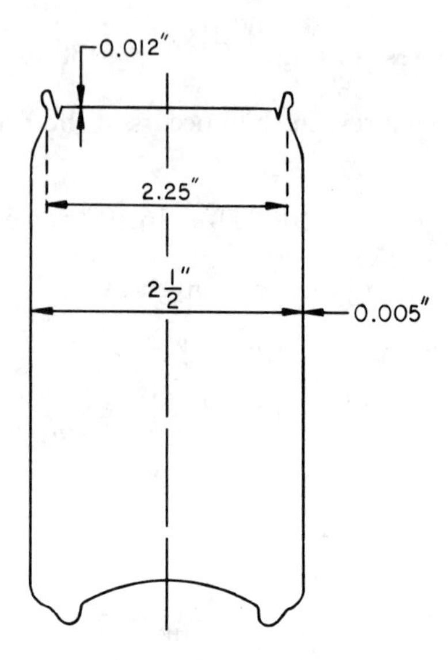

Figure P8-6.

8-7 Determine the maximum stress in the two cylindrical shells and the circular plate shown in Fig. P8-7 due to a pressure in the top compartment of 200 psi. Let $E = 25{,}000$ ksi and $\mu = 0.25$.

8-4 Short Cylindrical Shells

It was shown in the previous sections that the deflection due to applied edge shearing forces and bending moments dissipates rapidly as x increases and it becomes negligible at distances larger than $2.285/\beta$. This rapid reduction in deflection as x increased simplifies the solution of Eq. (8-16) by letting $C_1 = C_2 = 0$. When the length of the cylinder is less than about $2.285/\beta$, than C_1 and C_2 cannot be ignored and all four constants in Eq. (8-16) must be evaluated. Usually the alternate equation for the deflection, Eq. (8-17), results in a more accurate solution for short cylinders than Eq. (8-16). The calculations required in solving Eq. (8-17) are tedious because four constants are evaluated rather than two.

Example 8-5

Derive N_θ due to applied bending moment M_o at edge $x = 0$ for a short cylinder of length L.

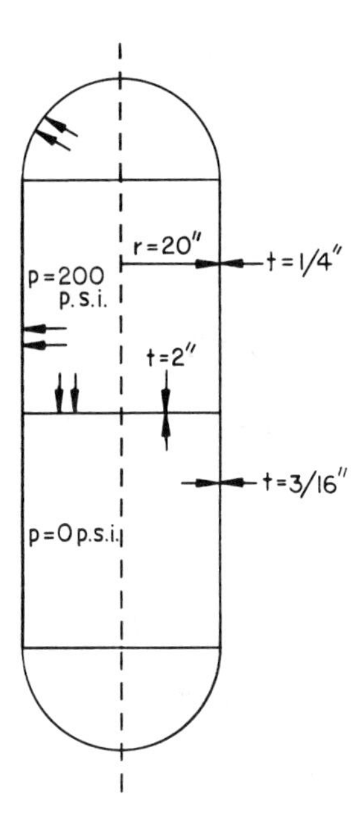

Figure P8-7.

Solution

The four boundary conditions are
At $x = 0$

$$\text{moment} = M_o = -D\left(\frac{d^2w}{dx^2}\right)$$

$$\text{shear} = 0 = -D\left(\frac{d^3w}{dx^3}\right).$$

At $x = L$

$$\text{moment} = 0 = -D\left(\frac{d^2w}{dx^2}\right)$$

$$\text{shear} = 0 = -D\left(\frac{d^3w}{dx^3}\right).$$

The second derivative of Eq. (8-17) is given by

$$\frac{d^2w}{dx^2} = 2\beta^2(C_1 \cos\beta x \cosh\beta x + C_2 \cos\beta x \sinh\beta x - C_3 \sin\beta x \cosh\beta x - C_4 \sin\beta x \sinh\beta x). \quad (1)$$

The third derivative of Eq. (8-17) is given by

$$\begin{aligned}\frac{d^3w}{dx^3} &= 2\beta^3[C_1(\cos\beta x \sinh\beta x - \sin\beta x \cosh\beta x)\\ &\quad + C_2(\cos\beta x \cosh\beta x - \sin\beta x \sinh\beta x)\\ &\quad - C_3(\sin\beta x \sinh\beta x + \cos\beta x \cosh\beta x)\\ &\quad - C_4(\sin\beta x \cosh\beta x + \cos\beta x \sinh\beta x)]. \end{aligned} \tag{2}$$

Substituting Eq. (1) into the first boundary condition gives

$$C_1 = \frac{-M_o}{2D\beta^2}.$$

Substituting Eq. (2) into the second boundary condition gives

$$C_2 = C_3.$$

From the third and fourth boundary conditions we obtain

$$C_3 = \frac{M_o}{2D\beta^2}\left(\frac{\sin\beta L \cos\beta L + \sinh\beta L \cosh\beta L}{\sinh\beta L - \sin\beta L}\right)$$

and

$$C_4 = \frac{M_o}{2D\beta^2}\left(\frac{\sin\beta L + \sinh\beta L}{\sinh\beta L - \sin\beta L}\right).$$

From Eq. (8-9)

$$\begin{aligned}N_\theta &= \frac{Etw}{r}\\ &= \frac{Et}{r}(C_1 \sin\beta x \sinh\beta x + C_2 \sin\beta x \cosh\beta x\\ &\quad + C_3 \cos\beta x \sinh\beta x + C_4 \cos\beta x \cosh\beta x)\end{aligned}$$

Problems

8-8 Solve problem 8-4 if the length of the cylinder is 2.0 inches.

8-9 Solve problem 8-3 assuming the total length of the cylinder is 3.0 inches and the loads are applied at the edges.

8-5 Stress Due to Thermal Gradients in the Axial Direction

Thermal temperature gradients in cylindrical shells occur either along the axial length or through the thickness of cylinders. Thermal gradients through the thickness are important only in thick cylindrical shells (Jawad and Farr 1989) and are beyond the scope of this book. Stress in a cylindrical shell due to temperature gradients in the axial direction can be obtained by subdividing the cylinder into infinitesimal rings of length dx. The thermal expansion in each ring due to a change of temperature T_x within the ring is given by

$$w = \alpha r T_x$$

where
α is the coefficient of thermal expansion. Some values of α are shown in Table 2-1.

Since adjacent cylindrical rings cannot have a mismatch in the deflection due to temperature T_x at their interface, a pressure p_x must be applied to eliminate the temperature deflection mismatch. Hence,

$$\frac{p_x r^2}{Et} = \alpha r T_x$$

and

$$p_x = Et\alpha T_x/r$$

$$\sigma_\theta = -p_x r/t = -E\alpha T_x. \qquad (8\text{-}33)$$

As the cylinder does not have any applied loads on it, the forces p_x must be eliminated by applying equal and opposite forces to the cylinder. Hence, Eq. (8-15) becomes

$$\frac{d^4w}{dx^4} + 4\beta^4 w = \frac{Et\alpha T_x}{rD}. \qquad (8\text{-}34)$$

The total stress in a cylinder due to axial thermal gradient distribution is obtained by adding the stresses obtained from Eqs. (8-33) and (8-34).

Example 8-6

The cylinder shown in Fig. 8-10a is initially at 0°F. The vessel is heated as shown in the figure. Determine the thermal stresses in the cylinder. Let $\alpha = 6.5 \times 10^{-6}$ inch/inch/°F, $E = 30{,}000$ ksi, $L = 10$ ft, $t = 0.25$ inch, $r = 30$ inches, and $\mu = 0.3$. The cylinder is fixed at point A and free at point B.

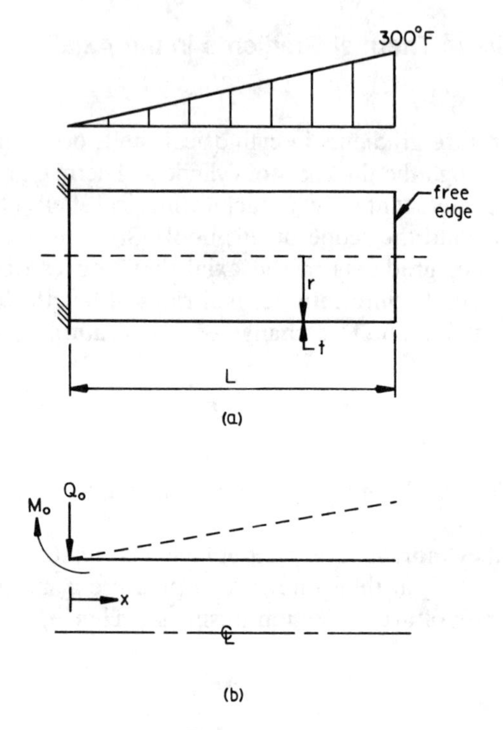

Figure 8-10.

Solution

The temperature gradient is expressed as $T_x = 300\,x/L$. The circumferential stress due to ring action is given by Eq. (8-33) as

$$\sigma_\theta = -E\alpha 300x/L \tag{1}$$

Equation (8-34) is written as

$$\frac{d^4w}{dx^4} + 4\beta^4 w = \frac{Et\alpha}{rD}(300x/L).$$

A particular solution of this equation in taken as

$$w = C_1 x + C_2$$

Substituting this expression into the differential equation gives

$$C_1 = 300r\alpha/L \qquad C_2 = 0$$

and

$$w = -r\alpha(300x/L).$$

From Eq. (8-9)

$$\sigma_\theta = N_\theta/t = Ew/r. \tag{2}$$

Adding Eqs. (1) and (2) gives

$$\sigma_\theta = 0$$

which indicates that the thermal stress in a cylinder due to linear axial thermal gradient is zero.

The slope (Fig. 8-10b) due to thermal gradient is

$$\theta = dw/dx = r\alpha(300/L)$$

At the fixed end, bending moments will occur due to the rotation θ caused by thermal gradients. The boundary conditions at the fixed edge are $w = 0$ and $\theta = 0$. From Eq. (8-24) and Fig. 8-10b,

$$0 = \beta M_o + Q_o.$$

From Eq. (8-25)

$$r\alpha(300/L) = \frac{1}{2\beta^2 D}(2\beta M_o + Q_o).$$

Solving these two equations gives

$$\begin{aligned} M_o &= 2r\alpha(300)\beta D/L \\ &= 19.63 \text{ inch-lbs/inch} \\ \sigma &= 6M_o/t^2 = 6 \times 19.63/0.25^2 = 1880 \text{ psi.} \end{aligned}$$

Problem

8-10 A coke drum in a refinery operates at 900°F. The ambient temperature is 100°F. The supporting cylinder is shown in Fig. P8-10. The temperature distribution in the cylinder, (Fig. P8-10) is parabolic between points A and B and linear between points B and C. Find the thermal stress in the cylinder if it is assumed fixed at both ends. Let $\alpha = 7.0 \times 10^{-6}$ inch/inch/°F, $\mu = 0.30$, and $E = 29{,}000$ ksi.

8-6 Discontinuity Stresses

The design of various components in a shell structure subjected to axisymmetric loads consists of calculating the thickness of the main components

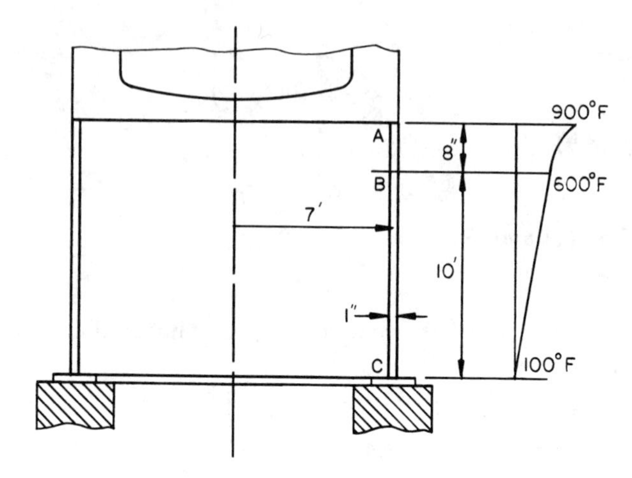

Figure P8-10.

first using the membrane theory and a given allowable stress. The forces due to various boundary conditions, such as those listed in Table 8-2, are then determined in accordance with the methods discussed in this chapter. In most cases, the magnitude of the discontinuity bending and membrane stresses at the junction is high. However, these high stresses are very local in nature and dissipate rapidly away from the junction as shown in the examples previously solved. Tests and experience have shown that these stresses are secondary in nature and are allowed to exceed the yield stress without affecting the structural integrity of the components.

Many Design Codes such as the ASME Pressure Vessel and Nuclear codes generally limit the secondary stresses at a junction to less than twice the yield stress at temperatures below the creep-rupture range. This stress level corresponds to approximately three times the allowable stress because the allowable stress is set at two-thirds the yield stress value. The justification for limiting the stress to twice the yield stress is best explained by referring to Fig. 8-11. The material stress–strain diagram is approximated by points ABO in Fig. 8-11. In the first loading cycle, the discontinuity stress, calculated elastically, at the junction increases from point A to B and then to C as the applied load is increased. The secondary stress is allowed to approach twice the yield stress indicated by point C. This point corresponds to point D on the actual stress–strain diagram which is in the plastic range.

When the applied loads are reduced, the local discontinuity stress at point D is also reduced along the elastic line DEF. The high discontinuity stress at the junction is very localized in nature and the material around

Table 8-2. Various discontinuity functions

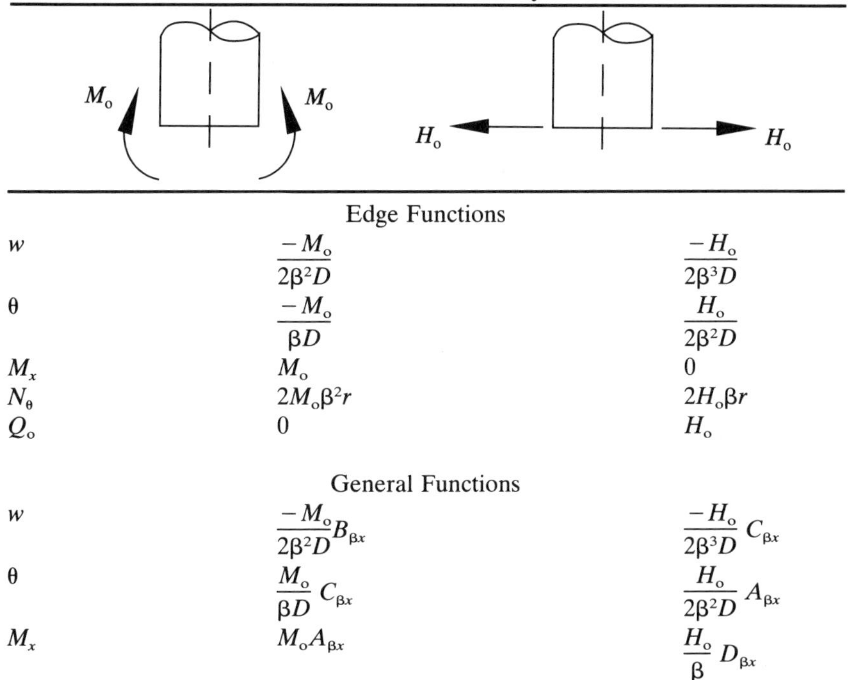

	Moment M_o	Force H_o
	Edge Functions	
w	$\dfrac{-M_o}{2\beta^2 D}$	$\dfrac{-H_o}{2\beta^3 D}$
θ	$\dfrac{-M_o}{\beta D}$	$\dfrac{H_o}{2\beta^2 D}$
M_x	M_o	0
N_θ	$2M_o\beta^2 r$	$2H_o\beta r$
Q_o	0	H_o
	General Functions	
w	$\dfrac{-M_o}{2\beta^2 D}B_{\beta x}$	$\dfrac{-H_o}{2\beta^3 D}C_{\beta x}$
θ	$\dfrac{M_o}{\beta D}C_{\beta x}$	$\dfrac{H_o}{2\beta^2 D}A_{\beta x}$
M_x	$M_o A_{\beta x}$	$\dfrac{H_o}{\beta}D_{\beta x}$
N_θ	$2M_o\beta^2 r B_{\beta x}$	$2H_o\beta r C_{\beta x}$
Q_o	$2\beta M_o D_{\beta x}$	$H_o B_{\beta x}$

the localized area is still elastic. Thus, when the applied loads are reduced, the elastic material in the vicinity of the plastic region tends to return to its original zero strain and causes the much smaller volume of plastic material with high discontinuity stress to move from points D to E and then to F. Accordingly, at the end of the first cycle after the structure is loaded and then unloaded, the highly stressed discontinuity area that was stressed to twice the yield stress in tension is now stressed in compression to the yield stress value.

On subsequent loading cycles, the discontinuity stress is permitted to have a magnitude of twice the yield stress. However, the high stressed area which is now at point F moves to point E and then to point D. The high stress with a magnitude of twice the yield stress in the junction remains within the elastic limit on all subsequent loading cycles.

If the secondary discontinuity stress at the junction is allowed to exceed twice the yield stress such as point G, then for the first loading cycle the strain approaches point H on the actual stress–strain diagram. Download-

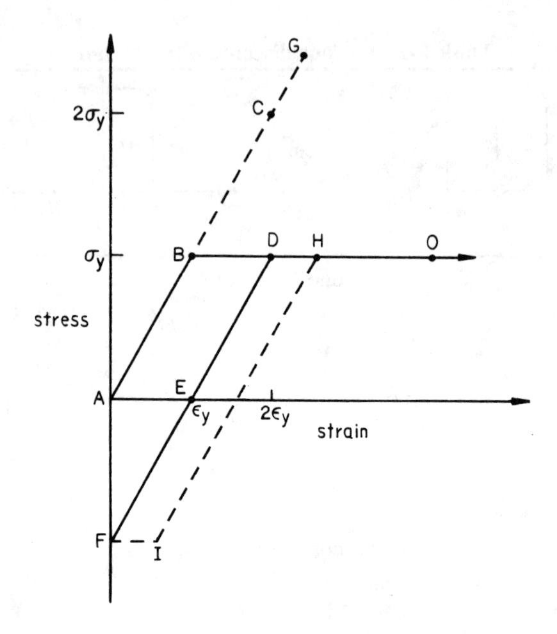

Figure 8-11.

ing will follow the path from H to I and then to F. Hence, yielding of the junction will occur both in the up and down cycles. Subsequent cycles will continue the yielding process which results in incremental plastic deformation at the junction that could lead to premature fatigue failure.

9

Bending of Shells of Revolution Due to Axisymmetric Loads

9-1 Basic Equations

In Chapter 6 the basic equations of equilibrium for the membrane forces in shells of revolution due to axisymmetric loads were developed. Referring to Fig. 6-4, it was shown that the two governing equations of equilibrium are given by Eqs. (6-9) and (6-11) as

$$\frac{d}{d\phi}(rN_\phi) - r_1 N_\theta \cos\phi + p_\phi r r_1 = 0 \tag{9-1}$$

and

$$\frac{N_\phi}{r_1} + \frac{N_\theta}{r_2} = p_r. \tag{9-2}$$

In many applications such as at a junction of a spherical to cylindrical shell subjected to axisymmetric loads, bending moments and shear forces are developed at the junction in order to maintain equilibrium and compatibility between the two shells. The effect of these additional moments and shears (Fig. 9-1) on a shell of revolution is the subject of this chapter. The axisymmetric moments and shears at the two circumferential edges of the infinitesimal element are shown in Fig. 9-1. Circumferential bending moments, which are constant in the θ-direction for any given angle ϕ, are applied at the meridional edges of the element. The shearing forces at the meridional edges must be zero in order for the deflections, which must be symmetric in the θ-direction because the loads are axisymmetric, to be constant in the θ-direction for any given angle ϕ.

Summation of forces in Fig. 9-1 parallel to the tangent at the meridian results in

$$Q(r\,d\theta)\sin d\phi/2 + \left(Q + \frac{dQ}{d\phi}\,d\phi\right)\left(r + \frac{dr}{d\phi}\,d\phi\right) d\theta \sin d\phi/2 = 0.$$

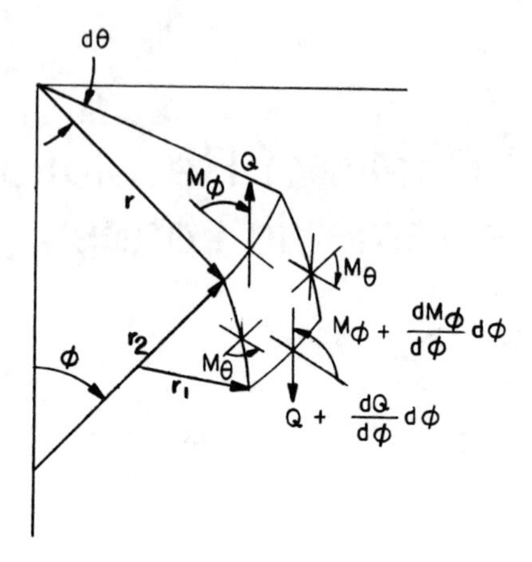

Figure 9-1.

Simplifying this equation gives

$$Qr\, d\phi\, d\theta = 0.$$

Adding this expression to Eq. (9-1) gives

$$\frac{d}{d\phi}(rN_\phi) - r_1 N_\theta \cos\phi - Qr + p_\phi r r_1 = 0. \tag{9-3}$$

Summation of forces perpendicular to the middle surface in Fig. 9-1 gives

$$-Q(r\, d\theta)\cos d\phi/2 + \left(Q + \frac{dQ}{d\phi}\, d\phi\right)\left(r + \frac{dr}{d\phi}\, d\phi\right) d\theta \cos d\phi/2 = 0.$$

Simplifying this expression and adding it to Eq. (9-2) gives

$$N_\phi r + N_\theta r_1 \sin\phi + \frac{d(Qr)}{d\phi} - p_r r r_1 = 0. \tag{9-4}$$

Summation of moments in the direction of a parallel circle gives

$$\frac{d}{d\phi}(M_\phi r) - M_\theta r_1 \cos\phi - Q r_1 r_2 \sin\phi = 0. \tag{9-5}$$

The second term involving M_θ in this equation is obtained from Fig. 9-2.

Equations (9-3) through (9-5) contain the five unknowns N_θ, N_ϕ, Q, M_θ, and M_ϕ. Accordingly, additional equations are needed and are obtained from the relationship between deflections and strains. The expressions for the meridional and circumferential strains were obtained in Chapter 7 as

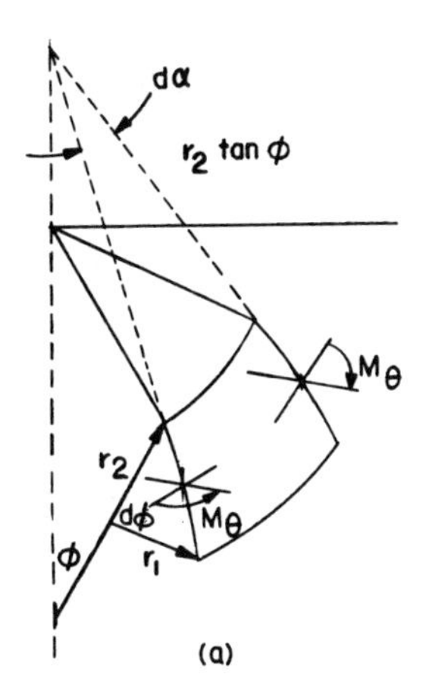

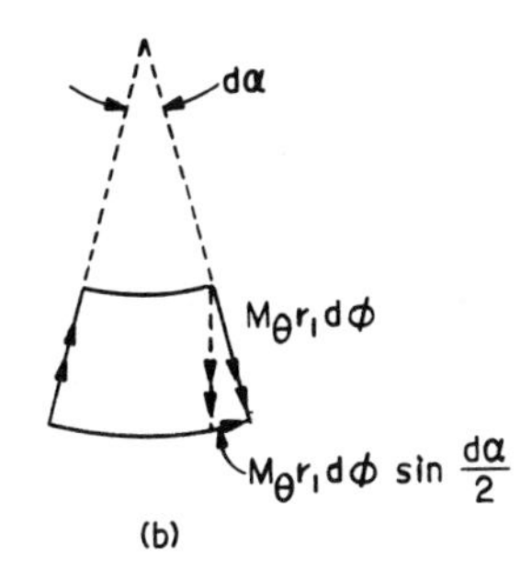

Figure 9-2.

Eqs. (7-15) and (7-16) and are given by

$$\varepsilon_\phi = \left(\frac{dv}{d\phi} - w\right) \Bigg/ r_1 \tag{9-6}$$

$$\varepsilon_\theta = (v \cot \phi - w)/r_2. \tag{9-7}$$

The expressions for N_θ and N_ϕ are obtained from Eq. (1-14) as

$$N_\phi = \frac{Et}{1 - \mu^2}(\varepsilon_\phi + \mu\varepsilon_\theta) \tag{9-8}$$

$$N_\theta = \frac{Et}{1 - \mu^2}(\varepsilon_\theta + \mu\varepsilon_\phi). \tag{9-9}$$

Substituting Eqs. (9-6) and (9-7) into Eqs. (9-8) and (9-9) gives

$$N_\phi = \frac{Et}{1-\mu^2}\left[\frac{1}{r_1}\left(\frac{dv}{d\phi} - w\right) + \frac{\mu}{r_2}(v\cot\phi - w)\right] \tag{9-10}$$

$$N_\theta = \frac{Et}{1-\mu^2}\left[\frac{1}{r_2}(v\cot\phi - w) + \frac{\mu}{r_1}\left(\frac{dv}{d\phi} - w\right)\right]. \tag{9-11}$$

The expression for change of curvature in the ϕ-direction is obtained from Fig. 9-3. The rotation α of point A in Fig. 9-3 is the summation of rotation α_1 due to deflection v and rotation α_2 due to deflection w (Fig. 9-4). From Fig. 9-4a,

$$\alpha_1 = v/r_1$$

and from Fig. 9-4b,

$$\alpha_2 = \frac{dw}{r_1\, d\phi}.$$

Hence,

$$\alpha = \alpha_1 + \alpha_2 = \left(v + \frac{dw}{d\phi}\right)\Big/ r_1. \tag{9-12}$$

Similarly the rotation β of point B is the summation of rotations due to deflection v and w. The rotation due to deflection v is expressed as

$$\left(v + \frac{dv}{d\phi}\, d\phi\right)\Big/ r_1$$

and the rotation due to w is

$$\frac{dw}{r_1\, d\phi} + \frac{d}{d\phi}\left(\frac{dw}{r_1\, d\phi}\right) d\phi.$$

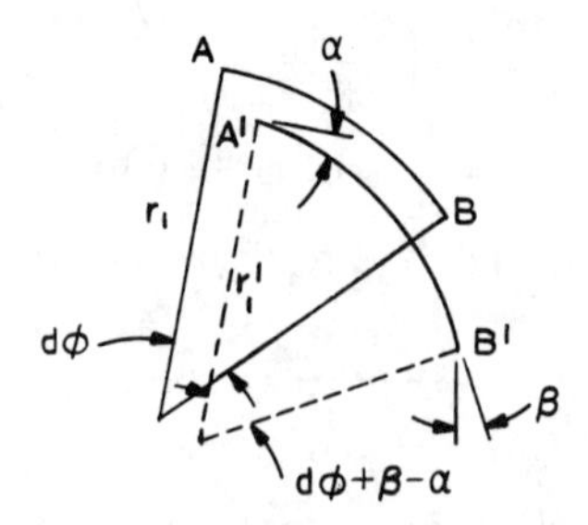

Figure 9-3.

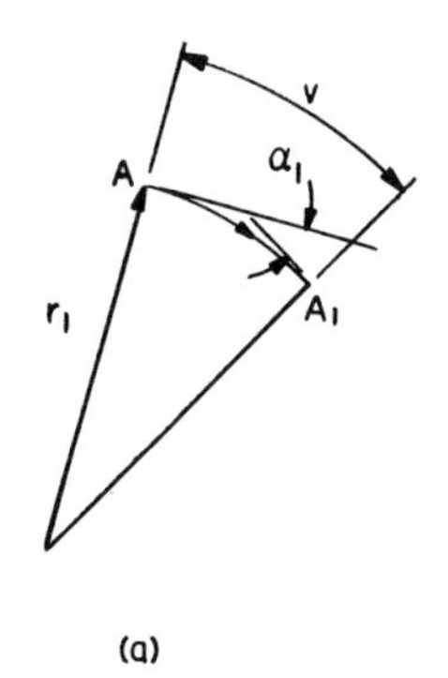

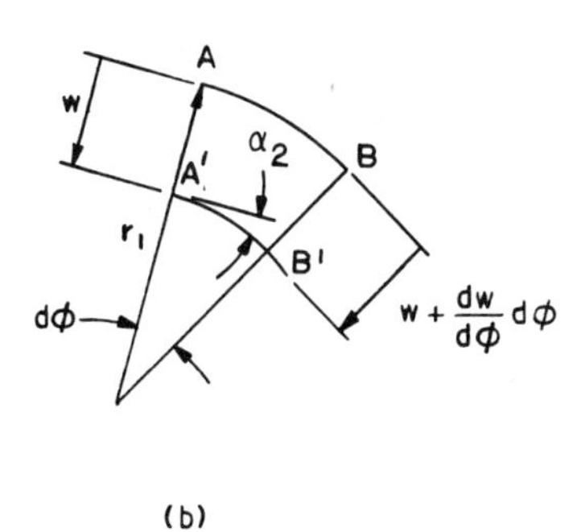

Figure 9-4.

Hence,

$$\beta = \left(v + \frac{dv}{d\phi}\, d\phi\right)\bigg/ r_1 + \frac{dw}{r_1\, d\phi} + \frac{d}{d\phi}\left(\frac{dw}{r_1\, d\phi}\right) d\phi. \qquad (9\text{-}13)$$

Due to rotation, the middle surface does not change in length. Thus, from Fig. 9-3

$$AB = A'B'$$
$$r_1\, d\phi = r_1(dQ + \beta - \alpha)$$

or,

$$\frac{1}{r_1'} = (d\phi + \beta - \alpha)/r_1\, d\phi.$$

Change in curvature

$$\chi_\phi = \frac{1}{r_1'} - \frac{1}{r_1}$$

$$= \frac{(\beta - \alpha)}{r_1\, d\phi}$$

$$\chi_\phi = \frac{d}{r_1\, d\phi}\left(\frac{v}{r_1} + \frac{dw}{r_1\, d\phi}\right). \tag{9-14}$$

The change of curvature in the θ-direction is obtained from Fig. 9-5 which shows the rotation of side AB due to the deformation of element $ABCD$. The original length AB is given by

$$AB = r_2 \sin\phi\, d\theta.$$

After rotation, AB is expressed by

$$AB = r_2' \sin(\phi + \alpha)\, d\theta.$$

Equating these two expressions and assuming small angle rotation yields

$$\frac{1}{r_2'} = \frac{1}{r_2}(1 + \alpha \cot\phi)$$

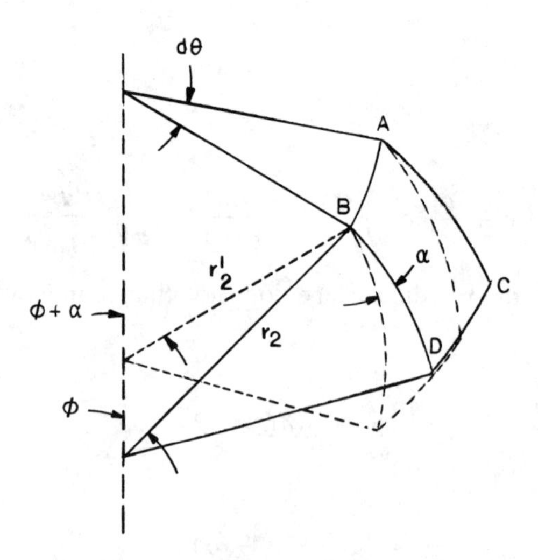

Figure 9-5.

and from

$$\chi_\theta = \frac{1}{r_2'} - \frac{1}{r_2}$$

we get

$$\chi_\theta = \frac{\alpha \cot \phi}{r_2}.$$

Substituting Eq. (9-12) into this expression gives

$$\chi_\theta = \frac{1}{r_2}\left(\frac{v}{r_1} + \frac{dw}{r_1\, d\phi}\right) \cot \phi. \tag{9-15}$$

The relationship between moment and rotation is obtained from Eqs. (1-6), (1-7), and (1-17) as

$$M_\phi = D(\chi_\phi + \mu\chi_\theta) \tag{9-16}$$

$$M_\theta = D(\chi_\theta + \mu\chi_\phi). \tag{9-17}$$

Substituting Eqs. (9-14) and (9-15) into these two expressions gives

$$M_\phi = D\left(\frac{1}{r_1}\frac{d\alpha}{d\phi} + \frac{\mu}{r_2}\,\alpha \cot \phi\right) \tag{9-18}$$

$$M_\theta = D\left(\frac{1}{r_2}\,\alpha \cot \phi + \frac{\mu}{r_1}\frac{d\alpha}{d\phi}\right). \tag{9-19}$$

The eight equations (9-3), (9-4), (9-5), (9-10), (9-11), (9-12), (9-18), and (9-19) contain eight unknowns. They are N_θ, N_ϕ, Q, M_ϕ, M_θ, v, w, and α. Solution of these equations is discussed next.

9-2 Spherical Shells

The forces and moments throughout a spherical shell due to edge shears and moments will be derived in this section. For spherical shells (Fig 9-1), the radii r_1 and r_2 are equal to R. Also, the pressures p_r and p_ϕ are set to zero for the case of applied edge loads and moments. The eight pertinent equations derived in Section 9-1 can now be reduced to two differential equations. The first equation is obtained by substituting the moment Eqs. (9-18) and (9-19) into Eq. (9-5). This gives

$$\frac{d^2\alpha}{d\phi^2} + \frac{d\alpha}{d\phi}\cot \phi - \alpha(\cot^2\phi + \mu) = -\frac{QR^2}{D}. \tag{9-20}$$

The second differential equation is more cumbersome to derive. We start by substituting Eq. (9-3) into Eq. (9-4) to delete N_θ. Then integrating the resultant equation with respect to ϕ gives

$$N_\phi = -Q \cot \phi. \tag{9-21}$$

Substituting this expression into Eq. (9-4) yields

$$N_\theta = -\frac{dQ}{d\phi}. \tag{9-22}$$

From Eqs. (9-10) and (9-11) we get

$$\frac{dv}{d\phi} - w = R(N_\phi - \mu N_\theta)/Et \tag{9-23}$$

and

$$v \cot \phi - w = R(N_\theta - \mu N_\phi)/Et. \tag{9-24}$$

Combining Eqs. (9-23) and (9-24) results in

$$\frac{dv}{d\phi} - v \cot \phi = R(1 + \mu)(N_\phi - N_\theta)/Et. \tag{9-25}$$

Differentiating Eq. (9-24) and combining it with Eq. (9-25) gives the expression for α

$$v + \frac{dw}{d\phi} = \frac{R}{Et}\left[(1 + \mu)(N_\phi - N_\theta) - \frac{dN_\theta}{d\phi} - R\frac{dN_\phi}{d\phi}\right].$$

Substituting Eqs. (9-21) and (9-22) into this expression results in

$$\frac{d^2Q}{d\phi^2} + \frac{dQ}{d\phi}\cot \phi - Q(\cot^2\phi - \mu) = Et\alpha. \tag{9-26}$$

Equations (9-20) and (9-26) must be solved simultaneously to determine Q and α. The exact solution of these two equations is too cumbersome to use for most practical problems and is beyond the scope of this book. Timoshenko (Timoshenko and Woinowsky-Krieger 1959) showed that a rigorous solution of Eqs. (9-20) and (9-26) results in expressions for α and Q that contain the terms $e^{\lambda\phi}$ and $e^{-\lambda\phi}$ where λ is a function of R/t. These terms have a large numerical value for thin shells with large R/t ratios. Substitution of these terms into Eqs. (9-20) and (9-26) for shells with large ϕ angles results in two equations with substantially larger numerical values for the higher derivatives $d^2Q/d\phi^2$ and $d^2\alpha/d\phi^2$ compared to the other terms in the equations. Hence, a reasonable approximation of Eqs. (9-20) and (9-26) is obtained by rewriting them as

$$\frac{d^2\alpha}{d\phi^2} = -\frac{QR^2}{D} \tag{9-27}$$

and

$$\frac{d^2Q}{d\phi^2} = Et\alpha. \tag{9-28}$$

Substituting Eq. (9-27) into Eq. (9-28) gives

$$\frac{d^4Q}{d\phi^4} + 4\lambda^4 Q = 0 \tag{9-29}$$

where

$$\left.\begin{aligned} \lambda^4 &= EtR^2/4D \\ &= 3(1 - \mu^2)(R/t)^2. \end{aligned}\right] \tag{9-30}$$

The solution of Eq. (9-29) is similar to that obtained for cylindrical shells and is given by either

$$\begin{aligned} Q &= e^{\lambda\phi}(C_1 \sin \lambda\phi + C_2 \cos \lambda\phi) \\ &\quad + e^{-\lambda\phi}(C_3 \sin \lambda\phi + C_4 \cos \lambda\phi) \end{aligned} \tag{9-31}$$

or

$$\begin{aligned} Q = K_1 \sin\lambda\phi \sinh\lambda\phi + K_2 \sin\lambda\phi \cosh\lambda\phi + K_3 \cos\lambda\phi \sinh\lambda\phi \\ + K_4 \cos\lambda\phi \cosh\lambda\phi. \end{aligned} \tag{9-32}$$

For continuous spherical shells without holes and subjected to edge forces, the constants C_3 and C_4 in Eq. (9-31) must be set to zero in order for Q to diminish as ϕ gets smaller. Hence, Eq. (9-31) becomes

$$Q = e^{\lambda\phi}(C_1 \sin \lambda\phi + C_2 \cos \lambda\phi). \tag{9-33}$$

This equation can be written in a more compact form by substituting for ϕ the quantity $\phi_0 - \zeta$ shown in Fig. 9-6. The new equation with redefined constants is

$$Q = C_1 e^{-\lambda\zeta} \sin(\lambda\zeta - C_2). \tag{9-34}$$

After obtaining Q, other forces and moments can be determined. Thus, from Eq. (9-21), we obtain the longitudinal membrane force

$$N_\phi = -C_1 e^{-\lambda\zeta} \sin(\lambda\zeta + C_2) \cot(\phi_o - \zeta). \tag{9-35}$$

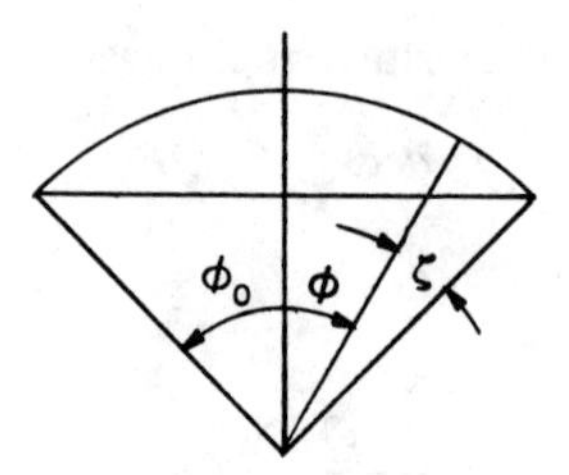

Figure 9-6.

The circumferential membrane force is determined from Eq. (9-22) as

$$N_\theta = -\sqrt{2}\, C_1 e^{-\lambda\zeta} \sin(\lambda\zeta + C_2 - \pi/4). \tag{9-36}$$

The rotation is determined from Eq. (9-28),

$$\alpha = -\frac{2\lambda^2}{Et} C_1 e^{-\lambda\zeta} \cos(\lambda\zeta + C_2). \tag{9-37}$$

The moments are obtained from Eqs. (9-18) and (9-19) using only higher order derivatives

$$\left.\begin{aligned} M_\phi &= -\frac{D}{R}\frac{d\alpha}{d\phi} \\ &= \frac{R}{\sqrt{2}\,\lambda} C_1 e^{-\lambda\zeta} \sin(\lambda\zeta + C_2 + \pi/4) \end{aligned}\right] \tag{9-38}$$

$$M_\theta = \mu M_\phi. \tag{9-39}$$

The horizontal displacement is obtained from Fig. 9-7 as

$$\begin{aligned} w_h &= v \cos\phi - w \sin\phi \\ &= (v \cot\phi - w) \sin\phi \end{aligned} \tag{9-40}$$

and from Eq. (9-24)

$$w_h = \frac{R}{Et}(N_\theta - \mu N_\phi) \sin\phi \tag{9-41}$$

$$= -\frac{\sqrt{2}\,\lambda R}{Et} C_1 e^{-\lambda\zeta} \sin(\phi_o - \zeta) \sin(\lambda\zeta + C_2 - \pi/4). \tag{9-42}$$

Table 9-1 gives various design values for spherical shells due to applied edge loads.

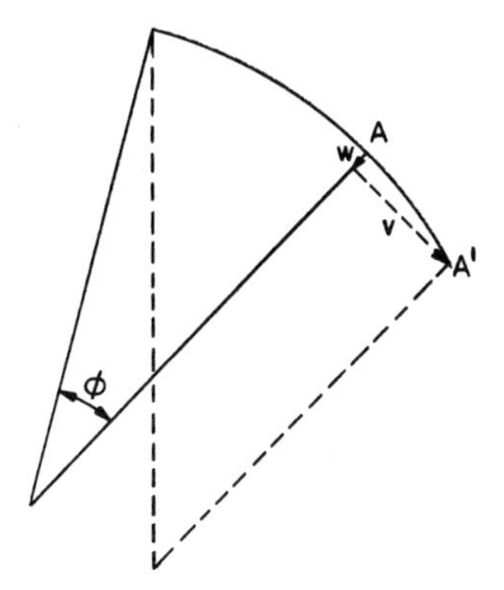

Figure 9-7.

Example 9-1

Find the location and maximum value of moment M_ϕ due to the horizontal force shown in Fig. 9-8.

Solution

The first boundary condition at the edge is

$$Q = H_o \sin \phi_o \quad \text{at } \zeta = 0$$

and from Eq. (9-34)

$$C_1 \sin(-C_2) = H_o \sin \phi_o. \tag{1}$$

From Eqs. (9-28) and (9-38)

$$M = \frac{-D}{R} \frac{1}{Et} \frac{d^3Q}{d\zeta^3}.$$

Substituting the third derivative of Eq. (9-34) into this expression gives

$$M_\phi = \frac{-2DC\lambda^3 e^{-\lambda\zeta}}{REt} [(\cos C_2)(\cos \lambda\zeta + \sin \lambda\zeta) + (\sin C_2)(\sin \lambda\zeta - \cos \lambda\zeta)].$$

The second boundary condition at the edge is

$$M_o = 0$$

or,

$$\cos C_2(1) + \sin C_2(-1) = 0. \tag{2}$$

Table 9-1. Edge Loads on Spherical Shells

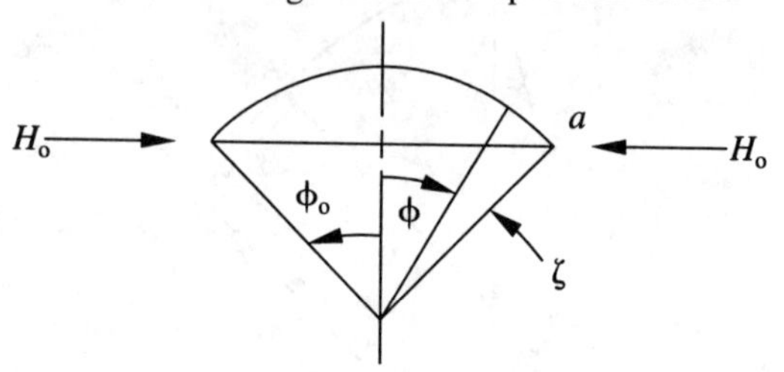

Edge Functions $\zeta = 0$ and $\phi = \phi_o$	
α	$-\dfrac{2H_o\lambda^2}{Et}\sin\phi_o$
w_h	$\dfrac{H_oR}{Et}\sin\phi_o(2\lambda\sin\phi_o - \mu\cos\phi_o)$

General Functions	
Q	$\sqrt{2}H_oe^{-\lambda\zeta}\sin\phi_o\cos\left(\lambda\zeta + \dfrac{\pi}{4}\right)$
N_θ	$-2H_o\lambda e^{-\lambda\zeta}\sin\phi_o\cos\lambda\zeta$
M_ϕ	$-\dfrac{H_oR}{\lambda}e^{-\lambda\zeta}\sin\phi_o\sin\lambda\zeta$
α	$-\dfrac{H_o}{Et}\left[2\sqrt{2}\,\lambda^2e^{-\lambda\zeta}\sin\phi_o\sin\left(\lambda\zeta + \dfrac{\pi}{4}\right)\right]$
w_h	$\dfrac{H_oR}{Et}e^{-\lambda\zeta}\sin\phi_o\left[2\lambda\sin\phi\cos\lambda\zeta - \sqrt{2}\,\mu\cos\phi\cos\left(\lambda\zeta + \dfrac{\pi}{4}\right)\right]$

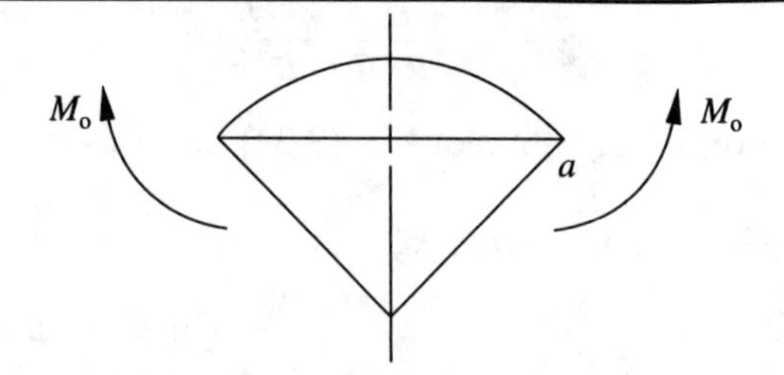

Edge Functions $\zeta = 0$ and $\phi = \phi_o$	
α	$\dfrac{4M_o\lambda^3}{EtR}$
w_h	$-\dfrac{2M_o\lambda^2}{Et}\sin\phi_o$

Table 9-1. (continued)

General Functions	
Q	$-\dfrac{2M_o\lambda}{R}\, e^{-\lambda\zeta} \sin \lambda\zeta$
N_θ	$2\sqrt{2}\,\dfrac{M_o\lambda^2}{R}\, e^{-\lambda\zeta} \cos\left(\lambda\zeta + \dfrac{\pi}{4}\right)$
M_ϕ	$\sqrt{2}\, M_o e^{-\lambda\zeta} \sin\left(\lambda\zeta + \dfrac{\pi}{4}\right)$
α	$\dfrac{4M_o\lambda^3}{EtR}\, e^{-\lambda\zeta} \cos \lambda\zeta$
w_h	$-\dfrac{2M_o\lambda}{Et}\, e^{-\lambda\zeta}\left[\sqrt{2}\,\lambda \sin\phi \cos\left(\lambda\zeta + \dfrac{\pi}{4}\right) + \mu \cos\phi \sin \lambda\zeta\right]$

Notation: $M_\theta = \mu M_\phi$; $N_\phi = -Q \cot \phi$; w_h = horizontal component of deflection; α = rotation; $\zeta = \phi_o - \phi$; $\lambda = \sqrt[4]{(R/t)^2\, 3(1 - \mu^2)}$; μ = Poisson's ratio. Inward deflection is positive. Positive rotation is in direction of positive moments. Tensile N_ϕ and N_θ are positive. Positive moments cause tension on the inside surface. Inward Q is positive.

From Eq. (2) we get

$$C_2 = \pi/4$$

and from Eq. (1) we get

$$C_1 = -\frac{2}{\sqrt{2}}\, H_o \sin \phi_o.$$

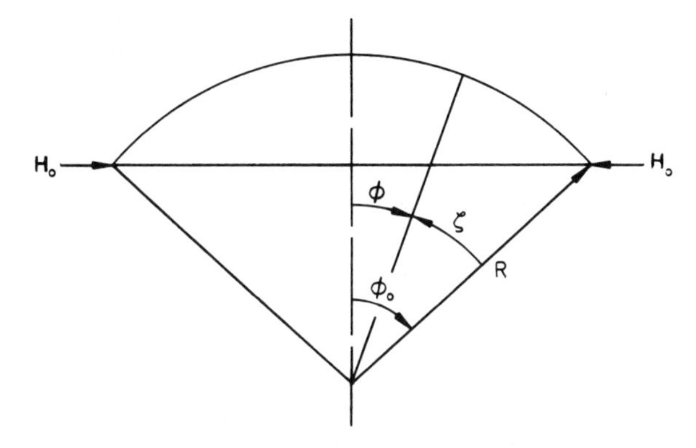

Figure 9-8.

The expression for M_ϕ becomes

$$M_\phi = \frac{4D\lambda^3 e^{-\lambda\zeta}}{REt} \sin\phi \sin\lambda\zeta.$$

Substituting

$$D = \frac{Et^3}{12(1 - \mu^2)}$$

into the moment expression results in

$$M_\phi = \frac{H_o R}{\lambda} e^{-\lambda\zeta} \sin\phi_o \sin\lambda\zeta \qquad (3)$$

which is the same as that given in Table 9-1.

The location of the maximum moment is obtained from

$$\frac{dM_\phi}{d\zeta} = 0$$

or

$$\zeta = \frac{\pi}{2\lambda}.$$

The maximum moment is obtained from Eq. (3) as

$$M_o = \frac{0.2079 H_o R}{\lambda} \sin\phi_o.$$

Example 9-2

Calculate the required thickness of the hemispherical and cylindrical shells in Fig. 9-9a and determine the discontinuity stress at the junction. Let $p = 200$ psi. Allowable membrane stress is 18 ksi and $\mu = 0.3$.

Solution

The required thickness of the cylindrical shell is

$$t = \frac{pr}{\sigma} = \frac{200 \times 36}{18{,}000} = 0.40 \text{ inch}.$$

The required thickness of the hemispherical shell is

$$t = \frac{pr}{2\sigma} = \frac{200 \times 36}{2 \times 18{,}000} = 0.20 \text{ inch}.$$

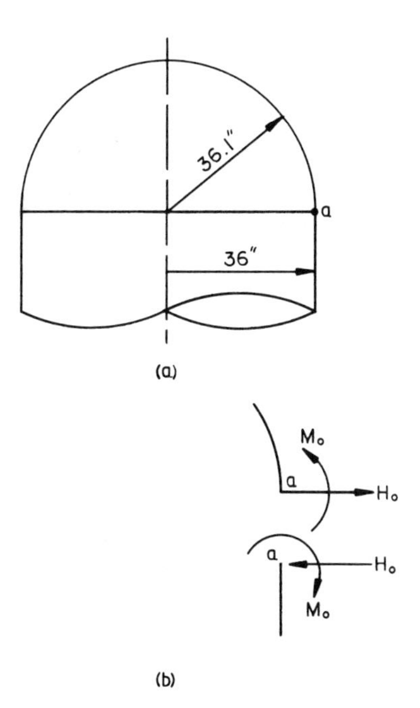

Figure 9-9.

The discontinuity forces are shown in Fig. 9-9b. The first compatibility equation at point a is

deflection of cylinder due to $p + H_o + M_o$

$$= \text{deflection of hemisphere due to } p + H_o + M_o \quad (1)$$

The expressions for the deflection of the cylinder are obtained from Eq. (8-32) and Table 8-2. For the hemisphere, the expressions are obtained from Eq. (1) of Example 7-3 and Table 9-1. Let outward deflection and clockwise rotation at point a be positive. Hence,

$$\beta = \sqrt[4]{\frac{3(1 - \mu^2)}{36^2 \times 0.4^2}} = 0.3387$$

$$D = \frac{E(0.4)^3}{12(1 - 0.3^2)} = \frac{5.8608E}{1000}$$

$$\lambda = \sqrt[4]{\left(\frac{36.1}{0.2}\right)^2 3(1 - 0.3^2)} = 17.2695.$$

Thus Eq. (1) becomes

$$\frac{(200)(36)^2}{E(0.4)}(0.85) - \frac{H_o}{(2)(0.3387)^3(5.8608E/1000)}$$
$$+ \frac{M_o}{(2)(0.3387)^2(5.8608E/1000)} = \frac{(200)(36.1)^2}{(2)(E)(0.2)}(1 - 0.3)$$
$$+ \frac{H_o(36.1)}{E(0.2)}(2 \times 17.2695) + \frac{2M_o(17.2695)^2}{E(0.2)}$$

$$550{,}800 - 2195.67H_o + 743.67M_o$$
$$= 456{,}123.5 + 6234.29H_o + 2982.36M_o$$

$$M_o + 3.77H_o = 42.29. \qquad (2)$$

The second compatibility equation at point a is given by rotation of cylinder due to $p + H_o + M_o$ = rotation of hemisphere due to $p + H_o + M_o$

$$0 - \frac{H_o}{(2)(0.3387)^2(5.8608E/1000)} + \frac{M_o}{(0.3387)(5.8608E/1000)}$$
$$= 0 - \frac{2H_o(17.2695)^2}{E(0.2)} - \frac{4M_o(17.2695)^3}{E(0.2)(36.1)}$$

$$1.50M_o + H_o = 0. \qquad (3)$$

Solving Eqs. (2) and (3) yields

$$M_o = -9.08 \text{ inch-lbs/inch} \qquad H_o = 13.63 \text{ lb/inch}$$

The negative sign for the moment indicates that the actual moment is opposite that assumed in Fig. 9-9b.

Cylindrical shell

$$\text{longitudinal bending stress} = \frac{6M_o}{t^2} = 340 \text{ psi}$$

$$\text{longitudinal membrane stress} = \frac{pr}{2t} = 9000 \text{ psi}$$

$$\text{total longitudinal stress} = 9340 \text{ psi}$$

From Eq. (8-13),

$$\text{hoop bending stress} = 0.3 \times 340 = 100 \text{ psi}$$

$$\text{deflection at point } a = (550{,}800 - 2195.67H_o + 743.67M_o)/E$$

$$= 514{,}120/E.$$

The hoop membrane force is obtained from Eq. (8-9) as

$$N_\theta = \frac{Etw}{r} = 5712 \text{ lbs/inch}$$

$$\text{hoop membrane stress} = \frac{5712}{0.4} = 14{,}300 \text{ psi}$$

$$\text{total hoop stress} = 14{,}400 \text{ psi}$$

Hemispherical shell

$$\text{longitudinal stress} = \frac{pr}{2t} + \frac{6M_o}{t^2}$$

$$= \frac{(200)(36.1)}{2(0.2)} + \frac{6(9.08)}{0.2^2}$$

$$= 19{,}410 \text{ psi}$$

From Table 9-1,

$$\text{hoop force} = \frac{pr}{2} + H_o\lambda - 2M_o\lambda^2/R$$

$$= \frac{(200)(36.1)}{2} + (13.63)(17.2695) - \frac{2(9.08)(17.2695)^2}{36.1}$$

$$= 18{,}480 \text{ psi}$$

If a spherical shell has an axisymmetric hole, and is subjected to edge loads, then Eq. (9-33) must be used to determine the constants K_1 through K_4. Other design functions are then established from the various equations derived. Table 9-2 lists various equations for open spherical shells.

Problems

9-1 Derive Eq. (9-35)

9-2 What is the maximum stress in the spherical shell and flat plate shown in Fig. P9-2 due to internal pressure of 600 psi? Let E = 16,000 ksi and μ = 0.0.

Table 9-2. Edge loads in open spherical shells (Baker et al. 1968)

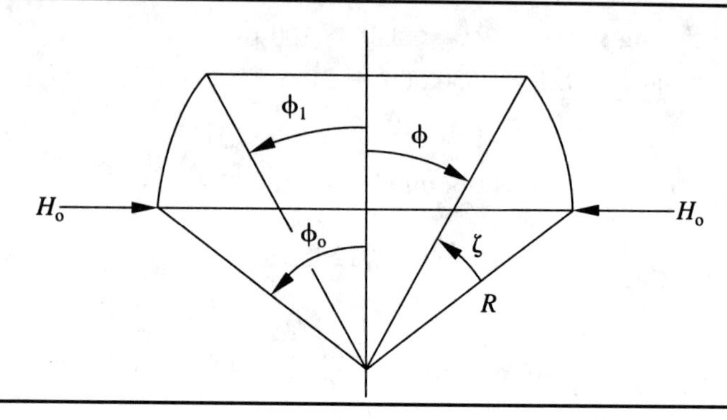

Q	$H_o \sin\phi_o \left[K_7(\zeta) - \dfrac{K_4}{K_1} K_{10}(\zeta) + \dfrac{K_2}{K_1} K_8(\zeta) \right]$
N_θ	$-H_o\lambda \sin\phi_o \left[-K_9(\zeta) - \dfrac{2K_4}{K_1} K_7(\zeta) + \dfrac{K_5}{K_1} K_{10}(\zeta) \right]$
M_ϕ	$-H_o \dfrac{R}{2\lambda} \sin\phi_o \left[-K_{10}(\zeta) + \dfrac{2K_4}{K_1} K_8(\zeta) - \dfrac{K_2}{K_1} K_9(\zeta) \right]$
α	$-H_o \dfrac{2\lambda^2}{Et} \sin\phi_o \left[-K_8(\zeta) - \dfrac{K_4}{K_1} K_9(\zeta) + \dfrac{K_2}{K_1} K_7(\zeta) \right]$
w_h	$H_o \dfrac{R\lambda}{Et} \sin\phi \sin\phi_o \left[-K_9(\zeta) - \dfrac{2K_4}{K_1} K_7(\zeta) + \dfrac{K_2}{K_1} K_{10}(\zeta) \right]$

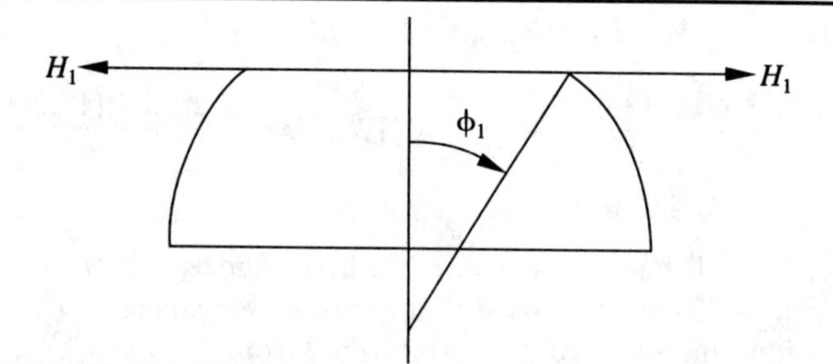

Q	$-H_1 \sin\phi_1 \left[-\dfrac{K_9}{K_1} K_{10}(\zeta) - \dfrac{2K_8}{K_1} K_8(\zeta) \right]$
N_θ	$H_1 2\lambda \sin\phi_1 \left[-\dfrac{K_9}{K_1} K_7(\zeta) + \dfrac{K_8}{K_1} K_{10}(\zeta) \right]$
M_ϕ	$H_1 \dfrac{R}{\lambda} \sin\phi_1 \left[\dfrac{K_9}{K_1} K_8(\zeta) - \dfrac{K_8}{K_1} K_9(\zeta) \right]$

Table 9-2. (continued)

α	$H_1 \dfrac{2\lambda^2}{Et} \sin \phi_1 \left[\dfrac{K_9}{K_1} K_9(\zeta) + \dfrac{2K_8}{K_1} K_7(\zeta) \right]$
w_h	$-H_1 \dfrac{2R\lambda}{Et} \sin \phi \sin \phi_1 \left[\dfrac{K_9}{K_1} K_7(\zeta) - \dfrac{K_8}{K_1} K_{10}(\zeta) \right]$

Q	$-M_o \dfrac{2\lambda}{R} \left[\dfrac{K_6}{K_1} K_{13}(\zeta) + \dfrac{K_5}{K_1} K_{14}(\zeta) - \dfrac{K_3}{K_1} K_8(\zeta) \right]$
N_θ	$M_o \dfrac{2\lambda^2}{R} \left[\dfrac{K_6}{K_1} K_{12}(\zeta) + \dfrac{K_5}{K_1} K_{11}(\zeta) - \dfrac{K_3}{K_1} K_{10}(\zeta) \right]$
M_ϕ	$M_o \left[\dfrac{K_6}{K_1} K_{11}(\zeta) - \dfrac{K_5}{K_1} K_{12}(\zeta) + \dfrac{K_3}{K_1} K_9(\zeta) \right]$
α	$M_o \dfrac{4\lambda^3}{EtR} \left[\dfrac{K_6}{K_1} K_{14}(\zeta) - \dfrac{K_5}{K_1} K_{13}(\zeta) - \dfrac{K_3}{K_1} K_7(\zeta) \right]$
w_h	$-M_o \dfrac{2\lambda^2}{Et} \sin \phi \left[\dfrac{K_6}{K_1} K_{12}(\zeta) + \dfrac{K_5}{K_1} K_{11}(\zeta) - \dfrac{K_3}{K_1} K_{10}(\zeta) \right]$

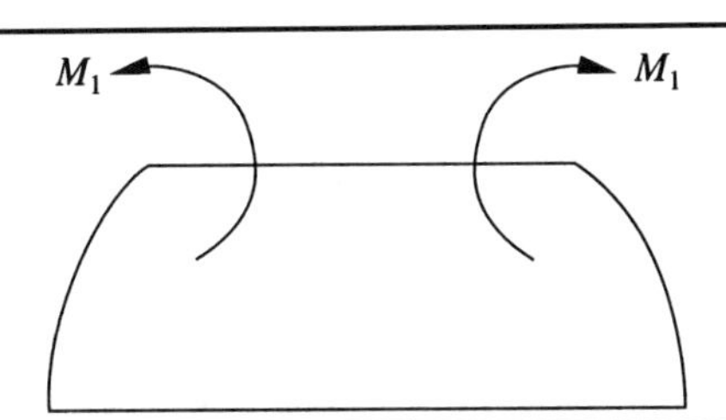

Q	$-M_1 \dfrac{2\lambda}{R} \left[-\dfrac{K_8}{K_1} K_{10}(\zeta) + \dfrac{K_{10}}{K_1} K_8(\zeta) \right]$
N_θ	$M_1 \dfrac{2\lambda^2}{R} \left[-\dfrac{2K_8}{K_1} K_7(\zeta) + \dfrac{K_{10}}{K_1} K_{10}(\zeta) \right]$
M_ϕ	$M_1 \left[\dfrac{2K_8}{K_1} K_8(\zeta) - \dfrac{K_{10}}{K_1} K_9(\zeta) \right]$

Table 9-2. (continued)

α	$M_1 \dfrac{4\lambda^3}{EtR}\left[\dfrac{K_8}{K_1} K_9(\zeta) + \dfrac{K_{10}}{K_1} K_7(\zeta)\right]$
w_h	$-M_1 \dfrac{2\lambda^2}{Et} \sin\phi \left[-\dfrac{2K_8}{K_1} K_7(\zeta) + \dfrac{K_{10}}{K_1} K_{10}(\zeta)\right]$

Notation: $M_\theta = \mu M_\phi$; $N_\phi = -Q \cot\phi$; w_h = horizontal component of deflection; α = rotation; $K_1 = \sinh^2\lambda - \sin^2\lambda$; $K_2 = \sinh^2\lambda + \sin^2\lambda$; $K_3 = \sinh\lambda\cosh\lambda + \sin\lambda\cos\lambda$; $K_4 = \sinh\lambda\cosh\lambda - \sin\lambda\cos\lambda$; $K_5 = \sin^2\lambda$; $K_6 = \sinh^2\lambda$; $K_7 = \cosh\lambda\cos\lambda$; $K_8 = \sinh\lambda\sin\lambda$; $K_8(\zeta) = \sinh\zeta\lambda\sin\zeta\lambda$; $K_9 = \cosh\lambda\sin\lambda - \sinh\lambda\cos\lambda$; $K_9(\zeta) = \cosh\zeta\lambda\sin\zeta\lambda - \sinh\zeta\lambda\cos\zeta\lambda$; $K_{10} = \cosh\lambda\sin\lambda + \sinh\lambda\cos\lambda$; $K_{10}(\zeta\lambda) = \cosh\zeta\lambda\sin\zeta\lambda + \sinh\zeta\lambda\cos\zeta\lambda$; $K_{11}(\zeta\lambda) = \cosh\zeta\lambda\cos\zeta\lambda - \sinh\zeta\lambda\sin\zeta\lambda$; $K_{12} = \cosh\lambda\cos\lambda + \sinh\lambda\sin\lambda$; $K_{12}(\zeta\lambda) = \cosh\zeta\lambda\cos\zeta\lambda + \sinh\zeta\lambda\sin\zeta\lambda$; $K_{13}(\zeta\lambda) = \cosh\zeta\lambda\sin\zeta\lambda$; $K_{14}(\zeta\lambda) = \sinh\zeta\lambda\cos\zeta\lambda$.

9-3 Determine the discontinuity stress at the spherical-to-cylindrical junction shown in Fig. P9-3. The dimensions of the stiffening ring at the junction are 4 inches × 3/4 inch. Let E = 30,000 ksi and μ = 0.30.
9-4 The heating compartment in Fig. P9-4 between the two spherical shells is subjected to 100 psi pressure. Find the discontinuity stress in the top and bottom spherical shells. E = 27,000 ksi and μ = 0.28.
9-5 Determine the length $L = R$ where the moment diminishes to 1% of moment M_o applied at the free edge in Fig. P9-5. How does this length compare with that in Example 8-1 for cylindrical shells?
9-6 Derive Eq. (9-42).

9-3 Conical Shells

The derivation of the expressions for the bending moments in conical shells is obtained from the general equations of Section 9-1. In this case the angle

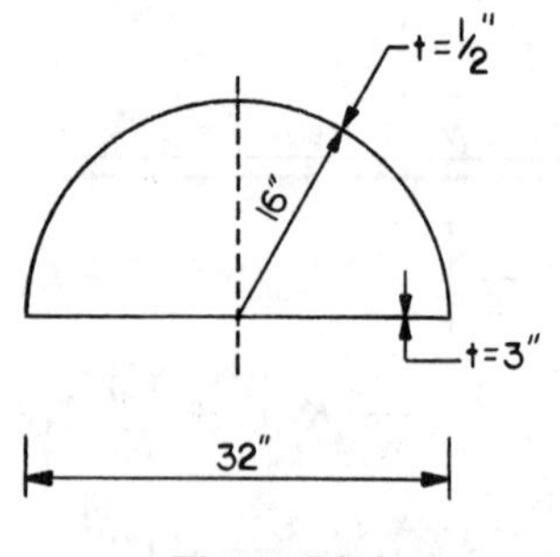

Figure P9-2.

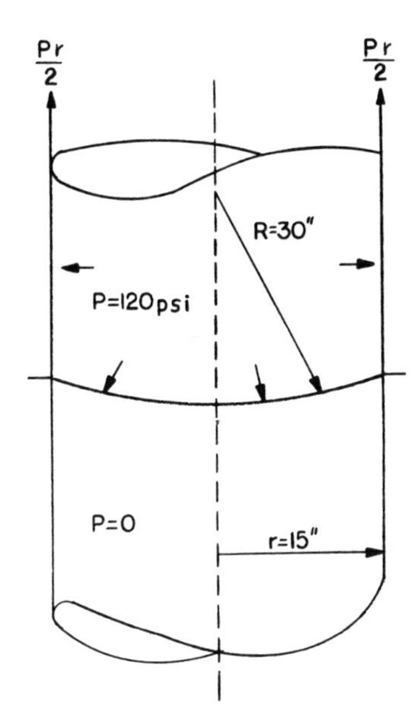

Figure 9-3.

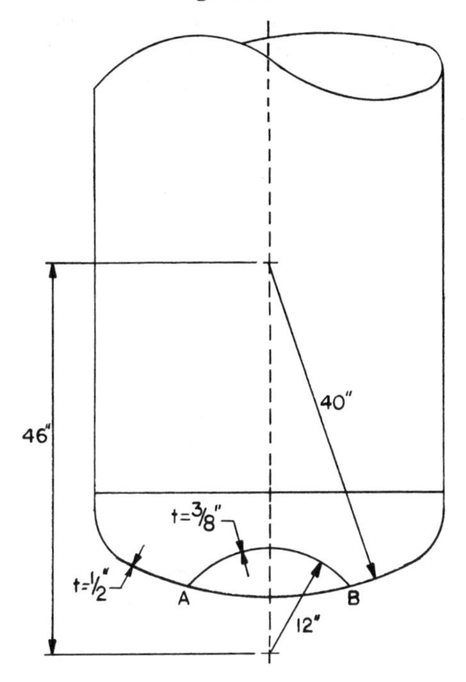

Figure P9-4.

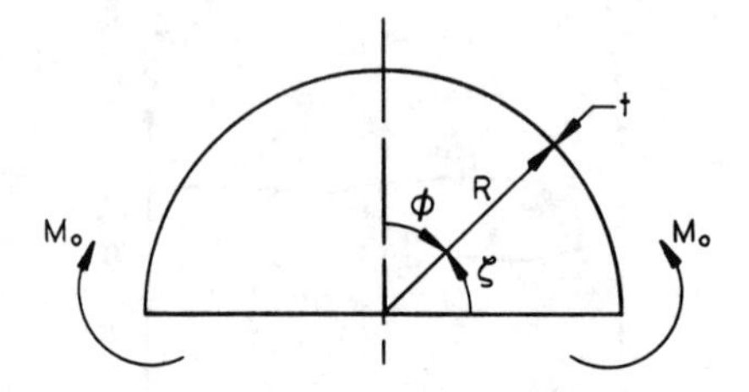

Figure P9-5.

is constant as shown in Fig. 9-10. Equations (9-5), (9-6), (9-7), (9-12), (9-13), (9-14), (9-20), and (9-21) have to be rewritten for conical shells with the following substitutions

$$\phi = \frac{\pi}{2} - \alpha, \qquad r_1 = \infty, \qquad r_2 = s \tan \alpha$$

$$ds = r_1 \, d\phi.$$

The solutions of the resulting eight equations (Flugge 1967) involve Bessel functions. These solutions are too cumbersome to use on a regular basis. However, simplified asymptotic solutions, similar to those developed for spherical shells, can be developed for the large end of conical shells with $\beta > 11$ where

$$\beta = 2\sqrt[4]{3(1 - \mu^2)} \sqrt{\frac{s \cot \alpha}{t}}. \tag{9-43}$$

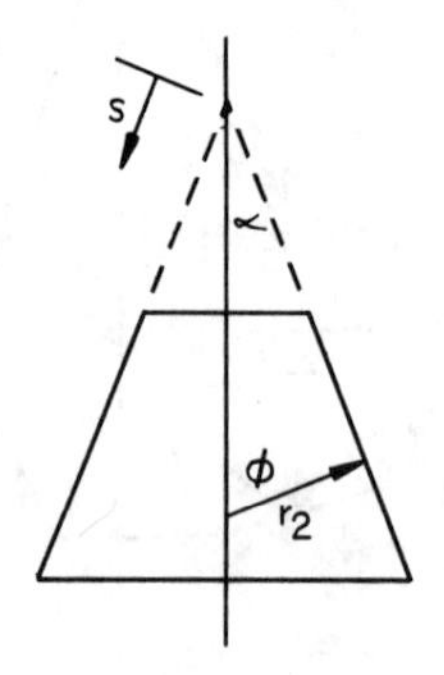

Figure 9-10.

This range of β is common for most conical shells encountered in industry.

$$Q_s = \frac{e^{\beta}}{2^{3/4} s\sqrt{\pi\beta}}\left[C_1 \cos\left(\beta - \frac{\pi}{8}\right) + C_2 \sin\left(\beta - \frac{\pi}{8}\right)\right] \tag{9-44}$$

$$M_s = \frac{e^{\beta}}{\sqrt[4]{2}\,\beta\,\sqrt{\pi\beta}}\left[C_1 \sin\left(\beta + \frac{\pi}{8}\right) - C_2 \cos\left(\beta + \frac{\pi}{8}\right)\right] \tag{9-45}$$

$$M_\theta = \mu M_s \tag{9-46}$$

$$N_\theta = \frac{-\sqrt{\beta}\,e^{\beta}\tan\alpha}{2\sqrt[4]{2}\,\sqrt{\pi}\,s}\left[C_1 \cos\left(\beta + \frac{\pi}{8}\right) + C_2 \sin\left(\beta + \frac{\pi}{8}\right)\right] \tag{9-47}$$

$$N_s = Q_s \tan\alpha \tag{9-48}$$

$$w = \frac{s\sin\alpha}{Et}(N_\theta - \mu N_s) \tag{9-49}$$

where C_1 and C_2 are obtained from the boundary conditions.

Application of Eqs. (9-43) to (9-49) to edge forces of a full cone is given in Table 9-3. Table 9-4 lists various equations for truncated cones.

Table 9-3. Edge loads in conical shells (Jawad and Farr 1989)

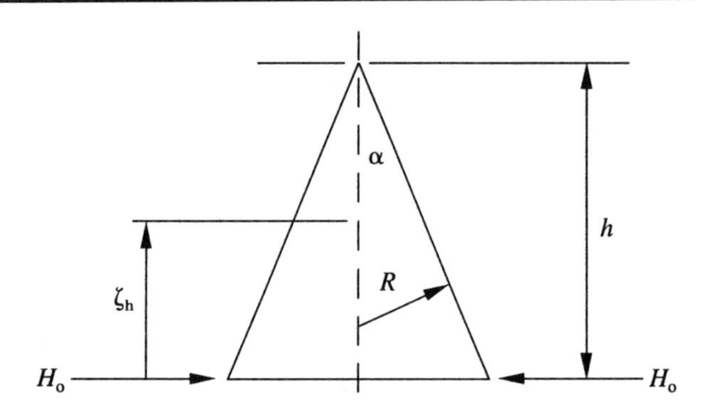

Edge Functions $\zeta = 0$	
θ	$-\dfrac{H_o h^2}{2D\beta^2 \cos\alpha}$
w_h	$\dfrac{H_o h^3}{2D\beta^3 \cos\alpha}\left[1 - \dfrac{\mu h \tan\alpha}{2R\beta\cos\alpha}\right]$

Table 9-3. (continued)

General Functions	
Q	$\sqrt{2}\, H_o e^{-\beta\zeta} \cos\alpha \cos\left(\beta\zeta + \frac{\pi}{4}\right)$
N_θ	$-\frac{2H_o R^2 \beta \cos^2\alpha}{h} e^{-\beta\zeta} \cos\beta\zeta$
M_ϕ	$-\frac{H_o h}{\beta} e^{-\beta\zeta} \sin\beta\zeta$
θ	$-\frac{H_o h^2}{\sqrt{2} D\beta^2 \cos\alpha} e^{-\beta\zeta} \sin\left(\beta\zeta + \frac{\pi}{4}\right)$
w_h	$\frac{H_o h^3 e^{-\beta\zeta}}{2D\beta^3 \cos\alpha}\left[\cos\beta\zeta - \frac{\mu h}{2R\beta} \tan\alpha \cos\left(\beta\zeta + \frac{\pi}{4}\right)\right]$

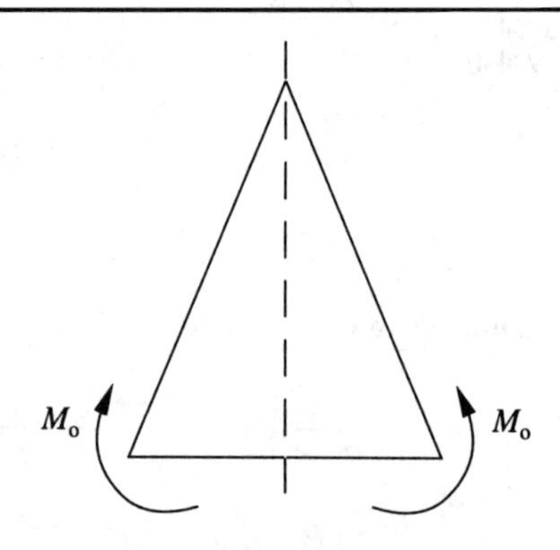

Edge Functions $\zeta = 0$	
θ	$\frac{M_o h}{D\beta \cos\alpha}$
w_h	$-\frac{M_o h^2}{2D\beta^2 \cos\alpha}$

General Functions	
Q	$-\frac{2M_o\beta}{h} \cos\alpha e^{-\beta\zeta} \sin\beta\zeta$
N_θ	$\frac{4R\beta^2 M_o}{\sqrt{2}\, h^2} \cos^2\alpha e^{-\beta\zeta} \cos\left(\beta\zeta + \frac{\pi}{4}\right)$

Table 9-3. (continued)

M_ϕ	$\sqrt{2}\, M_o e^{-\beta\zeta} \sin\left(\beta\zeta + \dfrac{\pi}{4}\right)$
θ	$\dfrac{M_o h}{D\beta \cos\alpha} e^{-\beta\zeta} \cos\beta\zeta$
w_h	$-\dfrac{M_o h^2}{2D\beta^2 \cos\alpha} e^{-\beta\zeta} \left[\sqrt{2}\cos\left(\beta\zeta + \dfrac{\pi}{4}\right) + \dfrac{\mu h}{\beta R \cos\alpha} \tan\alpha \sin\beta\zeta\right]$

Notation: $M_\theta = \mu M_\phi$; $N_\phi = -Q \cot\alpha$; w_h = horizontal component of deflection; θ = rotation; $\beta = (h/(\sqrt{Rt}\sin\alpha))\sqrt[4]{3(1-\mu^2)}$; μ = Poisson's ratio. Inward deflection is positive. Positive rotation is in direction of positive moments. Tensile N_ϕ and N_θ are positive. Positive moments cause tension on the inside surface. Inward Q is positive.

Table 9-4. Edge loads in open conical shells (Baker et al. 1968)

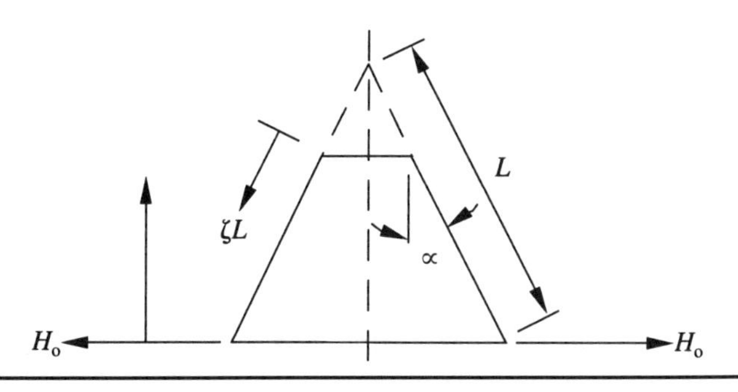

Q	$-H_o \cos\alpha \left[-\dfrac{K_9}{K_1} K_{10}(\zeta) + \dfrac{2K_8}{K_1} K_8(\zeta)\right]$
N_θ	$2H_o x_m \sin\alpha \left[-\dfrac{K_9}{K_1} K_7(\zeta) + \dfrac{K_8}{K_1} K_{10}(\zeta)\right]$
M_s	$H_o \dfrac{1}{\beta} \cos\alpha \left[\dfrac{K_9}{K_1} K_8(\zeta) - \dfrac{K_8}{K_1} K_9(\zeta)\right]$
θ	$H_o \dfrac{1}{2D\beta^2} \cos\alpha \left[\dfrac{K_9}{K_1} K_9(\zeta) + \dfrac{2K_8}{K_1} K_7(\zeta)\right]$
w_h	$-H_o \dfrac{1}{2D\beta^3} \cos^2\alpha \left[-\dfrac{K_9}{K_1} K_7(\zeta) + \dfrac{K_8}{K_1} K_{10}(\zeta)\right]$

Table 9-4. (continued)

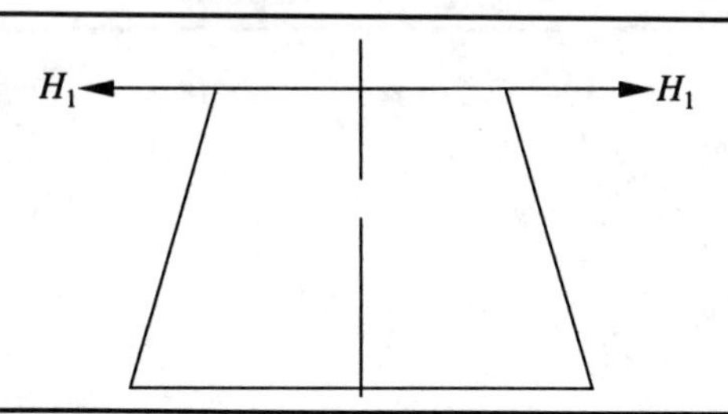

Q	$-H_1 \cos\alpha \left[K_7(\zeta) - \frac{K_4}{K_1} K_{10}(\zeta) + \frac{K_2}{K_1} K_8(\zeta) \right]$
N_θ	$H_1 x_m \beta \sin\alpha \left[K_9(\zeta) + \frac{2K_4}{K_1} K_7(\zeta) - \frac{K_2}{K_1} K_{10}(\zeta) \right]$
M_s	$H_1 \frac{1}{2} \cos\alpha \left[K_{10}(\zeta) - \frac{2K_4}{K_1} K_8(\zeta) + \frac{K_2}{K_1} K_9(\zeta) \right]$
θ	$H_1 \frac{\cos\alpha}{2D\beta^2} \left[-K_8(\zeta) + \frac{K_4}{K_1} K_9(\zeta) + \frac{K_2}{K_1} K_7(\zeta) \right]$
w_h	$-H_1 \frac{\cos^2\alpha}{4D\beta^3} \left[K_9(\zeta) + \frac{2K_4}{K_1} K_7(\zeta) - \frac{K_2}{K_1} K_{10}(\zeta) \right]$

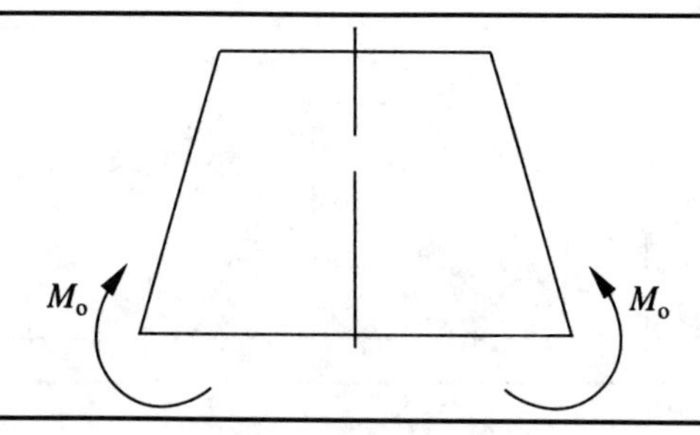

Q	$-M_o 2\beta \left[\frac{K_8}{K_1} K_{10}(\zeta) - \frac{K_{10}}{K_1} K_8(\zeta) \right]$
N_θ	$M_o 2\beta^2 x_m \tan\alpha \left[\frac{2K_8}{K_1} K_7(\zeta) - \frac{K_{10}}{K_1} K_{10}(\zeta) \right]$
M_s	$M_o \left[\frac{2K_8}{K_1} K_8(\zeta) - \frac{K_{10}}{K_1} K_9(\zeta) \right]$
θ	$M_o \frac{1}{D\beta} \left[\frac{K_8}{K_1} K_9(\zeta) + \frac{K_{10}}{K_1} K_7(\zeta) \right]$
w_h	$-M_o \frac{\cos\alpha}{2D\beta^2} \left[\frac{2K_8}{K_1} K_7(\zeta) - \frac{K_{10}}{K_1} K_{10}(\zeta) \right]$

Table 9-4. (continued)

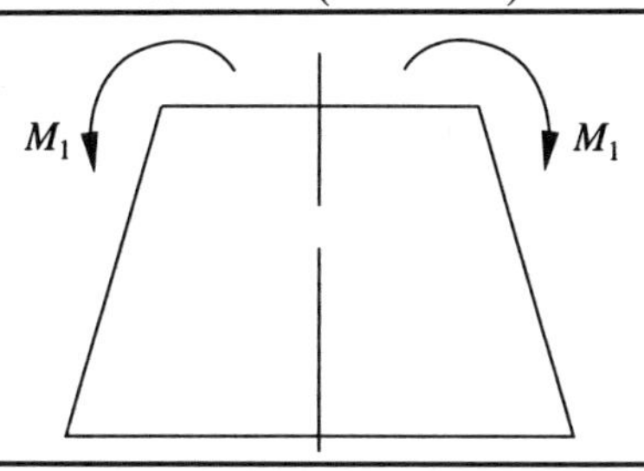

Q	$-M_1 2\beta\left[\frac{K_6}{K_1}K_{13}(\zeta) + \frac{K_5}{K_1}K_{14}(\zeta) - \frac{K_3}{K_1}K_8(\zeta)\right]$
N_θ	$M_1 2\beta^2 x_m \tan\alpha\left[\frac{K_6}{K_1}K_{12}(\zeta) + \frac{K_5}{K_1}K_{11}(\zeta) - \frac{K_3}{K_1}K_{10}(\zeta)\right]$
M_s	$M_1\left[\frac{K_6}{K_1}K_{11}(\zeta) - \frac{K_5}{K_1}K_{12}(\zeta) + \frac{K_3}{K_1}K_9(\zeta)\right]$
θ	$M_1\frac{1}{D\beta}\left[\frac{K_6}{K_1}K_{14}(\zeta) - \frac{K_5}{K_1}K_{13}(\zeta) - \frac{K_3}{K_1}K_7(\zeta)\right]$
w_h	$-M_1\frac{\cos\alpha}{2D\beta^2}\left[\frac{K_6}{K_1}K_{12}(\zeta) + \frac{K_5}{K_1}K_{11}(\zeta) - \frac{K_3}{K_1}K_{10}(\zeta)\right]$

Notation: $M_\theta = \mu M_\phi$; $N_\theta = -Q\cot\alpha$; w_h = horizontal component of deflection; θ = rotation; $\beta = \sqrt[4]{3(1-\mu^2)}/\sqrt{tx_m\tan\alpha}$; $K_1 = \sinh^2\beta L - \sin^2\beta L$; $K_2 = \sinh^2\beta L + \sin^2\beta L$; $K_3 = \sinh\beta L\cosh\beta L + \sin\beta L\cos\beta L$; $K_4 = \sinh\beta L\cosh\beta L - \sin\beta L\cos\beta L$; $K_5 = \sin^2\beta L$; $K_6 = \sinh^2\beta L$; $K_7 = \cosh\beta L\cos\beta L$; $K_7(\zeta) = \cosh\beta L\zeta\cos\beta L\zeta$; $K_8 = \sinh\beta L\sin\beta L$; $K_8(\zeta) = \sinh\beta L\zeta\sin\beta L\zeta$; $K_9 = \cosh\beta L\sin\beta L - \sinh\beta L\cos\beta L$; $K_9(\zeta) = \cosh\beta L\zeta\sin\beta L\zeta - \sinh\beta L\zeta\cos\beta L\zeta$; $K_{10} = \cosh\beta L\sin\beta L + \sinh\beta L\cos\beta L$; $K_{10}(\zeta) = \cosh\beta L\zeta\sin\beta L\zeta + \sinh\beta L\zeta\cos\beta L\zeta$; $K_{11}(\zeta) = \cosh\beta L\zeta\cos\beta L\zeta - \sinh\beta L\zeta\sin\beta L\zeta$; $K_{12}(\zeta) = \cosh\beta L\zeta\cos\beta L\zeta + \sinh\beta L\zeta\sin\beta L\zeta$; $K_{13}(\zeta) = \cosh\beta L\zeta\sin\beta L\zeta$; $K_{14}(\zeta) = \sinh\beta L\zeta\cos\beta L\zeta$.

Example 9-3

Find the discontinuity forces in the cone (Fig. 9-11a) due to an internal pressure of 60 psi. Let E = 30,000 ksi and $\mu = 0.30$.

Solution

The deflection w at point A due to internal pressure (Fig. 9-11b) is obtained from Eq. (8-32) by using the radius $r/\cos\alpha$ rather than r. Hence, radial deflection is expressed as

$$w = \frac{-pr^2}{Et\cos^2\alpha}(1 - \mu/2). \tag{1}$$

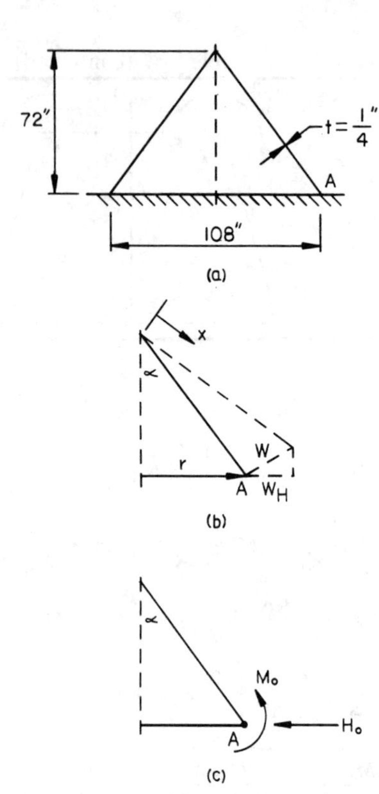

Figure 9-11.

The horizontal deflection is

$$
\begin{aligned}
w_H &= \frac{-pr^2}{Et\cos\alpha}(1 - \mu/2) \\
&= \frac{-(60)(54)^2(1 - 0.3/2)}{30{,}000{,}000(0.25)(0.8)} \\
w_H &= -0.0248 \text{ inch.}
\end{aligned}
$$

The rotation, θ, at point A is obtained by writing Eq. (1) as

$$
w = \frac{-px^2\sin^2\alpha}{Et\cos^2\alpha}(1 - \mu/2)
$$

and from Eq. (9-12) with $v = 0$, and $dx = r_1\,d\phi$

$$\theta = -\frac{dw}{r_1\,d\phi}$$

$$= -\frac{dw}{dx}$$

$$\theta = \frac{2px \sin^2 \alpha}{Et \cos^2 \alpha}(1 - \mu/2)$$

$$= \frac{2pr \sin \alpha}{Et \cos^2 \alpha}(1 - \mu/2)$$

$$= \frac{2(60)(54)(0.6)(0.85)}{30{,}000{,}000(0.25)(0.64)}$$

$$\theta = 0.000689 \text{ radians.}$$

From Table 9-3,

$$R = r/\cos \alpha = 54/0.8$$

$$= 67.50$$

$$\beta = \frac{72}{\sqrt{67.50 \times 0.25}\ 0.60}\sqrt[4]{3(1 - 0.3^2)}$$

$$= 37.549$$

$$D = 42{,}926.$$

From Fig. 9-11c,

$$\text{deflection due to } p + M_o + H_o = 0$$

or,

$$-0.0248 - \frac{hM_o}{2D\beta^2 \cos \alpha} + \frac{h^3 H_o}{2D\beta^3 \cos \alpha}\left(1 - \frac{\mu h \tan \alpha}{2R\beta \cos \alpha}\right) = 0$$

$$M_o - 1.910H_o = -463.21. \tag{2}$$

Similarly,

$$\text{rotation due to } p + M_o + H_o = 0$$

$$-0.000689 - \frac{hM_o}{D\beta \cos \alpha} + \frac{h^2 H_o}{2D\beta^2 \cos \alpha} = 0$$

or

$$M_o - 0.959H_o = -12.339. \tag{3}$$

Solving Eqs. (2) and (3) results in

$$M_o = 442.3 \text{ inch-lb/inch.}$$

$$H_o = 474.1 \text{ lb/inch.}$$

Problems

9-7 Find the discontinuity forces at points A, B, and C of Fig. P9-7 due to 500 psi pressure in the cone compartment ABC. Let $E = 25{,}000$ ksi and $\mu = 0.31$ in the cylinder and flat plate. Let $E = 30{,}000$ ksi and $\mu = 0.29$ in the cone.

9-8 Find the discontinuity forces at point A due to a 100 psi pressure in compartment AOB of Fig. P9-8. Assume the 40-inch head to be large enough so that the discontinuity forces at A are insignificant at the knuckle region. Let $E = 16{,}000$ ksi and $\mu = 0.30$.

9-9 Find the discontinuity forces at the cone to flat plate junction shown in Fig. P9-9. Let $p = 60$ psi, $E = 20{,}000$ ksi, $\mu = 0.25$ and $\alpha = 36.87°$.

9-4 Design Considerations

Numerous references are available for determining the bending stresses in spherical and conical shells due to various edge conditions. One of the

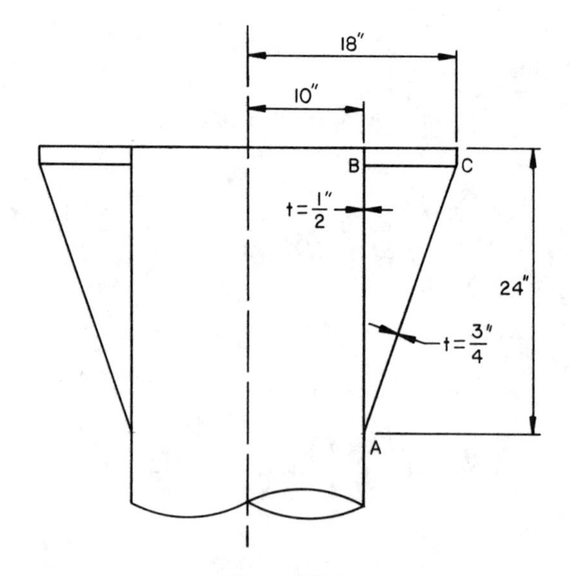

Figure P9-7.

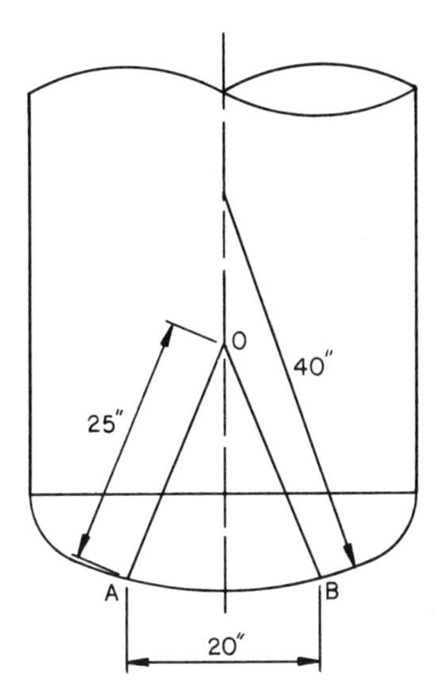

Figure P9-8.

most extensive coverages is given by Baker (Baker et al. 1968). Flugge (Flugge 1967) has extensive tables for solving the differential equations for spherical and conical shells.

When the configuration of the shell is other than spherical or conical, the classical methods discussed in this chapter become impractical to use and other more general methods such as finite element analysis are employed. It has also been shown by Baker that for some configurations such

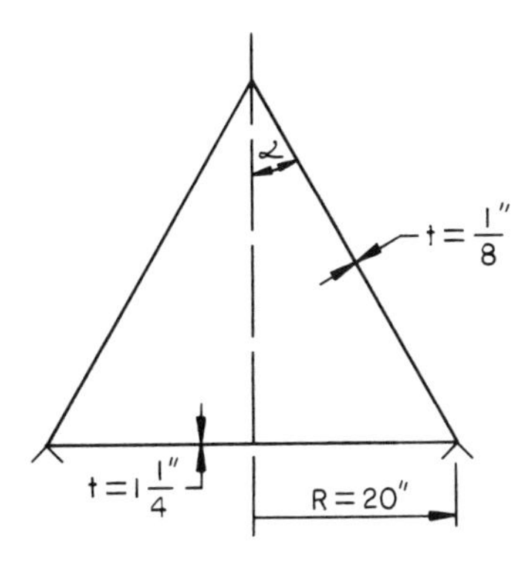

Figure P9-9.

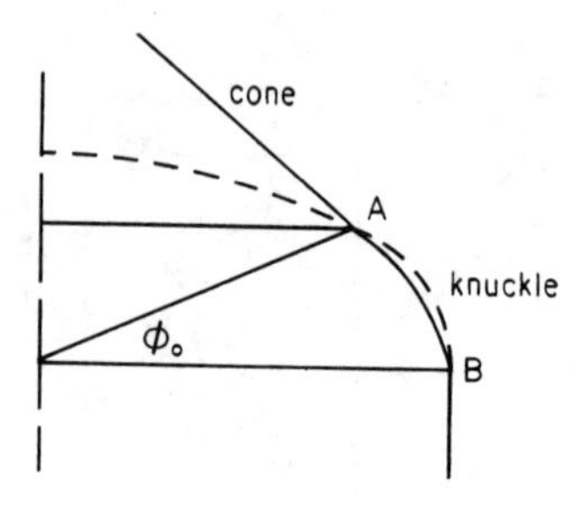

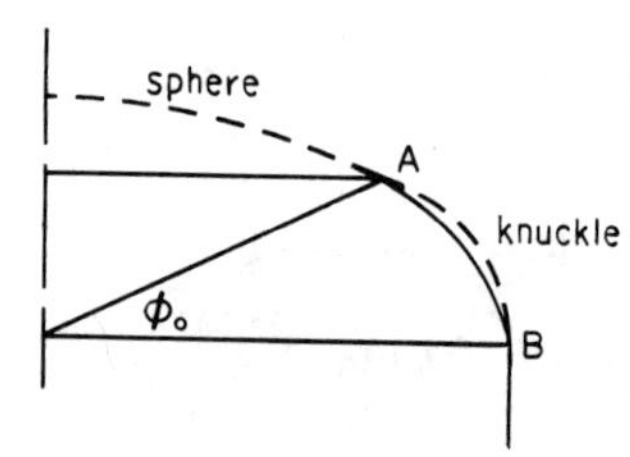

Figure 9-12.

as those shown in Fig. 9-12, the discontinuity forces in the region between points A and B can be approximated by using the spherical equations derived here for the knuckle portion as long as the angle ϕ_o is larger than about 20 degrees.

10

Buckling of Plates

10-1 Circular Plates

The differential equation for the bending of a circular plate subjected to lateral loads, p, is obtained from Eq. (3-11) as

$$r^2 \frac{d^2\phi}{dr^2} + r \frac{d\phi}{dr} - \phi = -\frac{Qr^2}{D}.$$

When in-plane forces N_r are applied as shown in Fig. 10-1, and the lateral loads, p, are reduced to zero, then the corresponding value of Q is

$$Q = N_r\phi.$$

Letting

$$A^2 = N_r/D \tag{10-1}$$

the differential equation becomes

$$r^2 \frac{d^2\phi}{dr^2} + r \frac{d\phi}{dr} - (r^2A^2 - 1)\phi = 0.$$

Defining

$$x = Ar \quad \text{and} \quad dx = A\,dr \tag{10-2}$$

we get

$$x^2 \frac{d^2\phi}{dx^2} + x \frac{d\phi}{dx} + (x^2 - 1)\phi = 0. \tag{10-3}$$

The solution of this equation is in the form of a Bessel function. From Eq. (B-3) of Appendix B,

$$\phi = C_1J_1(x) + C_2Y_1(x) \tag{10-4}$$

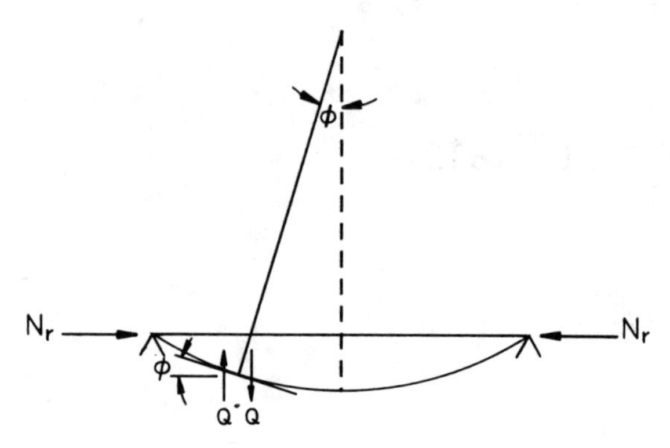

Figure 10-1.

at $r = 0$, $Y_1(x)$ approaches infinity. Hence, C_2 must be set to zero and Eq. (10-4) becomes

$$\phi = C_1 J_1(x). \tag{10-5}$$

For a fixed boundary condition, $\phi = 0$ at $r = a$ and a nontrivial solution of Eq. (10-5) is

$$J_1(x) = 0$$

or from Table B-1 of Appendix B, $x = 3.83$, and Eqs. (10-1) and (10-2) give

$$\sqrt{N_r/D}\,(a) = 3.83$$

or

$$N_{cr} = \frac{14.67D}{a^2}. \tag{10-6}$$

For a simply supported plate, the moment at the boundary $r = a$ is zero and Eqs. (10-5) and (3-4) give

$$N_{cr} = \frac{4.20D}{a^2}. \tag{10-7}$$

Example 10-1

What is the required thickness of a simply supported circular plate subjected to a lateral pressure of 2 psi and in-plane compressive force of 100 lb/inch if $a = 29$ inches, $\mu = 0.31$, $E = 30{,}000$ ksi, allowable stress in bending = 10,000 psi, and factor of safety ($F.S.$) for buckling = 3.0.

Solution

From Example 3-1,

$$M = \frac{Pa^2}{16}(3 + \mu) = 348.0 \text{ inch-lb/inch}$$

$$t = \sqrt{6 \times 348/10{,}000} = 0.46 \text{ inch.}$$

Try $t = 0.50$ inch.

$$\text{actual bending stress} = \frac{6 \times 348}{0.5^2} = 8350 \text{ psi}$$

$$\text{actual compressive stress} = 100/0.5 = 200 \text{ psi}$$

$$D = \frac{30{,}000 \times 0.5^3}{12(1 - 0.31^2)} = 345{,}720 \text{ inch-lbs.}$$

From Eq. (10-7),

$$\sigma_{cr} = \frac{4.20 \times 345{,}720}{29^2 \times 0.5} = 3450 \text{ psi.}$$

allowable compressive stress = 3450/3 = 1150 psi.
Using the interaction equation

$$\frac{\text{actual bending stress}}{\text{allowable bending stress}} + \frac{\text{actual compressive stress}}{\text{allowable compressive stress}} \leq 1.0$$

we get

$$\frac{8350}{10{,}000} + \frac{200}{1150} = 0.83 + 0.17 = 1.0 \quad \text{o.k.}$$

Use $t = 0.50$ inch.

Problems

10-1 Derive Eq. (10-7).
10-2 What is the required thickness of a fixed circular plate subjected to a lateral pressure of 3 psi and in-plane compressive force of 500 lb/inch if $a = 40$ inches, $\mu = 0.30$, $E = 25{,}000$ ksi, allowable stress in bending = 25,000 psi, and factor of safety (*F.S.*) for buckling = 4.0?
10-3 What is the effect of N_r on the bending moments if it were in tension rather than compression?

10-2 Rectangular Plates

The differential equation of a rectangular plate with lateral load, p, is given by Eq. (1-26) as

$$\nabla^4 w = p/D. \tag{10-8}$$

If the plate is additionally loaded in its plane (Fig. 10-2a) then summation of forces in the x-direction gives

$$N_x\,dy + N_{yx}\,dx - \left(N_x + \frac{\partial N_x}{\partial x}\,dx\right)dy - \left(N_{yx} + \frac{\partial N_{yx}}{\partial y}\,dy\right)dx = 0$$

or

$$\frac{\partial N_x}{\partial x} + \frac{\partial N_{yx}}{\partial y} = 0. \tag{10-9}$$

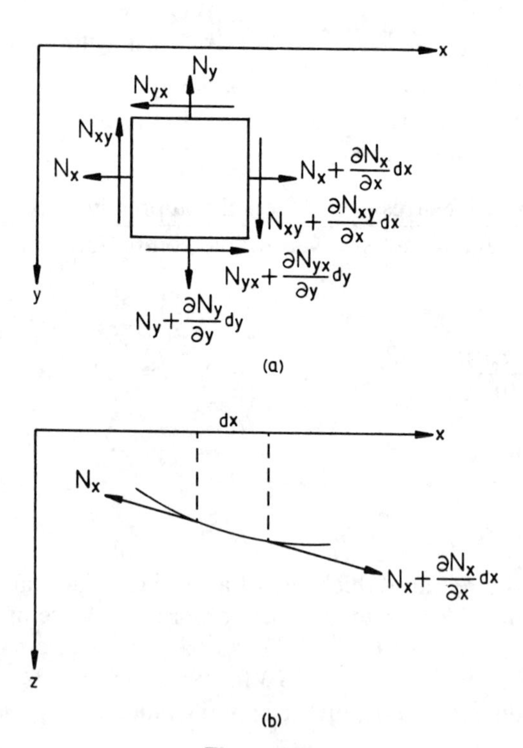

Figure 10-2.

Similarly, summation of forces in the y-direction gives

$$\frac{\partial N_y}{\partial y} + \frac{\partial N_{xy}}{\partial x} = 0. \tag{10-10}$$

Summation of forces in the z-direction (Fig. 10-2b) gives the following for N_x:

$$-N_x\,dy\,\frac{\partial w}{\partial x} + \left(N_x + \frac{\partial N_x}{\partial x}\,dx\right)\left(\frac{\partial w}{\partial x} + \frac{\partial}{\partial x}\left(\frac{\partial w}{\partial x}\right)dx\right)dy$$

which reduces to

$$N_x\,\frac{\partial^2 w}{\partial x^2}\,dx\,dy + \frac{\partial N_x}{\partial x}\,\frac{\partial w}{\partial x}\,dx\,dy. \tag{10-11}$$

Similarly, for N_y,

$$N_y\,\frac{\partial^2 w}{\partial y^2}\,dx\,dy + \frac{\partial N_y}{\partial y}\,\frac{\partial w}{\partial y}\,dx\,dy. \tag{10-12}$$

For N_{xy} from Fig. 10-3,

$$-\left(N_{xy}\,dy\,\frac{\partial w}{\partial y}\right) + \left(N_{xy} + \frac{\partial N_{xy}}{\partial x}\,dx\right)\left(\frac{\partial w}{\partial y} + \frac{\partial^2 w}{\partial x\,\partial y}\,dx\right)dy$$

or

$$N_{xy}\,\frac{\partial^2 w}{\partial x\,\partial y}\,dx\,dy + \frac{\partial N_{xy}}{\partial x}\,\frac{\partial w}{\partial y}\,dx\,dy. \tag{10-13}$$

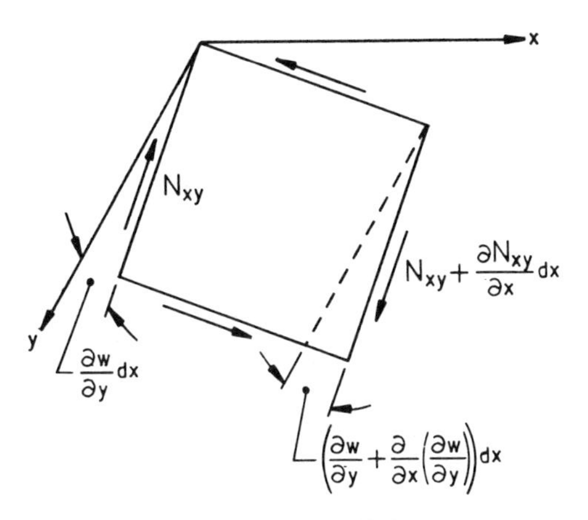

Figure 10-3.

Similarly for N_{yx}

$$N_{yx} \frac{\partial^2 w}{\partial x\, \partial y} dx\, dy + \frac{\partial N_{yx}}{\partial y} \frac{\partial w}{\partial x} dx\, dy. \tag{10-14}$$

The total sum of Eq. (10-8), which was obtained by summing forces in the z-direction, with Eqs. (10-11), (10-12), (10-13), and (10-14) gives the basic differential equation of a rectangular plate subjected to lateral and in-plane loads.

$$\nabla^4 w = \frac{1}{D}\left(p + N_x \frac{\partial^2 w}{\partial x^2} + N_y \frac{\partial^2 w}{\partial y^2} + 2N_{xy} \frac{\partial^2 w}{\partial x\, \partial y}\right) \tag{10-15}$$

It should be noted that Eqs. (10-9) and (10-10), which were obtained by summing forces in the x- and y-directions, were not utilized in Eq. (10-15). They are used to formulate large-deflection theory of plates which is beyond the scope of this book.

Another equation that is frequently utilized in buckling problems is the energy equation. It was shown in Chapter 4 that the strain energy expression due to lateral loads, p, is given by Eq. (4-1) as

$$U = \frac{D}{2}\int_A \left\{\left(\frac{\partial^2 w}{\partial x^2} + \frac{\partial^2 w}{\partial y^2}\right)^2 - 2(1-\mu)\left[\frac{\partial^2 w}{\partial x^2}\frac{\partial^2 w}{\partial y^2} - \left(\frac{\partial^2 w}{\partial x\, \partial y}\right)^2\right]\right\} dx\, dy. \tag{10-16}$$

The strain energy expression for the in-plane loads is derived from Fig. 10-4 which shows the deflection of a unit segment dx. Hence,

$$dx' = \sqrt{dx^2 - \left(\frac{\partial w}{\partial x} dx\right)^2}$$

$$\varepsilon_x = dx - dx' = \frac{1}{2}\left(\frac{\partial w}{\partial x}\right)^2 dx$$

or per unit length,

$$\varepsilon_x = \frac{1}{2}\left(\frac{\partial w}{\partial x}\right)^2.$$

Similarly,

$$\varepsilon_y = \frac{1}{2}\left(\frac{\partial w}{\partial y}\right)^2.$$

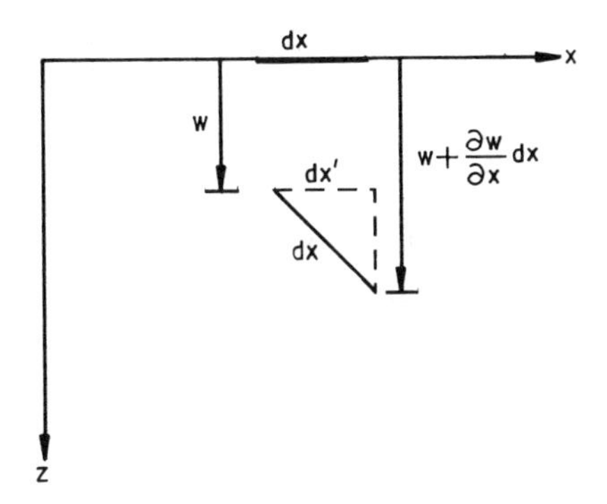

Figure 10-4.

It can also be shown that

$$\gamma_{xy} = \frac{\partial w}{\partial x}\frac{\partial w}{\partial y}.$$

Hence, the strain energy expression for the in-plane forces is given by

$$U = \int_A (N_x\varepsilon_x + N_y\varepsilon_y + N_{xy}\gamma_{xy})\, dx\, dy$$

$$U = \frac{1}{2}\int_A \left[N_x\left(\frac{\partial w}{\partial x}\right)^2 + N_y\left(\frac{\partial w}{\partial y}\right)^2 + 2N_{xy}\frac{\partial w}{\partial x}\frac{\partial w}{\partial y}\right] dx\, dy. \tag{10-17}$$

The total strain energy expression for rectangular plates loaded laterally and in-plane is the summation of expressions (10-16) and (10-17). Thus,

$$U = \frac{D}{2}\int_A \left\{\left(\frac{\partial^2 w}{\partial x^2} + \frac{\partial^2 w}{\partial y^2}\right)^2 - 2(1-\mu)\left[\frac{\partial^2 w}{\partial x^2}\frac{\partial^2 w}{\partial y^2} - \left(\frac{\partial^2 w}{\partial x\,\partial y}\right)^2\right]\right\} dx\, dy + \frac{1}{2}\int_A \left[N_x\left(\frac{\partial w}{\partial x}\right)^2 + N_y\left(\frac{\partial w}{\partial y}\right)^2 + 2N_{xy}\frac{\partial w}{\partial x}\frac{\partial w}{\partial y}\right] dx\, dy. \tag{10-18}$$

The total potential energy of a system is given by Eq. (4-3) as

$$\Pi = U - W \tag{10-19}$$

where W is external work. In order for the system to be in equilibrium, Eq. (10-19) must be minimized.

Example 10-2

Find the buckling stress of a simply supported rectangular plate (Fig. 10-5) subjected to forces N_x.

Solution

Let the deflection be expressed as

$$w = \sum_{m=1}^{\infty} \sum_{n=1}^{\infty} A_{mn} \sin \frac{m\pi x}{a} \sin \frac{n\pi y}{b},$$

which satisfies the boundary conditions.

Substituting this expression into Eq. (10-18), and noting that the expression

$$2(1 - \mu)\left[\frac{\partial^2 w}{\partial x^2}\frac{\partial^2 w}{\partial y^2} - \left(\frac{\partial^2 w}{\partial x\, \partial y}\right)^2\right] = 0,$$

gives

$$U = \frac{D}{2}\int_0^b \int_0^a \sum_{m=1}^{\infty} \sum_{n=1}^{\infty}\left[A_{mn}^2\left(\frac{m^2\pi^2}{a^2} + \frac{n^2\pi^2}{b^2}\right)^2 \sin^2\frac{m\pi x}{a} \sin^2\frac{n\pi y}{b}\right] dx\, dy$$
$$+ \frac{1}{2}\int_0^b \int_0^a \left[(-N_x) \sum_{m=1}^{\infty} \sum_{n=1}^{\infty} A_{mn}^2 \frac{m^2\pi}{a^2} \times \sin^2\frac{m\pi x}{a} \sin^2\frac{n\pi y}{b}\right] dx\, dy$$

or,

$$U = \frac{\pi^4 ab}{8} D \sum_{m=1}^{\infty} \sum_{n=1}^{\infty} A_{mn}\left(\frac{m^2}{a^2} + \frac{n^2}{b^2}\right)^2 - \frac{\pi^2 b}{8a} N_x \sum_{m=1}^{\infty} \sum_{n=1}^{\infty} m^2 A_{mn}^2. \tag{1}$$

Since there are no lateral loads, we can take the external work in the z-direction as zero and Eq. (10-19) becomes

$$\Pi = U.$$

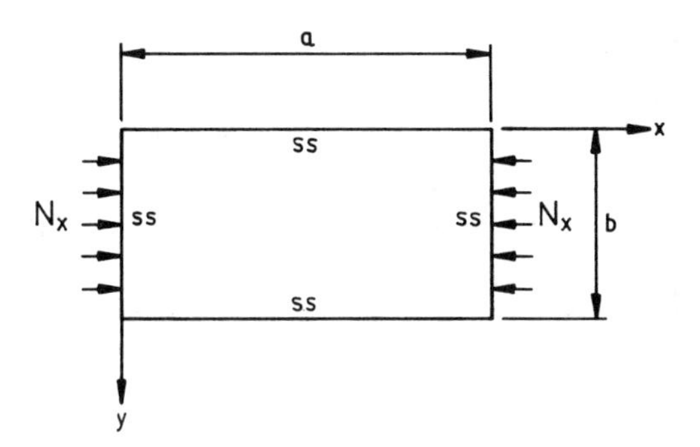

Figure 10-5.

Solving Eq. (1) for

$$\frac{\partial \Pi}{\partial A_{mn}} = 0$$

we get

$$N_{cr} = \frac{\pi^2 a^2 D}{m^2}\left(\frac{m^2}{a^2} + \frac{n^2}{b^2}\right)^2. \tag{2}$$

The smallest value of Eq. (2) is for $n = 1$. Also, if we substitute

$$\sigma_{cr} = N_{cr}/t$$

and

$$D = \frac{Et^3}{12(1 - \mu^2)}$$

we get

$$\sigma_{cr} = \frac{\pi^2 E}{12(1 - \mu^2)(b/t)^2} K \tag{3}$$

where

$$K = \left(\frac{m}{a/b} + \frac{a/b}{m}\right)^2. \tag{4}$$

A plot of Eq. (4) is shown in Fig. 10-6 and shows that the minimum value of K is 4.0.

Problems

10-4 Derive Eq. (10-15).

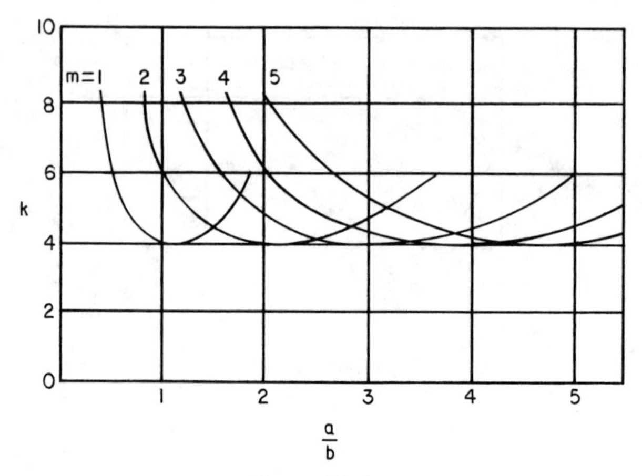

Figure 10-6.

10-5 What is the required thickness of the plate in Fig. 10-5 if $a = 40$ inches, $b = 15$ inches, $N_x = 400$ lb/inch, $E = 16{,}000$ ksi, and $\mu = 0.33$? Use a factor of safety of 4.0. Use increments of 1/16 inch for thickness.

10-3 Plates with Various Boundary Conditions

Figure 10-7 shows a plate simply supported on sides $x = 0$ and $x = a$ and subjected to force N_x. The differential Eq. (10-15) becomes

$$\nabla^4 w = -\frac{N_x}{D}\frac{\partial^2 w}{\partial x^2}. \tag{10-20}$$

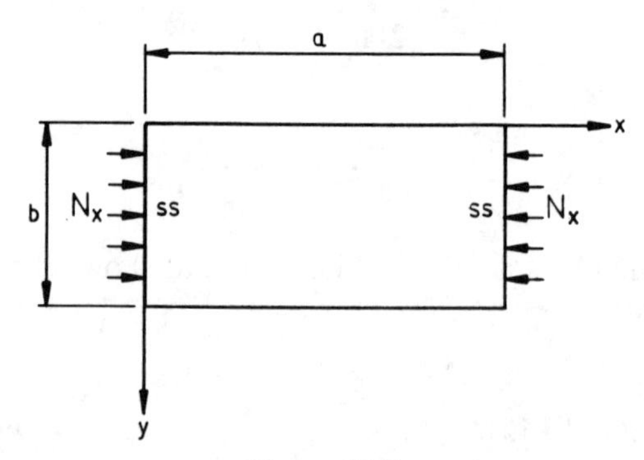

Figure 10-7.

Let the solution be of the form

$$w = \sum_{m=1}^{\infty} f(y) \sin \frac{m\pi x}{a}. \tag{10-21}$$

This solution satisfies the two boundary conditions $w = M_x = 0$ at $x = 0$ and $x = a$.

Substituting Eq. (10-21) into Eq. (10-20) results in

$$\frac{d^4 f}{dy^4} - A \frac{d^2 f}{dy^2} + Bf = 0 \tag{10-22}$$

where

$$A = \frac{2m^2\pi^2}{a^2}$$

$$B = \frac{m^4\pi^4}{a^4} - \frac{N_x}{D}\frac{m^2\pi^2}{a^2}.$$

The solution of Eq. (10-22) is

$$f(y) = C_1 e^{-\alpha y} + C_2 e^{\alpha y} + C_3 \cos \beta y + C_4 \sin \beta y \tag{10-23}$$

where

$$\alpha = \sqrt{\frac{m^2\pi^2}{a^2} + \sqrt{\frac{N_x}{D}\frac{m^2\pi^2}{a^2}}}$$

$$\beta = \sqrt{-\frac{m^2\pi^2}{a^2} + \sqrt{\frac{N_x}{D}\frac{m^2\pi^2}{a^2}}}$$

Values of the constants C_1 through C_4 are obtained from the boundary conditions $y = 0$ and $y = b$.

Case 1

Side $y = 0$ is fixed and side $y = b$ is free. The four boundary conditions are

$$\text{for } y = 0, \text{ deflection } w = 0$$

$$\text{rotation } \frac{\partial w}{\partial y} = 0$$

$$\text{for } y = b, \text{ moment } M_y = 0 = \frac{\partial^2 w}{\partial y^2} + \mu \frac{\partial^2 w}{\partial x^2}$$

$$\text{shear } Q = 0 = \frac{\partial^3 w}{\partial y^3} + (2 - \mu)\frac{\partial^3 w}{\partial x\, \partial y^2}.$$

From the first boundary condition we get

$$C_1 + C_2 + C_3 = 0$$

and from the second boundary condition we get

$$-\alpha C_1 + \alpha C_2 + \beta C_3 = 0$$

or,

$$C_1 = -\frac{C_3}{2} + \frac{\beta C_4}{2\alpha}$$

and

$$C_2 = -\frac{C_3}{2} - \frac{\beta C_4}{2\alpha}.$$

Substituting C_1 and C_2 into Eq. (10-23) gives

$$f(y) = C_3(\cos \beta y - \cosh \alpha y) + C_4\left(\sin \beta y - \frac{\beta}{\alpha}\sinh \alpha y\right)$$

With this expression and Eq. (10-21), we can solve the last two boundary conditions. This results in two simultaneous equations. The critical value of the compressive force, N_x, is determined by equating the determinant of these equations to zero. This gives

$$2gh(g^2 + h^2) \cos \beta b \cosh \alpha b = \frac{1}{\alpha\beta}(\alpha^2 h^2 - \beta^2 g^2) \sin \beta b \sinh \alpha b \qquad (10\text{-}24)$$

where

$$g = \alpha^2 - \mu \frac{m^2\pi^2}{a^2}$$

and

$$h = \beta^2 + \mu \frac{m^2\pi^2}{a^2}.$$

For $m = 1$, the minimum value of Eq. (10-24) can be expressed in terms of stress as

$$\sigma_{cr} = \frac{\pi^2 E}{12(1 - \mu^2)(b/t)^2} K \qquad (10\text{-}25)$$

where, for $\mu = 0.25$,

$$K_{\min} = 1.328. \qquad (10\text{-}26)$$

Case 2

Side $y = 0$ is simply supported and side $y = b$ is free. Again starting with Eq. (10-23) and satisfying the boundary conditions at $y = 0$ and $y = b$, we get a solution (Timoshenko and Gere 1961) identical to Eq. (10-25) with K given by

$$K = 0.456 + \frac{b^2}{a^2} \quad \text{for } \mu = 0.25. \tag{10-27}$$

Case 3

Sides $y = 0$ and $y = b$ are fixed. In this case, the minimum value of K in Eq. (10-25) is

$$K = 7.0 \quad \text{for } \mu = 0.25. \tag{10-28}$$

Example 10-3

Let the plate in Fig. 10-5 be simply supported at $x = 0$ and $x = a$, simply supported at $y = 0$, and free at $y = b$. Calculate the required thickness if $a = 22$ inches, $b = 17$ inches, $N_x = 300$ lb/in, factor of safety $(F.S.) = 2.0$, $\mu = 0.25$, $\sigma_y = 36$ ksi, and $E = 29{,}000$ ksi.

Solution

Assume $t = 0.25$ inch. From Eq. (10-27) the minimum value of $K = 0.456$ and Eq. (10-25) becomes

$$\sigma_{cr} = \frac{\pi^2\,(29{,}000{,}000)}{12\,(0.9375)\,(17/0.25)^2}\,0.456$$

$$= 2510 \text{ psi which is less than the yield stress.}$$

Allowable stress $= 2510/2 = 1255$ psi. Actual stress $= 300/0.25 = 1200$ psi. o.k.

Problems

10-6 Derive Eq. (10-22).
10-7 Derive Eq. (10-23).
10-8 Derive Eq. (10-24).
10-9 Determine the actual expression of K in Eq. (10-26). Plot K versus a/b for m values of 1, 2, and 3 and a/b values from 1.0 to 5.0.

10-4 Application of Buckling Expressions to Design Problems

Many codes utilize the expressions of Section 10-3 to establish buckling criteria for various members. The American Institute of Steel Construction Manual (AISC 1991) assumes the buckling stress of unsupported members in compression not to exceed the yield stress of the material. Thus, Eq. (10-25) can be written as

$$\sigma_y = \frac{\pi^2 E}{12\,(1 - \mu^2)\,(b/t)^2} K \tag{10-29}$$

or, for steel members with $\mu = 0.3$ and $E = 29{,}000$ ksi

$$\frac{b}{t} = 162 \sqrt{\frac{K}{\sigma_y}}. \tag{10-30}$$

Equation (10-29) is based on the assumption that the interaction between the buckling stress and the yield stress in designated by points ABC in Fig. 10-8. However, due to residual stress in structural members due to forming, the actual interaction curve is given by points ADC and Eq. (10-30) is modified by a factor of 0.7 as

$$\frac{b}{t} = 114 \sqrt{\frac{K}{\sigma_y}}. \tag{10-31}$$

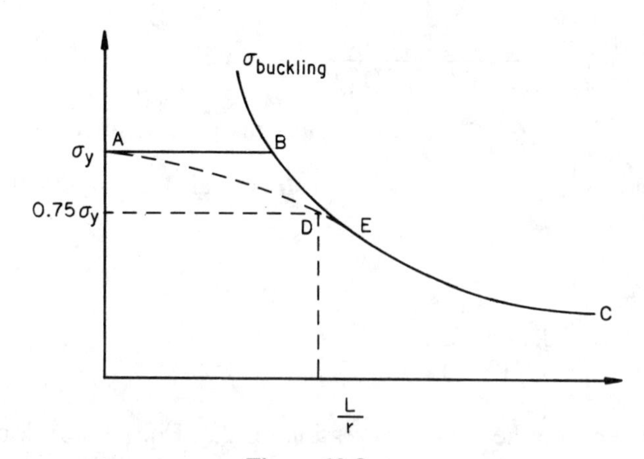

Figure 10-8.

Single-Angle Struts

Leg b of the angle shown in Fig. 10-9a is assumed free at point B. Point A is assumed simply supported because it can rotate due to deflections. Thus, Eq. (10-27) is applicable with a minimum value of $K = 0.456$. Thus Eq. (10-31) becomes

$$\frac{b}{t} = \frac{76}{\sqrt{\sigma_y}}.$$

Double Angles

Due to symmetry (Fig. 10-9b) the possibility of rotation is substantially reduced from that of case 1. Thus, the AISC uses the average of the simply supported–free case, Eq. (10-27), and the average of the simply supported–free and fixed–free, Eq. (10-26), cases.

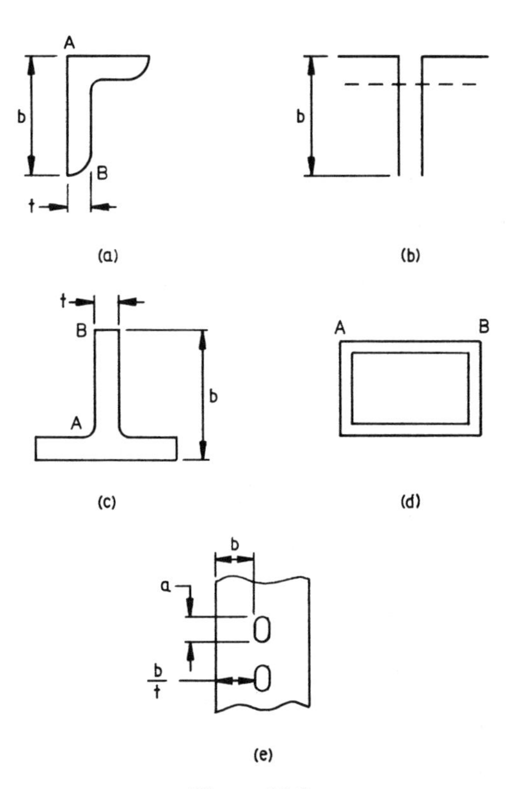

Figure 10-9.

$$K = \frac{0.456 + \dfrac{0.456 + 1.328}{2}}{2} = 0.674$$

and Eq. (8.31) is approximated by AISC as

$$\frac{b}{t} = \frac{95}{\sqrt{\sigma_y}}.$$

Stems of T's

For this case, point A in Fig. 10-9c is assumed fixed due to the much thicker flange, and point B is taken as free. The K value is taken from Eq. (10-26) and expression (10-31) gives

$$\frac{b}{t} = \frac{132}{\sqrt{\sigma_y}}.$$

The AISC reduces this value further down to

$$\frac{b}{t} = \frac{127}{\sqrt{\sigma_y}}.$$

Flanges of Box Sections

Points A and B in Fig. 10-9d are conservatively taken as simply supported. In this case, $k = 4.0$ and Eq. (10-31) becomes

$$\frac{b}{t} = \frac{228}{\sqrt{\sigma_y}}.$$

The AISC increases this value to match experimental data and it becomes

$$\frac{b}{t} = \frac{238}{\sqrt{\sigma_y}}.$$

Perforated Cover Plates

In Fig. 10-9e, the plate between the perforation and edge is assumed fixed because additional rigidity is obtained from the continuous areas between the perforations. The dimension of the perforated plate is assumed to be $a/b = 1.0$. This ratio results in a K value of about 7.69. This is higher than that given by Eq. (10-28) which is based on the smallest possible value.

Equation (10-31) becomes

$$\frac{b}{t} = \sqrt{\frac{317}{\sigma_y}}.$$

Other Compressed Members

Other members are assumed to have a K value that varies between 4.0 for simply supported edges and 7.0 for fixed edges. The AISC uses a value of $K = 4.90$. This gives

$$\frac{b}{t} \sqrt{\frac{253}{\sigma_y}}.$$

Another standard that uses the plate buckling equations to set a criterion is the American Association of State Highway and Transportation Officials (AASHTO 1992). The equations are very similar to those of AISC. AASHTO uses the square root of the actual compressive stress in the b/t equations rather than the square root of the yield stress. Also, limits are set on the maximum b/t values for various strength steels.

A theoretical solution of the buckling of rectangular plates due to various loading and boundary conditions is available in numerous references. Two such references are Timoshenko and Gere (1961) and Iyengar (1988). Timoshenko discusses mainly isotropic plates whereas Iyengar handles composite plates.

Various NASA publications are also available for the solution of the buckling of rectangular plates with various loading and boundary conditions. NASA's Handbook of Structural Stability consists of five parts and contains numerous theoretical background and design aids. Part I, edited by Gerard and Becker, includes various classical buckling solutions for flat plates. Part II, edited by Becker, is for buckling of composite elements. Part III is for buckling of curved plates and shells and Part IV discusses failure of plates and composite elements. Parts III and IV were edited by Gerard and Becker. Compressive strength of flat stiffened panels is given in Part V which is edited by Gerard.

11

Buckling of Cylindrical Shells

11-1 Basic Equations

In this chapter, we derive the equations for buckling of cylindrical shells (Fig. 11-1) subjected to compressive forces. Two approaches for developing the buckling equations will be discussed. The first is that of Sturm (Sturm 1941) which is well suited for designing cylindrical shells at various temperatures using actual stress–strain curves as discussed in Section 11-5. This approach is used in many pressure vessel codes for the design of cylindrical shells.

The second approach for analyzing buckling of cylindrical shells is that of Donnell (Gerard 1962). This method is discussed in Section 11-6 and is used extensively in the aerospace industry.

We begin Sturm's derivation by taking an infinitesimal element of a cylindrical shell with applied forces and moments as shown in Fig. 11-2. Summation of forces and moments in the x-, y-, and z-directions results in the following six equations of equilibrium:

$$\frac{\partial N_x}{\partial x} + \frac{\partial N_{\theta x}}{\partial y} - Q_x \frac{\partial^2 w}{\partial x^2} = 0 \tag{11-1}$$

$$\frac{\partial N_\theta}{\partial y} + \frac{\partial N_{x\theta}}{\partial x} + Q_\theta \frac{\partial \psi}{\partial y} = 0 \tag{11-2}$$

$$\frac{\partial Q_\theta}{\partial y} + \frac{\partial Q_x}{\partial x} = p + N_\theta \frac{\partial \psi}{\partial y} - N_x \frac{\partial^2 w}{\partial x^2} - N_{x\theta} \frac{\partial^2 w}{\partial y\, \partial x} - N_{\theta x} \frac{\partial^2 w}{\partial x\, \partial y} \tag{11-3}$$

$$Q_x = \frac{\partial M_x}{\partial x} + \frac{\partial M_{\theta x}}{\partial y} \tag{11-4}$$

$$Q_\theta = \frac{\partial M_\theta}{\partial y} + \frac{\partial M_{x\theta}}{\partial x} \tag{11-5}$$

$$N_{\theta x} - N_{x\theta} + M_{\theta x}\frac{\partial \psi}{\partial y} + M_{x\theta}\frac{\partial^2 w}{\partial x^2} = 0 \tag{11-6}$$

where w is the deflection in the z-direction.

The above equations cannot be solved directly because there are more unknowns than available equations. Accordingly, additional equations are needed. We can utilize the stress–strain relationship of Eq. (1-14) and rewrite it in terms of force–strain relationship as

$$N_x = \frac{Et}{1 - \mu^2}(\varepsilon_x + \mu\varepsilon_\theta) \tag{11-7}$$

$$N_\theta = \frac{Et}{1 - \mu^2}(\varepsilon_\theta + \mu\varepsilon_x) \tag{11-8}$$

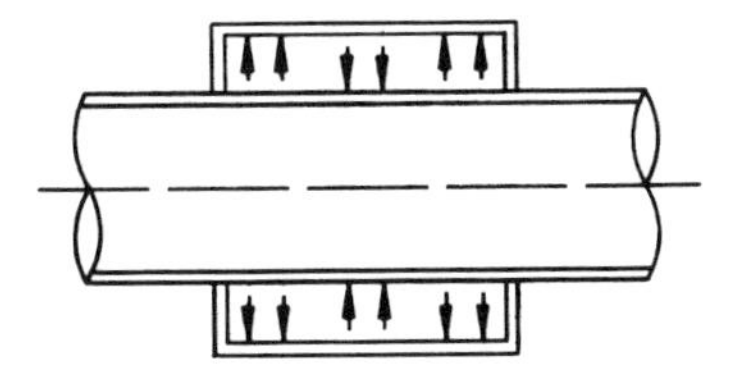

(a) inside cylinder under lateral external pressure

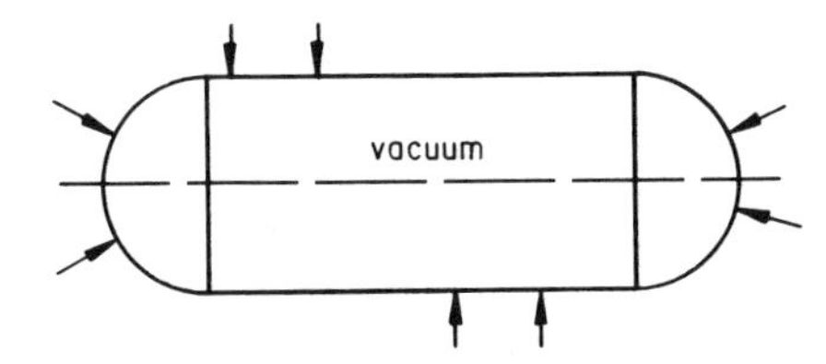

(b) cylinder under external lateral and end pressure

(c) cylinder under axial compressive force

Figure 11-1.

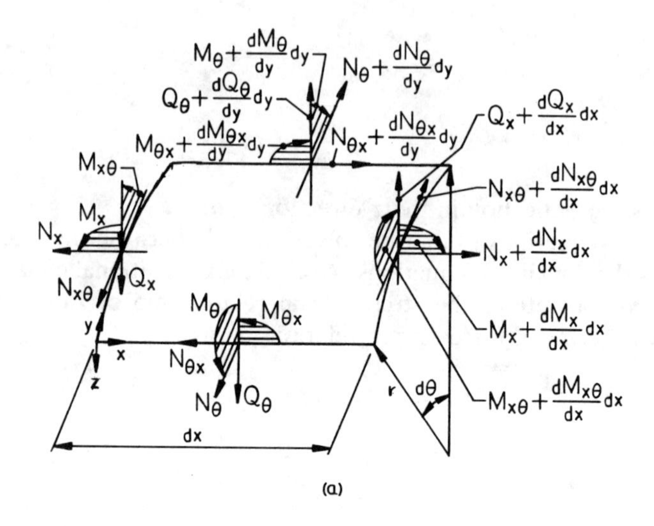

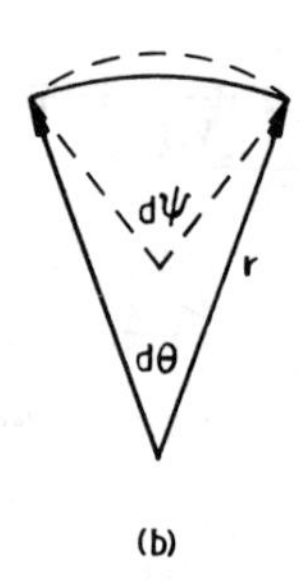

Figure 11-2.

$$N_{\theta x} = \frac{Et}{2(1 + \mu)} \gamma_{\theta x} \tag{11-9}$$

where ε_x and ε_θ are the strains in the x- and y-directions and $\gamma_{\theta x}$ is the shearing strain.

Similarly, the moment–deflection equations are expressed as

$$M_x = -D\left(\frac{\partial^2 w}{\partial x^2} + \frac{\mu}{r^2}\frac{\partial^2 w}{\partial \theta^2} + \mu\frac{w}{r^2}\right) \tag{11-10}$$

$$M_\theta = -D\left(\mu\frac{\partial^2 w}{\partial x^2} + \frac{1}{r^2}\frac{\partial^2 w}{\partial \theta^2} + \frac{w}{r^2}\right) \tag{11-11}$$

$$M_{\theta x} = -D(1 - \mu)\frac{1}{r}\frac{\partial^2 w}{\partial \theta\, \partial x}. \tag{11-12}$$

The relationship between strain and deformation is given by

$$\varepsilon_x = \frac{\partial u}{\partial x} \tag{11-13}$$

$$\varepsilon_\theta = \frac{\partial v}{\partial y} + \frac{v}{r} \tag{11-14}$$

$$\gamma_{\theta x} = \frac{\partial u}{\partial y} + \frac{\partial v}{\partial x} \tag{11-15}$$

where u and v are the deflections in the x- and y-directions, respectively.

Equation (11-14) is based on the fact that the radial deflection, w, for thin shells produces bending as well as stretching of the middle surface. Hence, from Fig. 11-3

$$\frac{ds_2 - ds_1}{ds_1} = \frac{(r + w)\, d\theta + (\partial v/\partial\theta)\, d\theta - r\, d\theta}{r\, d\theta}$$

$$= \frac{\partial v}{r\, \partial\theta} + \frac{w}{r} = \frac{\partial v}{\partial y} + \frac{w}{r}.$$

The change in angle $\delta\psi$ shown in Fig. 11-2 is expressed as

$$\frac{\partial\psi}{\partial\theta} = 1 - \frac{1}{r}\frac{\partial^2 w}{\partial\theta^2} - \frac{w}{r} + \varepsilon_\theta. \tag{11-16}$$

This equation is obtained from Fig. 11-2 where the angle $\delta\psi$ is the total sum of

(1) $d\theta$ which is the original angle

(2) $-\dfrac{1}{r}\dfrac{\partial^2 w}{\partial\theta^2}\, d\theta$ which is the change in slope of length ds

(3) $-\dfrac{w}{r}\, d\theta$ which is due to radial deflection

(4) $\varepsilon_\theta\, d\theta$ which is due to circumferential strain.

The derivative of these four expressions results in Eq. (11-16).

Assuming $M_{xy} = M_{yx}$, the above 16 equations contain the following unknowns: N_x, N_θ, $N_{x\theta}$, $N_{\theta x}$, Q_x, Q_θ, M_x, M_θ, $M_{x\theta}$, ε_x, ε_θ, $\gamma_{x\theta}$, w, u, v, and ψ. These 16 equations can be reduced to four by the following various substitutions.

From Eqs. (11-13), (11-14), and (11-15) we get

$$\frac{\partial^2\varepsilon_\theta}{\partial x^2} + \frac{\partial^2\varepsilon_x}{\partial y^2} - \frac{\partial^2\gamma_{\theta x}}{\partial x\, \partial y} - \frac{1}{r}\frac{\partial^2 w}{\partial x^2} = 0. \tag{11-17}$$

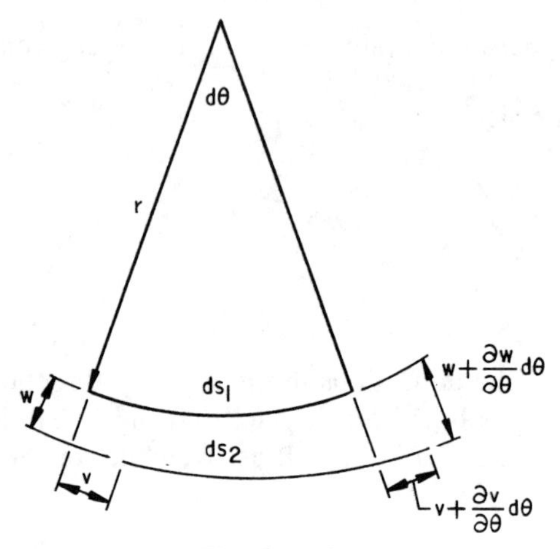

Figure 11-3.

Substituting Eqs. (11-7), (11-8), and (11-9) into Eq. (11-17) gives

$$\frac{\partial^2 N_\theta}{\partial x^2} - \mu \frac{\partial^2 N_x}{\partial x^2} + \frac{1}{r^2}\frac{\partial^2 N_x}{\partial \theta^2} - \frac{\mu}{r^2}\frac{\partial^2 N_\theta}{\partial \theta^2} - \frac{2(1+\mu)}{r}\frac{\partial^2 N_{\theta x}}{\partial \theta\, \partial x} = \frac{Et}{r}\frac{\partial^2 w}{\partial x^2}. \quad (11\text{-}18)$$

The shearing forces in Eqs. (11-1) and (11-2) are eliminated by combining these two equations with Eqs. (11-4), (11-5), and (11-6). This gives

$$\frac{\partial^2 N_\theta}{\partial y^2} - \frac{\partial^2 N_x}{\partial x^2} + \frac{1+\varepsilon_\theta}{r}\frac{\partial^2 M_\theta}{\partial y^2} + \frac{2(1+\varepsilon_\theta)}{r}\frac{\partial^2 M_{\theta x}}{\partial y\, \partial x} = 0. \quad (11\text{-}19)$$

Equations (11-4) and (11-5) are combined with Eq. (11-3) to give

$$\frac{\partial^2 M_x}{\partial x^2} + 2\frac{\partial^2 M_{\theta x}}{\partial y\, \partial x} + \frac{\partial^2 M_\theta}{\partial y^2} = p + N_\theta \frac{\partial \psi}{\partial y} - N_x \frac{\partial^2 w}{\partial x^2} - N_{x\theta}\frac{\partial^2 w}{\partial y\, \partial x} - N_{\theta x}\frac{\partial^2 w}{\partial y\, \partial x}. \quad (11\text{-}20)$$

Substituting Eqs. (11-10), (11-11), and (11-12) into Eq. (11-20) results in the first of the four basic equations we are seeking:

$$-D\left[\frac{\partial^4 w}{\partial x^4} + \frac{\mu}{r^2}\frac{\partial^2 w}{\partial x^2} + \frac{2}{r^2}\frac{\partial^4 w}{\partial \theta^2\, \partial x^2} + \frac{1}{r^4}\frac{\partial^4 w}{\partial \theta^4} + \frac{1}{r^4}\frac{\partial^2 w}{\partial \theta^2}\right] = p + N_\theta \frac{1}{r}\frac{\partial \psi}{\partial \theta} - N_x \frac{\partial^2 w}{\partial x^2} - N_{x\theta}\frac{1}{r}\frac{\partial^2 w}{\partial \theta\, \partial x} - N_{\theta x}\frac{1}{r}\frac{\partial^2 w}{\partial \theta\, \partial x}. \quad (11\text{-}21)$$

The second basic equation is obtained by combining Eqs. (11-11), (11-12), and (11-19):

$$\frac{1}{r^2}\frac{\partial^2 N_\theta}{\partial \theta^2} - \frac{\partial^2 N_x}{\partial x^2} - \frac{(1+\varepsilon_\theta)}{r} D \left[\frac{1}{r^4}\frac{\partial^4 w}{\partial \theta^4} + \frac{1}{r^4}\frac{\partial^2 w}{\partial \theta^2} + \frac{(2-\mu)}{r^2}\frac{\partial^4 w}{\partial \theta^2\,\partial x^2} \right] = 0. \quad (11\text{-}22)$$

Combining Eqs. (11-18) and (11-1) yields the third basic equation:

$$\frac{\partial^2 N_\theta}{\partial x^2} + (2+\mu)\frac{\partial^2 N_x}{\partial x^2} + \frac{1}{r^2}\frac{\partial^2 N_x}{\partial \theta^2} - \frac{\mu}{r^2}\frac{\partial^2 N_\theta}{\partial \theta^2} = \frac{Et}{r}\frac{\partial^2 w}{\partial x^2}. \quad (11\text{-}23)$$

Solving for ε_θ from Eqs. (11-7) and (11-8) and differentiating twice with respect to x gives the fourth basic equation:

$$\frac{\partial^2 N_\theta}{\partial x^2} - \mu\frac{\partial^2 N_x}{\partial x^2} = \frac{Et}{r}\left(\frac{\partial^3 v}{\partial x^2\,\partial \theta} + \frac{\partial^2 w}{\partial x^2}\right). \quad (11\text{-}24)$$

Equations (11-21) through (11-24) are the four basic equations needed to develop a solution for the buckling of cylindrical shells.

11-2 Lateral Pressure

When the external pressure is applied only to the side of the cylinder as shown in Fig. 11-1a, then the solution can be obtained as follows. Let

$$N_\theta = -pr + f(x, y) \quad (11\text{-}25)$$

where $f(x, y)$ is a function of x and y which expresses the variation of N_θ from the average value. When the deflection, w, is small, then the function $f(x,y)$ is also very small. Similarly, the end force N_x is expressed as

$$N_x = 0 + g(x, y). \quad (11\text{-}26)$$

Since $N_{\theta x} = N_{x\theta} = 0$, then

$$N_{\theta x} = 0 + h(x, y) \quad (11\text{-}27)$$

$$N_{x\theta} = 0 + j(x, y). \quad (11\text{-}28)$$

Substituting Eqs. (11-25) through (11-28) into Eqs. (11-21) through (11-24) and neglecting higher order terms such as

$$g\frac{\partial^2 w}{\partial \theta\,\partial x},\quad h\frac{\partial^2 w}{\partial \theta\,\partial x},\quad j\frac{\partial^2 w}{\partial \theta\,\partial x},\quad \text{and } f \text{ with terms in } \frac{\partial \psi}{\partial \theta}$$

other than unity results in the following four equations:

$$D\left[\frac{\partial^4 w}{\partial x^4} + \frac{\mu}{r^2}\frac{\partial^2 w}{\partial x^2} + \frac{2}{r^2}\frac{\partial^4 w}{\partial \theta^2\,\partial x^2} + \frac{1}{r^4}\left(\frac{\partial^2 w}{\partial \theta^2} + \frac{\partial^4 w}{\partial \theta^4}\right)\right] + \frac{1}{r}f(x, y) = -p\left[\frac{1}{r}\frac{\partial^2 w}{\partial \theta^2} + \frac{w}{r} + \varepsilon_\theta\right] \quad (11\text{-}29)$$

$$\frac{1}{r^2}\frac{\partial^2 f}{\partial \theta^2} - \frac{\partial^2 g}{\partial x^2} - \frac{(1+\varepsilon_\theta)}{r^3}D\left[\frac{1}{r^2}\frac{\partial^4 w}{\partial \theta^4} + \frac{1}{r^2}\frac{\partial^2 w}{\partial \theta^2} + (2-\mu)\frac{\partial^4 w}{\partial \theta^2\,\partial x^2}\right] = 0 \quad (11\text{-}30)$$

$$\frac{\partial^2 f}{\partial x^2} + (2+\mu)\frac{\partial^2 g}{\partial x^2} + \frac{1}{r^2}\frac{\partial^2 g}{\partial \theta^2} - \frac{\mu}{r^2}\frac{\partial^2 f}{\partial \theta^2} = \frac{Et}{r}\frac{\partial^2 w}{\partial x^2} \quad (11\text{-}31)$$

$$\frac{\partial^2 f}{\partial x^2} - \mu\frac{\partial^2 g}{\partial x^2} = \frac{Et}{r}\left(\frac{\partial^3 v}{\partial x^2\,\partial \theta} + \frac{\partial^2 w}{\partial x^2}\right). \quad (11\text{-}32)$$

Equations (11-29) through (11-32) can be solved for various boundary conditions. For a simply supported cylinder, the following conditions are obtained from Fig. 11-4:

$$\text{at } x = \pm L/2, \quad w = \frac{\partial^2 w}{\partial x^2} = \frac{\partial^2 w}{\partial \theta^2} = 0.$$

Also, because of symmetry,

$$\frac{\partial w}{\partial \theta} = 0 \text{ for all values of } \theta \text{ when } x = 0$$

and

$$\frac{\partial w}{\partial \theta} = 0 \text{ for all values of } x \text{ when } \theta = 0.$$

Similarly, $\dfrac{\partial v}{\partial \theta} = 0$ for all values of θ at $x = \pm L/2$.

These boundary conditions suggest a solution of the form

$$w = A\cos n\theta \cos\frac{\pi x}{L}$$

$$v = B\sin n\theta \cos\frac{\pi x}{L}$$

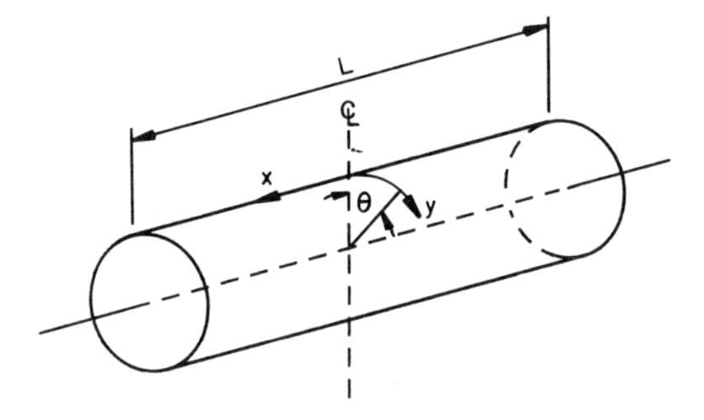

Figure 11-4.

where n is the number of lobes as defined in Fig. 11-5. Substituting these two expressions into Eqs. (11-29) through (11-32) gives

$$D\left[\frac{\pi^4}{L^4} + \frac{(2n^2 - \mu)\pi^2}{r^2L^2} - \frac{n^2}{r^4} + \frac{n^4}{r^4}\right] A \cos n\theta \cos \frac{\pi x}{L}$$

$$= \frac{p}{r}(An^2 + Bn) \cos n\theta \cos \frac{\pi x}{L} - \frac{1}{r} f(x, y) \quad (11\text{-}33)$$

$$\frac{1}{r^2}\frac{\partial^2 f}{\partial \theta^2} - \frac{\partial^2 g}{\partial x^2} = \frac{(1 - \varepsilon_\theta)}{r^3} D\left[\frac{n^4}{r^2} - \frac{n^2}{r^2} + (2 - \mu)\frac{n^2\pi^2}{L^2}\right] A \cos n\theta \cos \frac{\pi x}{L} \quad (11\text{-}34)$$

$$(2 + \mu)\frac{\partial^2 g}{\partial x^2} + \frac{1}{r^2}\frac{\partial^2 g}{\partial \theta^2} + \frac{\partial^2 f}{\partial x^2} - \frac{\mu}{r^2}\frac{\partial^2 f}{\partial \theta^2} = \frac{-Et}{r}\frac{\pi^2}{L^2} A \cos n\theta \cos \frac{\pi x}{L} \quad (11\text{-}35)$$

$$\frac{\partial^2 f}{\partial x^2} - \mu \frac{\partial^2 g}{\partial x^2} = -\frac{Et}{r}\frac{\pi^2}{L^2}(Bn + A) \cos n\theta \cos \frac{\pi x}{L}. \quad (11\text{-}36)$$

From Eq. (11-33) if follows that

$$f(x, y) = C \cos n\theta \cos \frac{\pi x}{L} \quad (11\text{-}37)$$

and from Eqs. (11-34) and (11-35) that

$$g(x, y) = G \cos n\theta \cos \frac{\pi x}{L}. \quad (11\text{-}38)$$

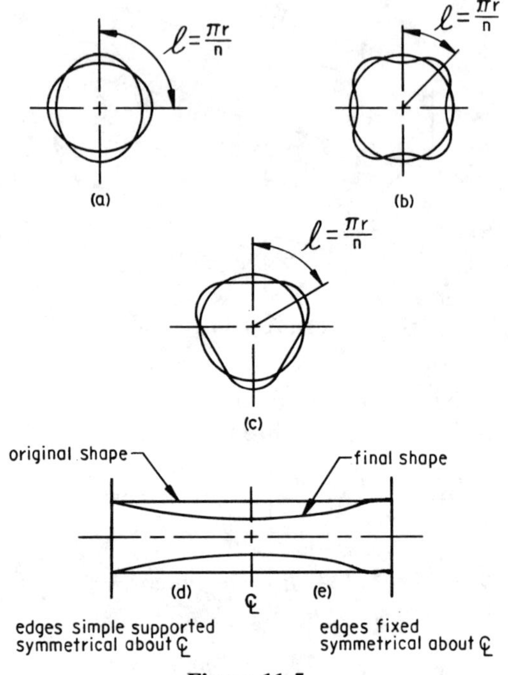

Figure 11-5.

From Eq. (11-36)

$$Bn + A = \frac{r}{Et}(C - \mu G) \tag{11-39}$$

and the values of C and G are found to be

$$\frac{C}{A} = \frac{Et}{r\alpha^2} - \frac{Et}{r^3(1 - \mu^2)}(H - 1)\left(1 - \frac{Pr}{Et}\right)\left(\frac{\alpha + 1 - \mu}{\lambda\alpha}\right) \tag{11-40}$$

$$\frac{G}{A} = \frac{Et}{r\alpha\lambda} + \frac{Et}{r^3(1 - \mu^2)}(H - 1)\left(1 - \frac{pr}{Et}\right)\left(\frac{1 - \mu(\alpha - 1)}{\lambda\alpha}\right) \tag{11-41}$$

where

$$H = n^2[1 + (\lambda - 1)(2 - \mu)]$$

$$\lambda = \frac{\pi^2 r^2}{n^2 L^2} + 1$$

$$\alpha = \frac{n^2 L^2}{\pi^2 r^2} + 1.$$

From Eq. (11-33),

$$\frac{Et}{r^2\alpha^2} - \frac{D}{r^4}\frac{\alpha + 1 + \mu}{\alpha\lambda}(H - 1)A\cos n\theta\cos\frac{\pi x}{L}$$
$$+ \frac{D}{r^4}n^2(n^2\lambda^2 - \mu(\lambda - 1) - 1)A\cos n\theta\cos\frac{\pi x}{L}$$
$$= \frac{p}{r}FA\cos n\theta\cos\frac{\pi x}{L} \quad (11\text{-}42)$$

where

$$F = n^2 - 1 + \frac{1}{\alpha} - \frac{\mu}{\alpha\lambda} - \frac{D}{r^2Et\alpha\lambda}\left\{(H - 1)\left(1 - \frac{pr}{Et}\right)[\alpha(1 - \mu^2) + (1 + \mu^2)] + \alpha + 1 + \mu\right\}. \quad (11\text{-}43)$$

Equation (11-42) indicates that solutions different from zero exist only if

$$p_{cr} = \frac{X + Y - Z}{F} \quad (11\text{-}44)$$

where

$$X = \frac{Et}{r\alpha^2}$$
$$Y = \frac{D}{r^3}\{n^2[n^2\lambda^2 - \mu(\lambda - 1) - 1]\}$$
$$Z = \frac{D}{r^3}\frac{\alpha + 1 + \mu}{\alpha\lambda}(H - 1).$$

Equation (11-44) can be written as

$$p_{cr} = \frac{1}{8}KE\left(\frac{t}{r}\right)^3 \quad (11\text{-}45)$$

where

$$K = K_1 + 4K_2(r/t)^2 \quad (11\text{-}46)$$
$$K_1 = \frac{2}{3}\frac{n^2[n^2\lambda^2 - \mu(\lambda - 1)] - U(H - 1)}{F(1 - \mu^2)} \quad (11\text{-}47)$$

$$K_2 = \frac{2}{\alpha^2 F} \tag{11-48}$$

where

$$U = \frac{\alpha + 1 + \mu}{\alpha\lambda}.$$

Equation (11-45) is the basic equation for the buckling of cylindrical shells subjected to lateral pressure. A plot of K in Eq. (11-45) is shown in Fig. 11-6.

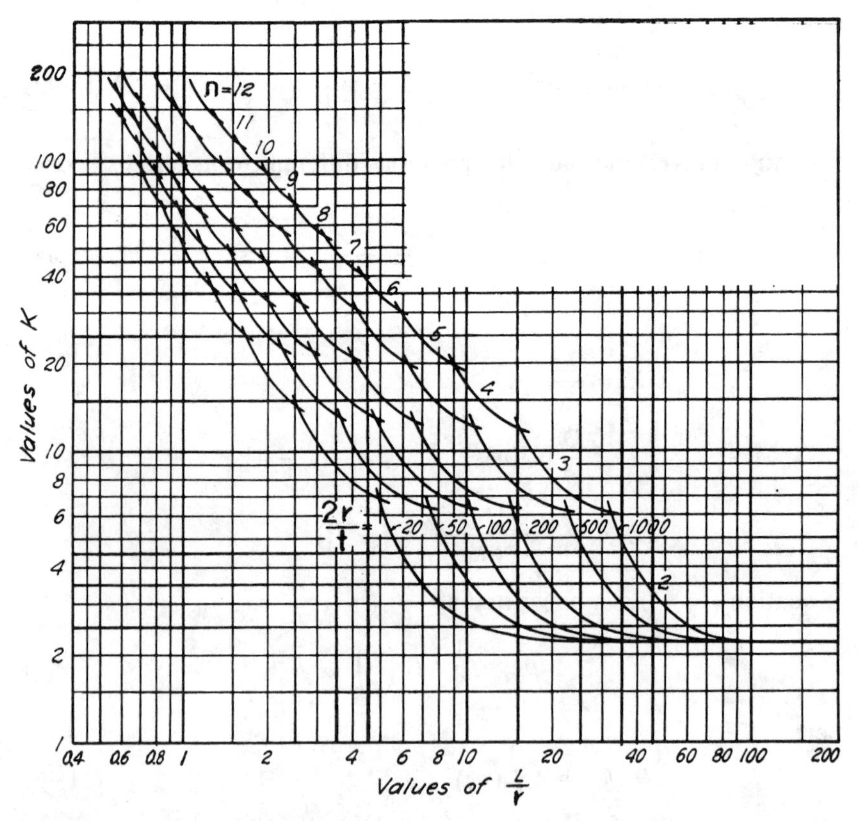

Figure 11-6. Buckling coefficient K for cylinders with pressure on sides only, edges simply supported; $\mu = 0.30$. (Sturm 1941.)

Example 11-1

Find the allowable external pressure for the inner cylinder shown in Fig. 11-7. Let $L = 10$ ft, $r = 2$ ft, $E = 29{,}000$ ksi, $t = 1/2$ inch, factor of safety $(F.S.) = 2.5$, and $\mu = 0.3$. Assume the inner cylinder to be simply supported.

Solution

$$L/r = 5, \qquad 2r/t = 96$$

From Fig. 11-6, $K = 11$. Hence, from Eq. (11-45),

$$\begin{aligned} p_{cr} &= \frac{11}{8}(29{,}000{,}000)(0.5/24)^3 \\ &= 360 \text{ psi} \\ p &= p_{cr}/F.S. = 360/2.5 = 144 \text{ psi.} \end{aligned}$$

Equation (11-45) assumes the end of the cylindrical shell to be simply supported. A similar equation can be derived for the case of a cylindrical shell with fixed ends. In this case the slope and deflection at the ends are zero. Proceeding in a similar fashion as for the simply supported case, a buckling equation is obtained. The derivation is more complicated than that for the simply supported cylinder. The resulting buckling equation is the same as Eq. (11-45) with the exception of the value of K. A plot of K for the fixed end condition is shown in Fig. 11-8.

Problems

11-1 Derive Eq. (11-40).
11-2 Derive Eq. (11-42).

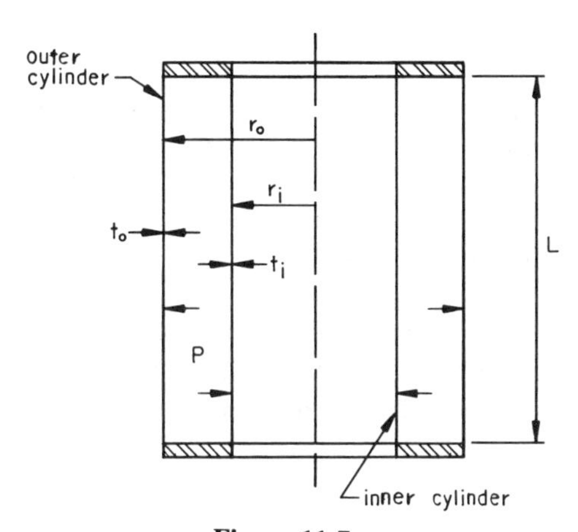

Figure 11-7.

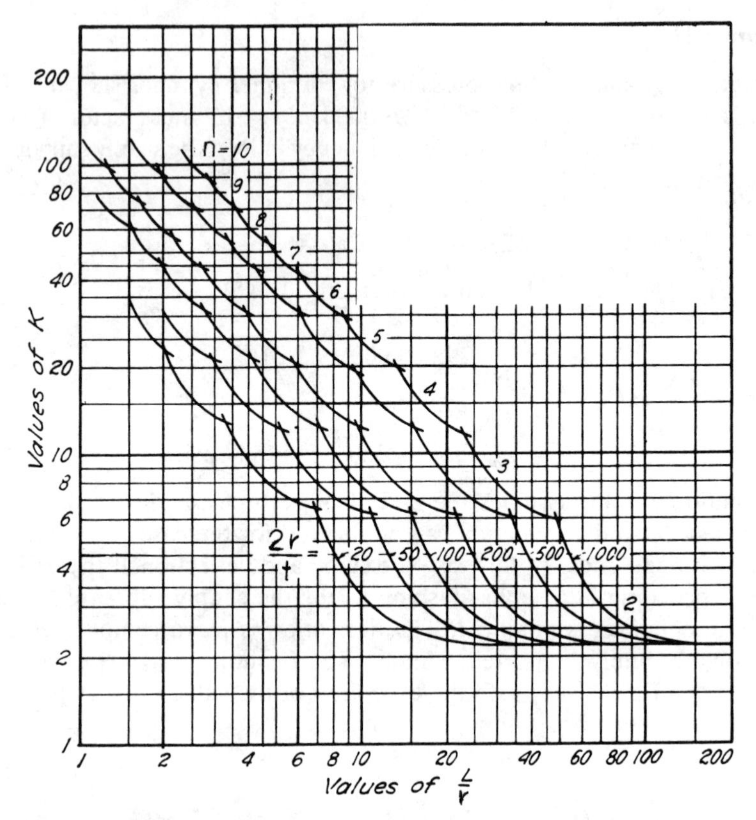

Figure 11-8. Buckling coefficient K for cylinders with pressure on sides only, edges fixed; $\mu = 0.30$. (Sturm 1941.)

11-3 What is the required thickness of the inner cylindrical shell shown in Fig. P11-3? Let $p = 15$ psi, $L = 20$ ft, $r = 20$ inches, $F.S. = 3.0$, $E = 15{,}000$ ksi, and $\mu = 0.3$. Assume ends to be fixed.

11-3 Lateral and End Pressure

Many cylindrical shells are subjected to axial forces in the lateral and axial directions (Fig. 11-1b) or to vacuum. The governing equations are very similar to those derived for the lateral condition. For lateral and axial loads, Eq. (11-26) is written as

$$N_x = -\frac{pr}{2} + g(x, y) \tag{11-49}$$

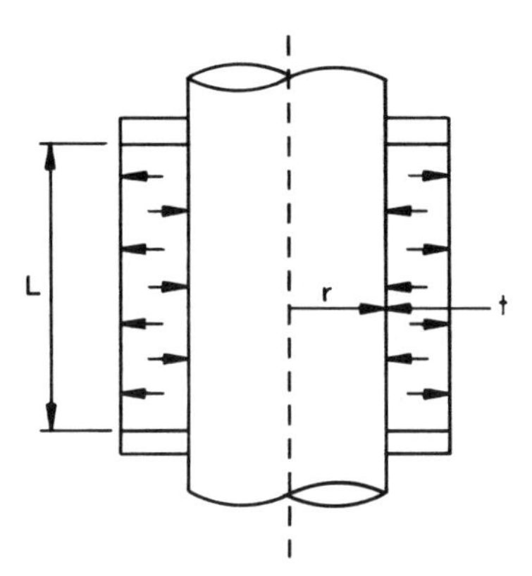

Figure P11-3.

while Eqs. (11-25), (11-27), and (11-28) remain the same. Equation (11-29) becomes

$$D\left[\frac{\partial^4 w}{\partial x^4} + \frac{\mu}{r}\frac{\partial^4 w}{\partial x^4} + \frac{2}{r^2}\frac{\partial^4 w}{\partial \theta^2\,\partial x^2} + \frac{1}{r^4}\frac{\partial^2 w}{\partial \theta^2} + \frac{\partial^4 w}{\partial \theta^4}\right] + \frac{1}{r}f(x, y)$$

$$= -p\left[\frac{1}{r}\frac{\partial^2 w}{\partial \theta^2} + \frac{w}{r} + \varepsilon_\theta\right] - \frac{pr}{2}\frac{\partial^2 w}{\partial x^2} \qquad (11\text{-}50)$$

while Eqs. (11-30), (11-31), and (11-32) remain the same. Using the boundary conditions for a simply supported cylinder, the governing Eq. (11-45) can be written as

$$p_{cr} = \frac{1}{8}K'E(t/r)^3 \qquad (11\text{-}51)$$

where

$$K' = K_1' + 4K_2'(r/t)^2 \qquad (11\text{-}52)$$

$$K_1' = K_1 \frac{F}{F + \dfrac{\pi^2 r^2}{2L^2}}$$

$$K_2' = K_2 \frac{F}{F + \dfrac{\pi^2 r^2}{2L^2}}.$$

A plot of Eq. (11-52) for a simply supported cylinder is shown in Fig. 11-9.
A similar equation can be derived for a cylinder with fixed ends. The resulting K' value is plotted in Fig. 11-10.

Example 11-2

Determine if the cylinder shown in Fig. 11-11 is adequate for full vacuum condition. Let $E = 16{,}000$ ksi, $\mu = 0.3$ and factor of safety $(F.S.) = 4$.

Solution

$$L/r = \frac{96}{24} = 4.00$$

$$2r/t = 192.$$

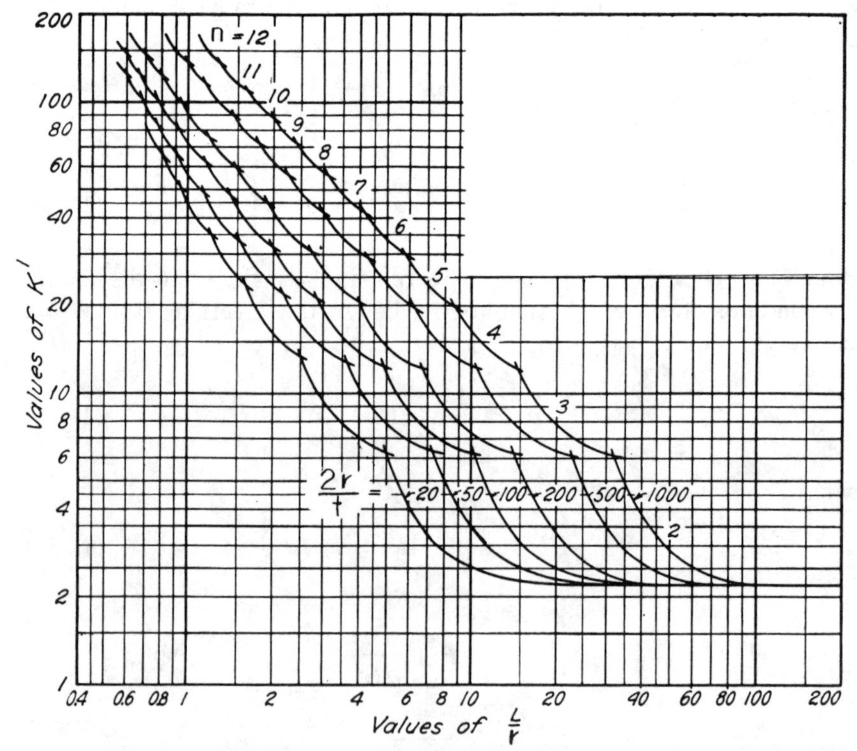

Figure 11-9. Buckling coefficient K' for cylinders with pressure on sides and ends, edges simply supported; $\mu = 0.30$. (Sturm 1941.)

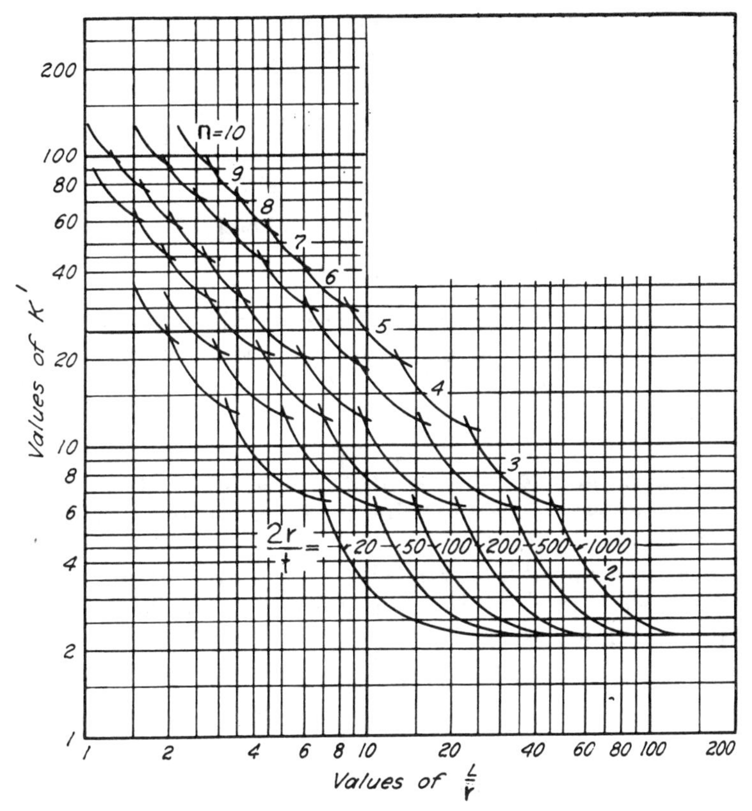

Figure 11-10. Buckling coefficient K' for cylinders with pressure on sides and ends, edges fixed; $\mu = 0.30$. (Sturm 1941.)

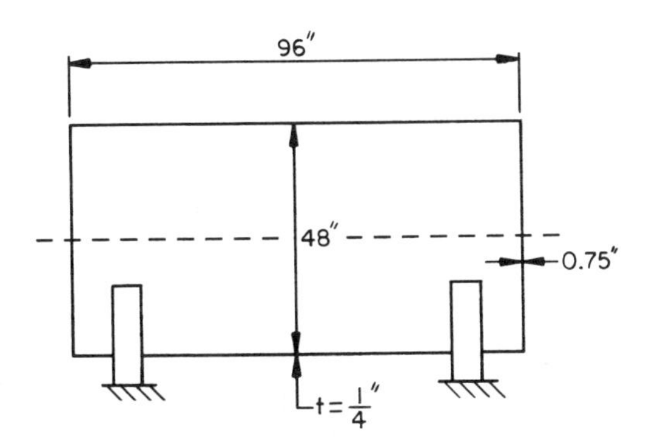

Figure 11-11.

Assume the ends of the cylinder to be simply supported due to the flexibility of the end plates. From Fig. 11-9, $K' = 20$ and from Eq. (11-51),

$$p = (1/4)\left[\frac{20}{8} \times 16{,}000{,}000\ (0.25/24)^3\right]$$
$$= 11.3 \text{ psi} < 14.7 \text{ psi.} \quad \text{no good.}$$

For $t = 5/16$ inch, we get $K' = 16$ and $p = 17.66$ psi.

Problems

11-4 Find the required thickness of the inner cylinder shown in Fig. P11-4 due to a full vacuum condition, 15 psi, in the inner cylinder. Use Eq. (11-51). Let $E = 29{,}000$ ksi and $\mu = 0.30$. Use multiples of 1/16 inch in determining the cylinder thickness.
11-5 In Problem 11-4, find the required thickness in the outer cylinder due to a full vacuum condition in the annular jacket space. Use Eq. (11-51) even though the axial force on the outer cylinder is substantially less than that given by Eq. (11-51).
11-6 Solve Problem 11-5 using Eq. (11-45) as a more appropriate equation due to the small end load. What is the difference between this thickness and that obtained in Problem 11-5?

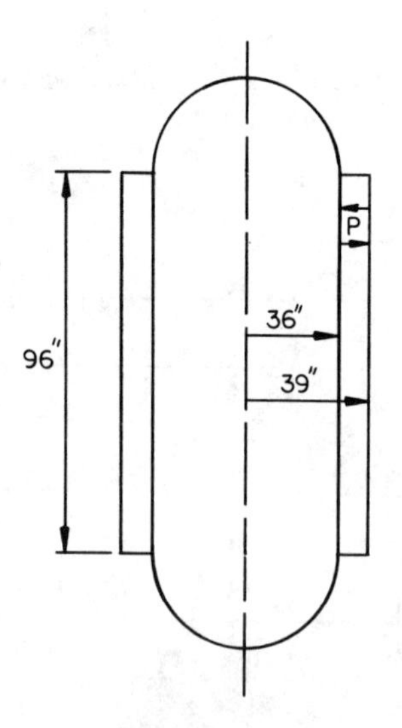

Figure P11-4.

11-4 Axial Compression

When an axial force is applied at the end of a cylinder, the buckling strength is developed as follows. Let W be the applied end load per unit length of circumference. Equations (11-25) and (11-26) become

$$N_\theta = 0 + f(x, y) \tag{11-53}$$

$$N_x = -W + g(x, y). \tag{11-54}$$

For a simply supported cylinder, the buckling equation can be expressed as

$$W_{cr} = \frac{X + Y - Z}{n^2 + V(H - 1)} \tag{11-55}$$

where

$$X = \frac{Et}{\alpha_0 \lambda_0}$$

$$Y = \frac{D(\alpha_0 - 1)}{r^2} \{n^2[n^2\lambda_0^2 - \mu(\lambda_0 - 1)] - 1\}$$

$$Z = \frac{D(\alpha_0 - 1)}{r^2} \frac{\alpha_0 + 1 + \mu}{\alpha_0 \lambda_0} (H - 1)$$

$$V = \frac{\mu D(\alpha + 1 + \mu)}{rEt^2\lambda_0^2}$$

$$\alpha_0 = \frac{n^2 L_0^2}{\pi^2 r^2} + 1$$

$$\lambda_0 = \frac{\pi^2 r^2}{n^2 L_0^2} + 1$$

L_0 = length of one buckle wave ($L_0 << L$).

Equation (11-55) can be simplified as follows.

Let $\sigma_{cr} = W_{cr}/t$, and for long cylinders with small values of r/t, Eq. (11-55) becomes

$$\sigma_{cr} = 0.6Et/r. \tag{11-56}$$

Tests have shown that for actual cylinders, which have a slight out-of-roundness, the stress magnitude given by Eq. (11-56) is unconservative. Figure 11-12 shows the scatter of some of the data obtained from various tests. An empirical equation that defines the lower bound of available test

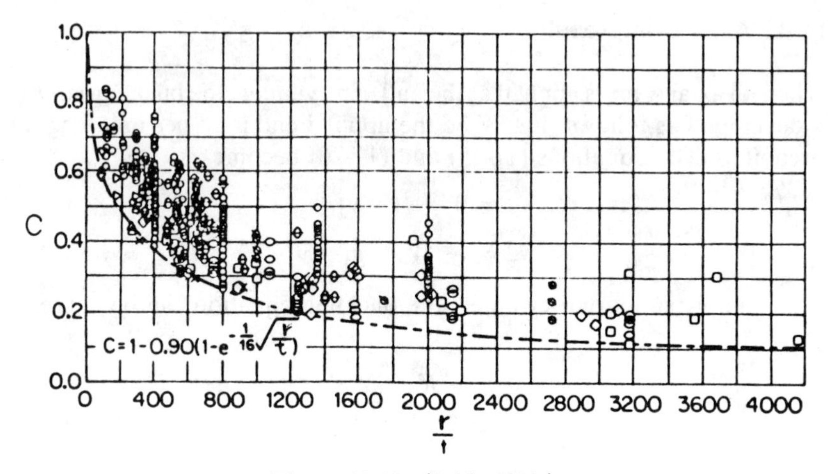

Figure 11-12. (Seide 1981.)

data (ACI 1989) is given by

$$\sigma_{cr} = 0.6CEt/r \tag{11-57}$$

where

$$C = 1 - 0.9(1 - e^{(-1/16)\sqrt{(r/t)}}).$$

Example 11-3

Determine the thickness of the lower cylinder shown in Fig. 11-13 needed to support a reactor of weight 400,000 lbs. Use Eq. (11-57). Let $E =$ 30,000 ksi and factor of safety $(F.S.) = 4.0$.

Solution

Try $t = 0.25$ inch. Then $r/t = 192$.
From Eq. (11-57), $C = 0.48$
and

$$\begin{aligned}\sigma &= (1/4)(0.48 \times 0.6 \times 30{,}000{,}000 \times 0.25/48) \\ &= 11{,}250 \text{ psi.}\end{aligned}$$

Actual stress is given by

$$= \frac{400{,}000}{\pi \times 96 \times 0.25} = 5300 \text{ psi.}$$

Theoretically, the thickness can be reduced further. However, consideration must be given to handling and erection procedures.

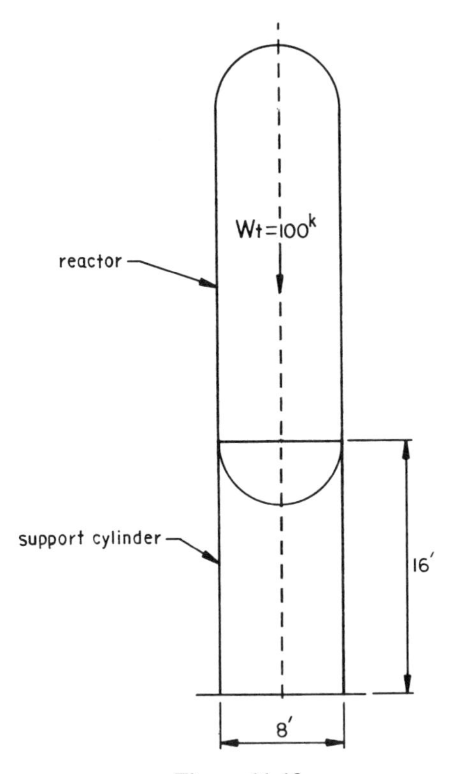

Figure 11-13.

Problems

11-7 What is the required thickness of the cylindrical shell shown in Fig. P11-7? The weight of the spherical tank is 100,000 lbs. Let $E = 30{,}000$ ksi, $\mu = 0.30$, factor of safety, $(F.S.) = 5.0$. Use increments of 1/16 inch to determine thickness.

11-8 An oil storage standpipe is shown in Fig. P11-8. Is the thickness adequate for axial compression due to dead weight? Let $E = 30{,}000$ ksi, $\mu = 0.30$, factor of safety $(F.S.) = 5.0$, and weight of steel $= 490$ pcf.

11-5 Donnell's Equations

Donnell in 1933 was able to combine Eqs. (11-1) through (11-16) by means of ingenious substitutions to obtain one equation for the buckling of cylindrical shells. The eighth-order differential equation he obtained is a

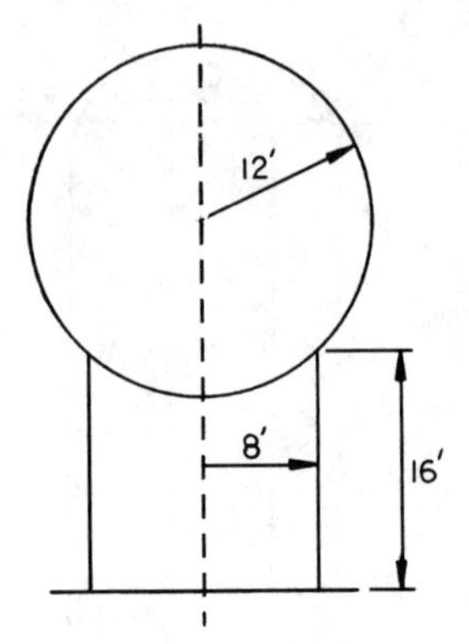

Figure P11-7.

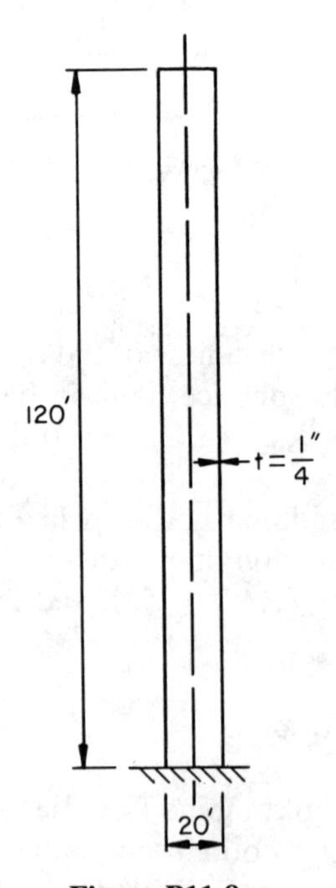

Figure P11-8.

function of deflection (Gerard 1962) and is expressed as

$$D\nabla^8 w + \frac{Et}{r^2}\frac{\partial^4 w}{\partial x^4} + \nabla^4\left(N_x \frac{\partial^2 w}{\partial x^2} + 2N_{xy}\frac{\partial^2 w}{\partial x\,\partial\theta} + N_\theta \frac{\partial^2 w}{\partial\theta^2}\right) = 0. \quad (11\text{-}58)$$

This equation can be used for various loading conditions. Let us take the simple case of applied lateral pressure, where

$$N_x = N_{xy} = 0 \quad \text{and} \quad N_\theta = pr = \sigma_{cr} t. \quad (11\text{-}59)$$

Equation (11-58) then becomes

$$D\nabla^8 w + \frac{Et}{r^2}\frac{\partial^4 w}{\partial x^4} + \nabla^4\left(N_\theta \frac{\partial^2 w}{\partial\theta^2}\right) = 0. \quad (11\text{-}60)$$

An expression for the deflection that satisfies the simply supported boundary condition of a cylinder can be expressed as

$$w = w_{mn} \sin\frac{m\pi x}{L} \sin\frac{n\pi y}{\pi r}. \quad (11\text{-}61)$$

Substituting this expression into Eq. (11-60) gives the nontrivial solution

$$\sigma_{cr} = \frac{\pi^2 KE}{12(1-\mu^2)}(t/L)^2 \quad (11\text{-}62)$$

where

$$K = \frac{(m^2+\beta^2)^2}{\beta^2} + \frac{12Z^2}{\pi^4\beta^2(1+\beta^2/m^2)^2} \quad (11\text{-}63)$$

$$\beta = \frac{nL}{\pi r} \quad (11\text{-}64)$$

$$Z = \text{curvature parameter} = \frac{L^2}{rt}\sqrt{1-\mu^2}. \quad (11\text{-}65)$$

The minimum value of Eq. (11-63) is obtained when $m = 1$. Equation (11-63) then becomes

$$K = \frac{(1+\beta^2)^2}{\beta^2} + \frac{12Z^2}{\pi^4\beta^2(1+\beta^2)^2}. \quad (11\text{-}66)$$

Short Cylinders

For short cylinders, the curvature parameter Z can be set to zero and Eq. (11-66) becomes

$$K = (\beta + 1/\beta)^2. \quad (11\text{-}67)$$

Intermediate-Length Cylinders

As the ratio L/r increases, the quantity $\beta + 1$ can be approximated by β. Hence, Eq. (11-66) becomes

$$K = \beta^2 + \frac{12Z^2}{\pi^4\beta^6}. \qquad (11\text{-}68)$$

Minimizing this quantity with respect to β gives

$$K = 1.038\sqrt{Z}. \qquad (11\text{-}69)$$

Long Cylinders

For long cylinders, the buckling mode is similar to that of a circular ring and is elliptic in shape. Hence, $n = 2$ and

$$\beta = \frac{2L}{\pi r}$$

and Eq. (11-68) reduces to

$$K = \beta^2. \qquad (11\text{-}70)$$

A plot of K defined by Eqs. (11-67), (11-69), and (11-70) is shown in Fig. 11-14.

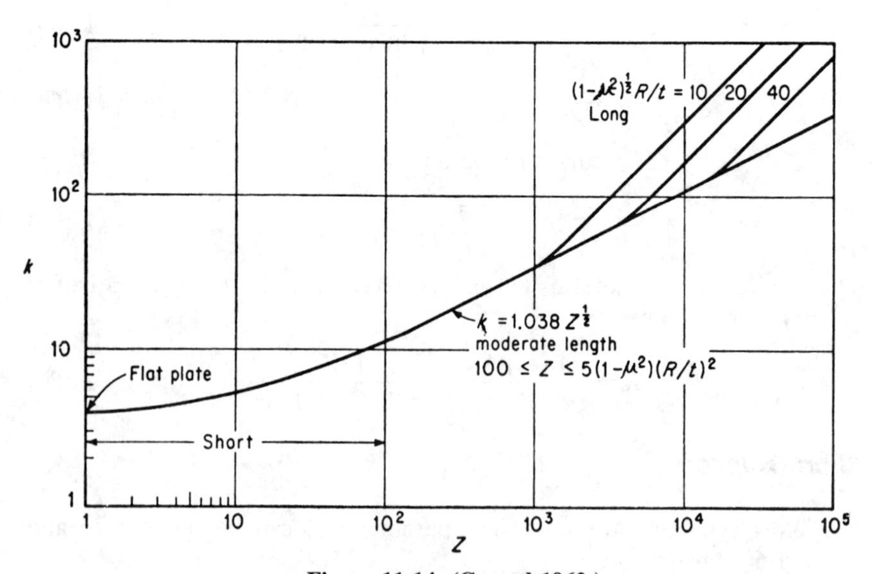

Figure 11-14. (Gerard 1962.)

Example 11-4

The aluminum pipe shown in Fig. 11-15 is subjected to a vacuum pressure of 10 psi. The pipe has stiffeners spaced at 8-ft intervals. Let E = 10,000 ksi, factor of safety $(F.S.)$ = 3.0, and μ = 0.25. (a) Find the required thickness based on Eq. (11-62). (b) Compare the result with Sturm's Eq. (11-51).

(a)

$$\text{Let } t = 3/16 \text{ inch.}$$

Then actual stress is

$$\begin{aligned}\sigma &= pr/t \\ &= 10 \times 12/0.1875 = 640 \text{ psi.}\end{aligned}$$

From Eq. (11-65)

$$\begin{aligned}Z &= \frac{L^2}{rt}\sqrt{1 - \mu^2} = \frac{96^2}{12 \times 0.1875}\sqrt{1 - 0.25^2} \\ &= 3966.\end{aligned}$$

From Eq. (11-69), $K = 1.038\sqrt{Z} = 65.4$

The allowable stress from Eq. (11-62) with a factor of safety of 3.0 is

$$\begin{aligned}\sigma_{\text{all}} &= \frac{\pi^2 \times 65.4 \times 10{,}000{,}000}{3 \times 12(1 - 0.25^2)}(0.1875/96)^2 \\ &= 730 \text{ psi} > 640 \text{ psi.} \quad \text{ok.}\end{aligned}$$

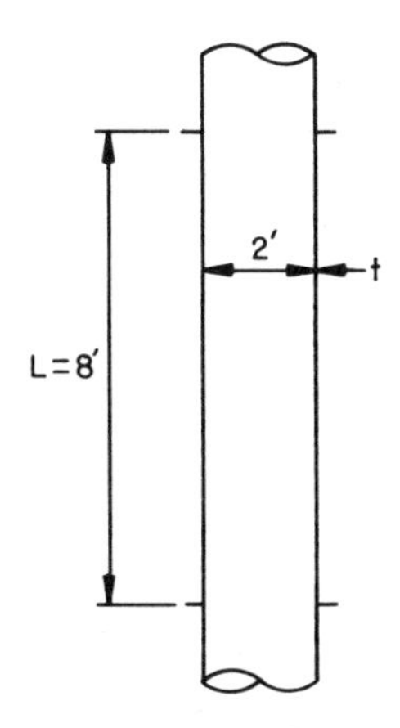

Figure 11-15.

(b)

$$L/r = 96/12 = 8, \qquad D/t = 128.$$

From Fig. 11-9, $K = 7.2$.
From Eq. (11-51) with a factor of safety of 3,

$$p = \frac{1}{3}\frac{7.2}{8}(10{,}000{,}000)\,(0.1875/12)^3$$
$$= 11.4 \text{ psi}$$
$$\sigma = pr/t = 11.4 \times 12/0.1875 = 732 \text{ psi}.$$

Problems

11-9 Solve Problem 11-4 by using Eq. (11-62). Is there any difference in the results? If so, why?
11-10 Solve Problem 11-5 by using Eq. (11-62).

11-6 Design Equations

The equations derived in Sections 11-1 through 11-5 are used in numerous codes and standards for design purposes. Some of these design equations are given in this section.

External Pressure

The ASME Power Boiler, Pressure Vessel, and Nuclear Reactor Codes use Eq. (11-51) as a basis for establishing design rules for external loads. This method permits the use of stress–strain curves of actual materials of construction to obtain allowable external pressure. This procedure prevents the possibility of calculating an allowable external pressure that results in a stress value that is above the yield stress of the material. Equation (11-51) for lateral and end pressure can be written as

$$\sigma_{cr} = p_{cr}r/t = \frac{K'}{2}E(t/2r)^2. \tag{11-71}$$

Define

$$\varepsilon_{cr} = \sigma_{cr}/E.$$

Equation (11-71) becomes

$$\varepsilon_{cr} = \frac{K'}{2}(t/2r)^2. \tag{11-72}$$

A plot of this equation as a function of ε_{cr}, $L/2r$, and $t/2r$ is shown in Fig. 11-16. This figure is normally used by entering the values of $L/2r$ and $t/2r$ for a given shell and determining the critical strain ε_{cr}.

In order to determine the critical stress, plots of stress–strain diagrams are made for different materials at various temperatures. These diagrams are plotted by ASME on a log–log scale with the ordinate plotted as $\sigma_{cr}/2$ and abscissa as ε_{cr}. The value of $\sigma_{cr}/2$ in these diagrams is referred to by ASME as B. A sample of a stress–strain diagram for carbon steel is shown in Fig. 11-17. The ASME procedure consists of determining critical strain ε_{cr} from Fig. 11-16 and the B value from Fig. 11-17. The allowable external pressure is then calculated from

$$p = \frac{(2B)t}{r(F.S.)}. \tag{11-73}$$

If the value of ε_{cr} falls to the left of the material curves in Fig. 11-17, then the allowable external pressure is given by

$$p = \frac{\varepsilon_{cr}E}{(r/t)(F.S.)}. \tag{11-74}$$

The ASME Codes use a factor of safety ($F.S.$) of 3.0 in Eqs. (11-73) and (11-74) for external pressure design.

The ability of a cylindrical shell to resist external pressure increases with a reduction in its effective length. The ASME code gives rules for adding stiffening rings to reduce the effective length. Rules for the design of stiffening rings (Jawad and Farr 1989) are also given in the ASME code.

An empirical equation developed by the U.S. Navy (Raetz 1957) for the buckling of cylindrical shells under lateral and axial pressure in the elastic range is given by

$$p_{cr} = \frac{2.42E}{(1 - \mu^2)^{3/4}} \frac{(t/2r)^{2.5}}{[L/2r - 0.45(t/2r)^{1/2}]}. \tag{11-75}$$

This equation is a good approximation of Eq. (11-51).

Tests that led to the development of Eq. (11-75) also showed that the effective cylindrical length for cylinders with end closures in the form of hemispherical or elliptical shape is equal to the length of the actual cylinder plus one-third the depth of the end closures as illustrated in Fig. 11-18.

For structures with large r/t ratios, Eq. (11-75) can be simplified to

$$p_{cr} = \frac{2.42E}{(1 - \mu^2)^{3/4}} \frac{(t/2r)^{2.5}}{L/2r}.$$

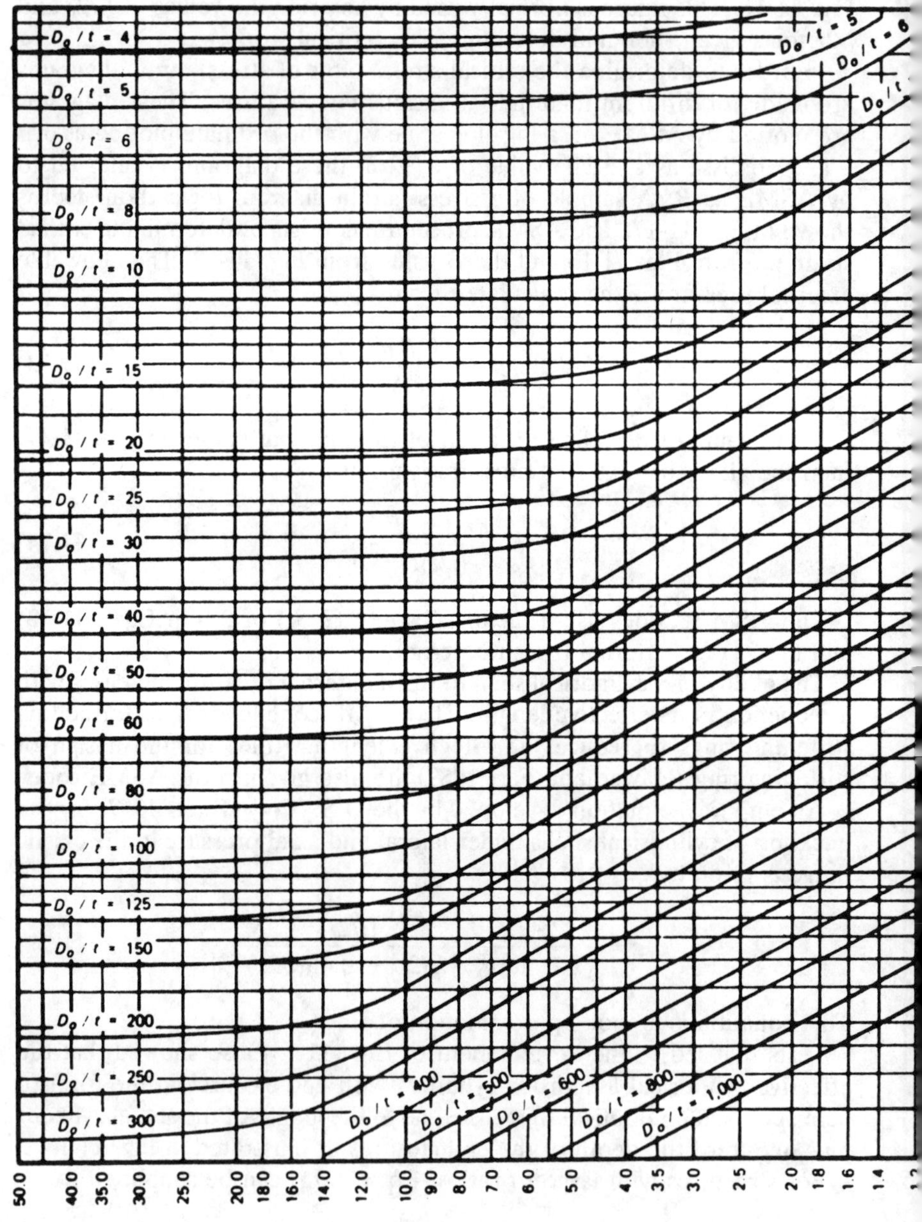

D_o/t = 4
D_o/t = 5
D_o/t = 6
D_o/t = 8
D_o/t = 10
D_o/t = 15
D_o/t = 20
D_o/t = 25
D_o/t = 30
D_o/t = 40
D_o/t = 50
D_o/t = 60
D_o/t = 80
D_o/t = 100
D_o/t = 125
D_o/t = 150
D_o/t = 200
D_o/t = 250
D_o/t = 300
D_o/t = 400
D_o/t = 500
D_o/t = 600
D_o/t = 800
D_o/t = 1,000
D_o/t = 5
D_o/t = 6
50.0
40.0
35.0
30.0
25.0
20.0
18.0
16.0
14.0
12.0
10.0
9.0
8.0
7.0
6.0
5.0
4.0
3.5
3.0
2.5
2.0
1.8
1.6
1.4
Length ÷ Outside Diameter = L/D_o

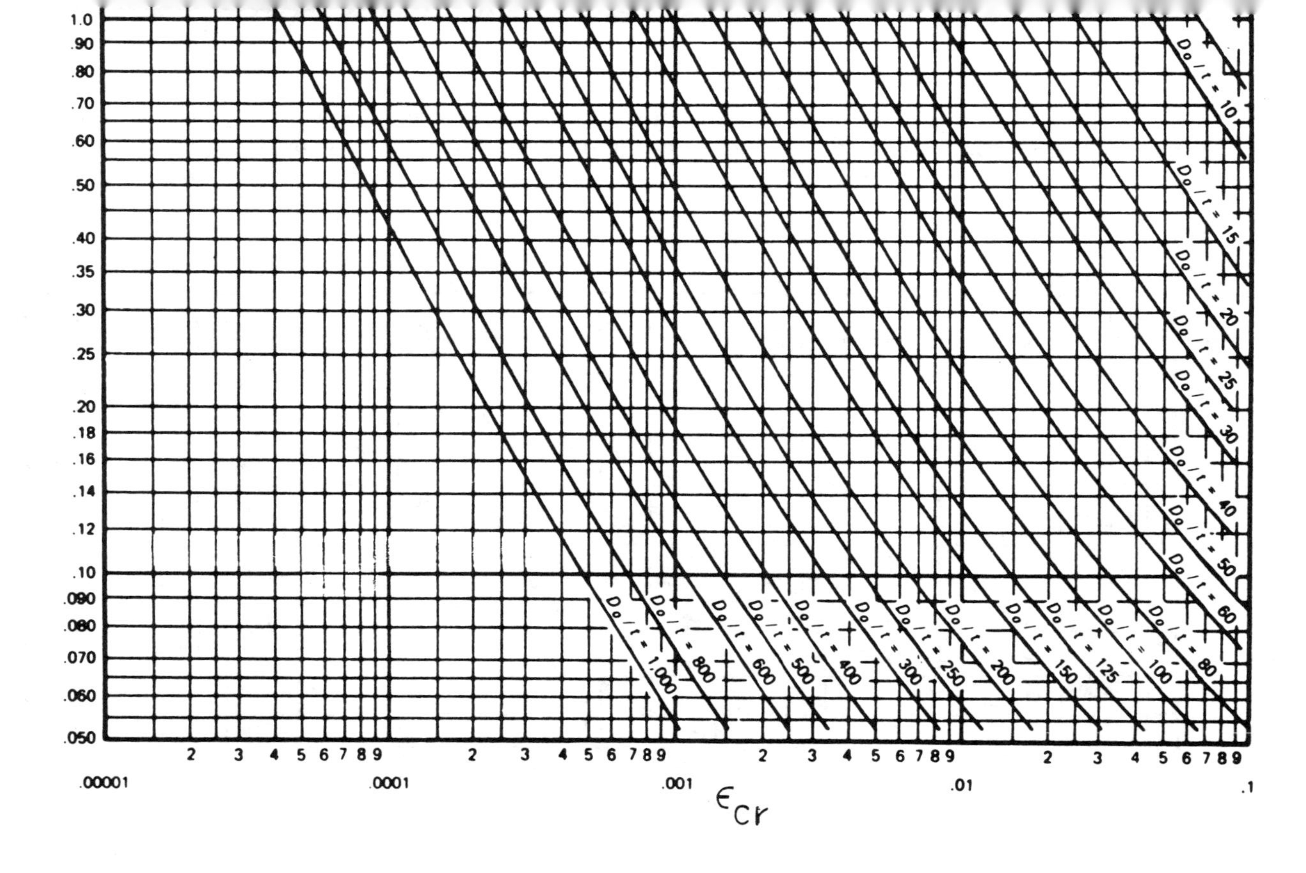

Figure 11-16. A plot of Eq. (11-71). (Courtesy of ASME.)

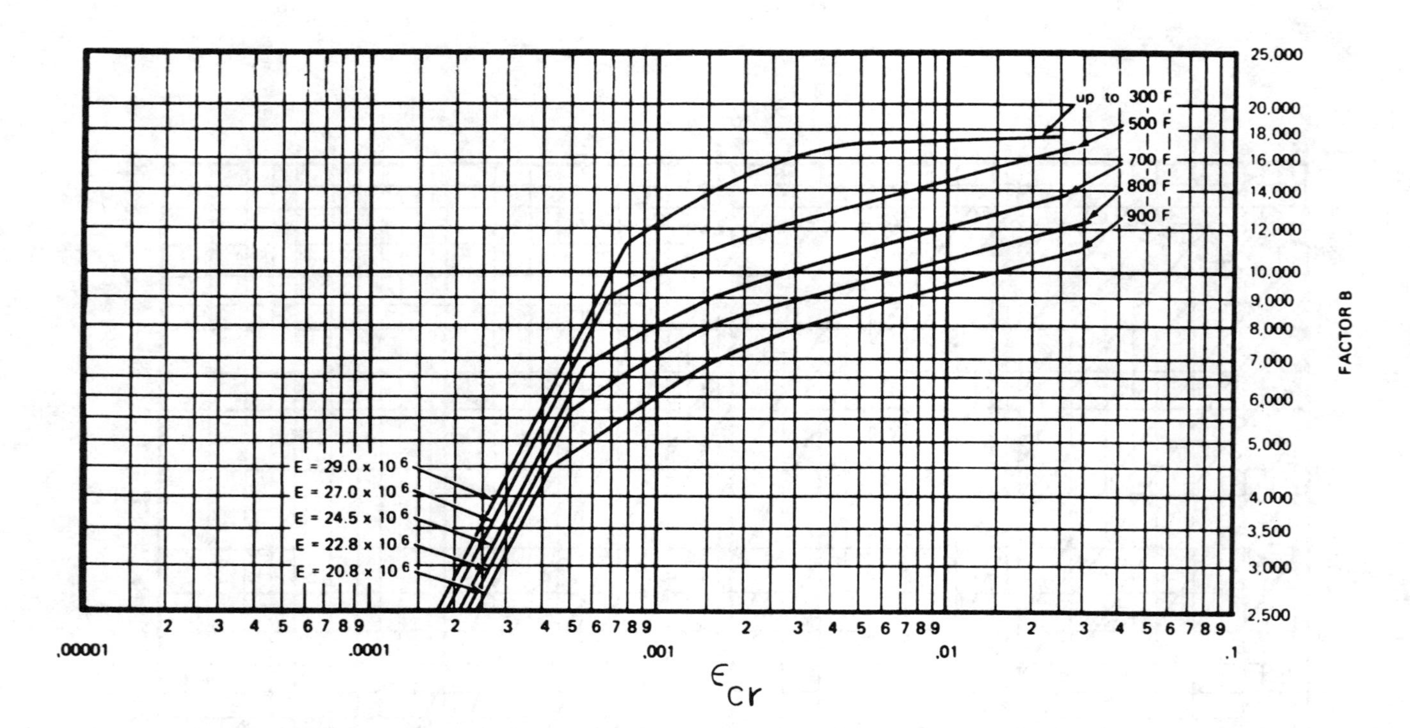

Figure 11-17. Chart for determining thickness of cylindrical and spherical shells under external pressure when constructed of carbon or low alloy steels and type 405 and type 410 stainless steels. (Courtesy of ASME.)

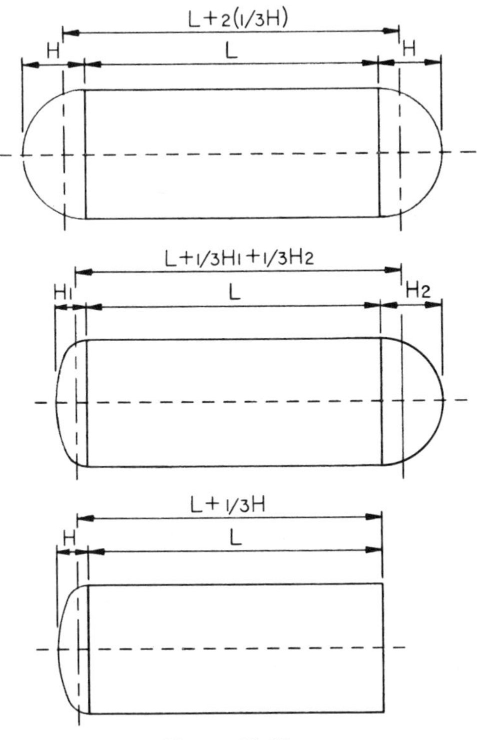

Figure 11-18.

Substituting into this equation the value of $\mu = 0.30$ for metallic structures and $\mu = 0.15$ for concrete structures results in

$$\left.\begin{aligned} p_{cr} &= \frac{0.92E(t/r)^{2.5}}{L/r}, \quad \text{for metallic structures} \\ p_{cr} &= \frac{0.87E(t/r)^{2.5}}{L/r} \quad \text{for concrete structures} \end{aligned}\right] \tag{11-76}$$

Example 11-5

Determine the allowable pressure for the cylinder in Fig. 11-19 based on Eq. (11-75). Compare the result with that obtained from Eq. (11-51). Let $E = 16{,}000$ ksi, $\mu = 0.3$ and factor of safety $(F.S.) = 4$.

Solution

$$L/r = \frac{96 + 2 \times 9/3}{24} = 4.25$$

$$2r/t = 192.$$

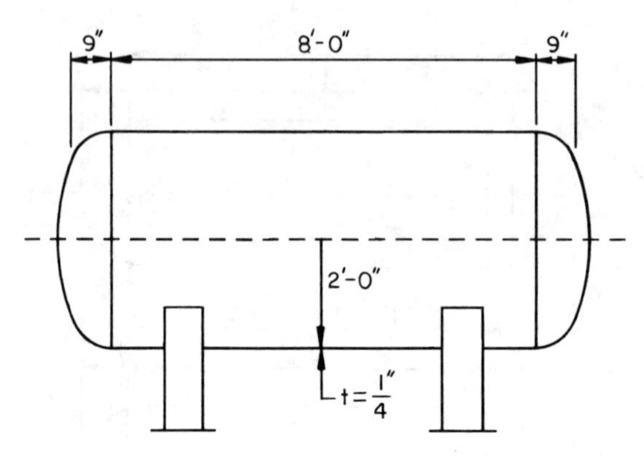

Figure 11-19.

From Eq. (11-75),

$$p = (1/4)\left[\frac{2.42 \times 16{,}000{,}000}{(0.91)^{0.75}} \frac{(0.25/48)^{2.5}}{2.125 - 0.45(0.25/48)^{1.2}}\right]$$
$$= 9.74 \text{ psi.}$$

From Fig. 11-9, $K' = 18$. And from Eq. (11-51),

$$p = \frac{1}{4 \times 8} \times 18 \times 16{,}000{,}000\ (0.25/24)^3$$
$$= 10.17 \text{ psi.}$$

Axial Compression

The allowable axial compressive stress for cylindrical shells in the ASME Code is obtained by using a factor of safety of 10 in Eq. (11-56) to account for the reduction in strength as shown in Fig. 11-12. This gives the following approximate equation for the elastic buckling:

$$\sigma = \frac{0.0625E}{r/t} \tag{11-77}$$

or in terms of strain

$$\varepsilon = \frac{0.0625}{r/t}.$$

In the inelastic region, the material charts (Fig. 11-17) must be used. Accordingly, the above strain equation must be multiplied by a factor of 2 to compensate for the fact that the ASME material charts are plotted with $B = \sigma/2$ which is half of the critical stress. Thus, we enter the charts with

$$\varepsilon = \frac{0.125}{r/t} \tag{11-78}$$

and read a value of B from the charts. This B value is the allowable compressive stress in the cylinder with a theoretical factor of safety of 10.

An empirical equation that is often used in the design of stacks and other self-supporting cylindrical structures made of low carbon steel was developed by the Chicago Bridge and Iron Company (Roark and Young 1975) as

$$\text{Allowable stress} = (X)\,(Y) \tag{11-79}$$

where

$$X = [1{,}000{,}000\; t/r]\left[2 - \frac{2}{3}(100\; t/r)\right] \quad \text{for } t/r < 0.015$$

$$X = 15{,}000 \text{ psi} \quad \text{for } t/r > 0.015$$

$$Y = 1 \quad \text{for } 2L/r \leq 60$$

$$Y = \frac{21{,}600}{18{,}000 + 2(L/r)} \quad \text{for } 2L/r > 60$$

Minimum t = 1/4 inch.

Another expression that is often used in determining the allowable compressive stress of steel cylindrical shells of flat-bottom oil storage tanks was developed by the API (API Standard 620). From Eq. (11-56) with E = 30,000,000 psi and a factor of safety of 10, the allowable compressive stress becomes

$$\sigma = 1{,}800{,}000\,(t/r). \tag{11-80}$$

This allowable compressive stress is limited to 15,000 psi. A stress transition equation between this limit and Eq. (11-56) was developed by API as shown in Fig. 11-20. The API standard also includes curves for allowable stress of cylindrical shells subjected to biaxial stress combinations.

It should be noted that Eqs. (11-77) and (11-80) do not include any terms for the length of the cylinder. Hence, for extremely large L/r ratios, Euler's equation for column buckling may control the allowable stress rather than shell buckling and should be checked. In this case, the expression to be considered is

$$\sigma_{cr} = \frac{\pi^2 EI}{(KL)^2 A} \tag{11-81}$$

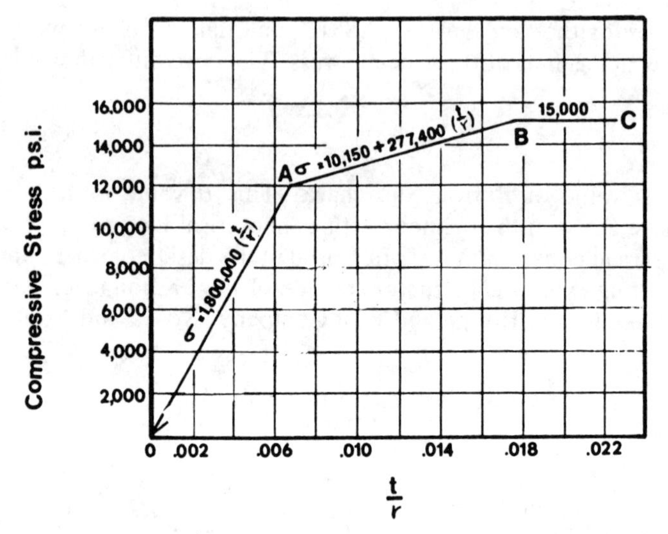

Figure 11-20. (Courtesy of American Petroleum Institute.)

where

A = material cross sectional area;
I = material moment of inertia;
K = 1.0 for cylinders with simply supported ends
= 0.5 for cylinders with fixed ends
= 2.0 for cantilever cylinders.

The critical axial compressive stress in reinforced concrete structures is obtained from Eqs. (11-56) and (11-57). Buckling stress for other loading conditions such as torsion and shear as well as various loading combinations are given in various publications such as that from NACA (Gerard and Becker 1957).

Example 11-6

The reactor shown in Fig. 11-21a is constructed of carbon steel with yield stress of 38 ksi and E = 30,000 ksi. The design temperature is 100°F. (a) Determine by the ASME method the required thickness due to vacuum. (b) Check the thickness due to the wind loading shown in Fig. 11-21b.

Solution

(a) Vacuum condition

$$\text{Try } t = 3/8 \text{ inch.}$$

$$L = 30 + (2)(1)/3 = 30.67 \text{ ft} = 368 \text{ inch.}$$

$$L/2r = 7.7 \quad \text{and} \quad 2r/t = 128.$$

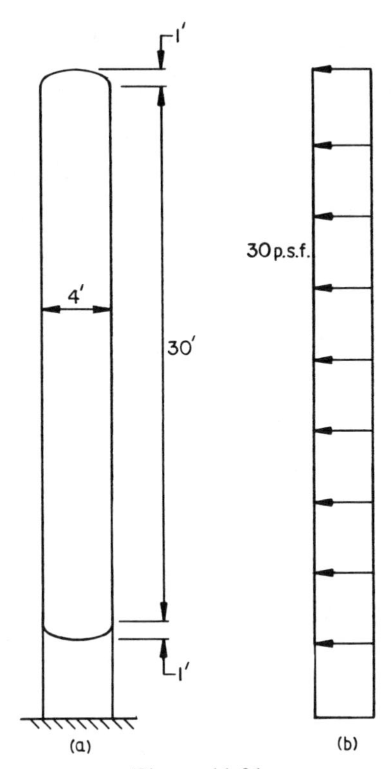

Figure 11-21.

From Fig. 11-16, $\varepsilon_{cr} = 0.00012$ inch/inch
From Fig. 11-17, ε_{cr} falls to the left of the material curve.
Hence, from Eq. (11-74)

$$p = \frac{(0.00012)(29{,}000{,}000)(0.375)}{3 \times 24} = 18.1 \text{ psi.} \quad \text{ok}$$

(b) The bending moment at the bottom of the cylinder due to wind load is given by

$$M = (30)(4)(31)^2/2 = 57{,}660 \text{ ft-lbs}$$

$$\text{stress } \sigma = \frac{M}{\pi r^2 t} = \frac{57{,}660 \times 12}{\pi(24)^2(0.375)} = 1020 \text{ psi.}$$

From Eq. (11-78),

$$\varepsilon = \frac{0.125}{24/0.375} = 0.002 \text{ inch/inch}$$

and from Fig. 11-17, the allowable compressive stress is determined as

$$B = 15{,}000 \text{ psi} > 1020 \text{ psi.} \quad \text{ok.}$$

Check Euler's buckling ($K = 2.0$) by calculating

$$A = 2\pi rt = 56.55 \text{ in.}^2 \quad \text{and} \quad I = \pi r^3 t = 16{,}286 \text{ in.}^4$$

$$\sigma_{cr} = \frac{\pi^2 \times 30{,}000{,}000 \times 16{,}286}{(2 \times 360)^2(56.55)}$$

$$\sigma_{cr} = 164{,}500 \text{ psi.}$$

Use 38,000 psi yield stress.

$$F.S. = 38{,}000/1020 = 37.$$

Hence, Euler's buckling does not control.

Example 11-7

The stack shown in Fig. 11-22 is subjected to an effective wind pressure of 40 psf. What is the required thickness? Use Eq. (11-79).

Solution

Maximum bending moment at bottom

$$M = (40)(8)(50)^2/2 = 400{,}000 \text{ ft-lbs}$$

Try $t = 1/4$ inch (minimum allowed by equation)

$$\sigma = Mc/I = \frac{M}{\pi tr^2} = \frac{400{,}000}{\pi \times 0.25 \times 48^2}$$

$$= 2660 \text{ psi.}$$

From Eq. (11-79),

$$2(L/r) = 25$$

Hence,

$$Y = 1.0$$

$$t/r = 0.0052.$$

Hence,

$$X = [1{,}000{,}000 \times 0.25/48][2 - (0.667)(100 \times 0.25/48)]$$

$$X = 8600 \text{ psi.}$$

Allowable stress $= XY = 8600 \times 1.0 = 8600 \text{ psi} > 2660 \text{ psi.}$

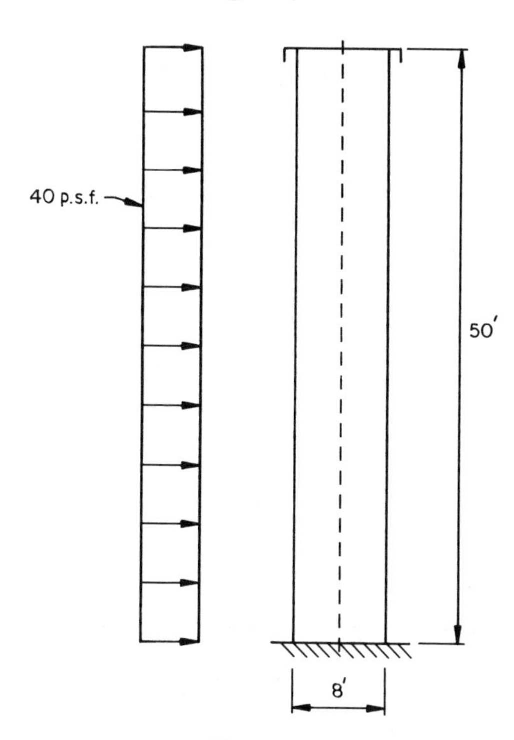

Figure 11-22.

Example 11-8

The flat-bottom gasoline tank shown in Fig. 11-23 is subjected to a snow load of 30 psf. Calculate the actual stress in the cylinder due to snow load and compare it to the allowable stress.

Solution

Actual stress at bottom of cylindrical shell

$$\sigma = \frac{p(\pi r^2)}{2\pi r t}$$

$$= \frac{(30/144)(240)}{2(0.25)} = 100 \text{ psi.}$$

From Fig. 11-20 with $t/r = 0.00104$,

$$\sigma = 1{,}800{,}000(0.00104) = 1870 \text{ psi} > 100 \text{ psi.}$$

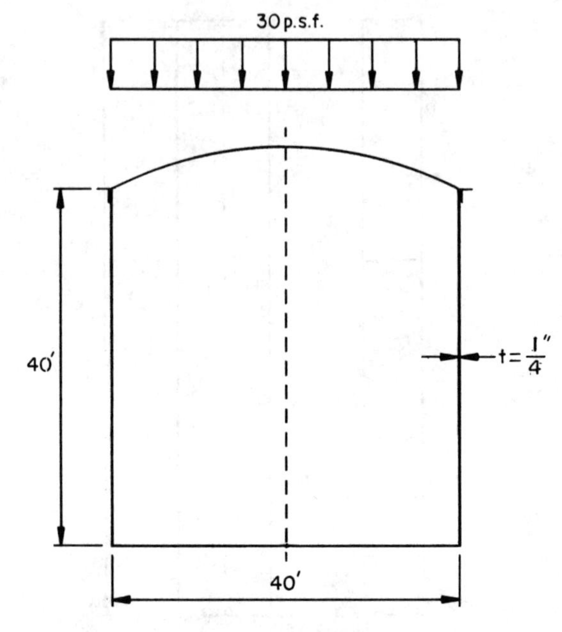

Figure 11-23.

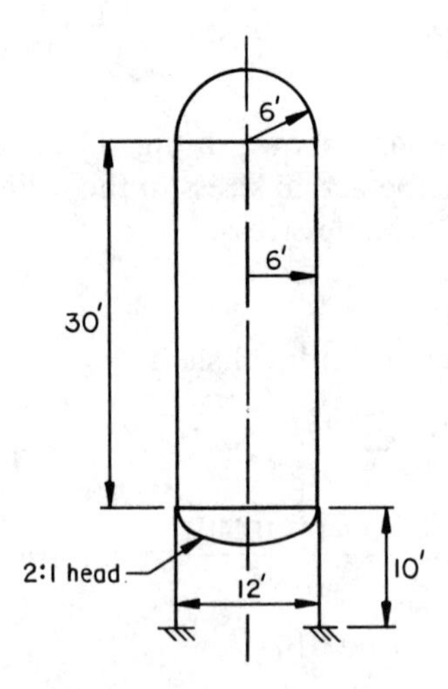

Figure P11-11.

Problems

11-11 Find the required thickness of the cylinder shown in Fig. P11-11 due to a full vacuum condition. Use Eq. (11-73) or (11-74).
11-12 Solve Problem 11-11 using Eq. (11-75).
11-13 What is the required thickness of the supporting cylinder shown in Fig. P11-11? The weight of the contents is 100,000 lbs.
11-14 What is the required thickness of the steel stack shown in Fig. P11-14 due to an effective wind load of 50 psf? Add the weight of the steel in the calculations. The thickness of the top 25 ft may be made different than the thickness of the bottom 25 ft. Weight of steel is 490 pcf. Use Eq. (11-79).
11-15 Calculate the required cylindrical shell thickness of the flat-bottom tank shown in Fig. 11-23 due to a snow load of 30 psf plus the dead weight of the roof and cylinder. Use Eq. (11-80).

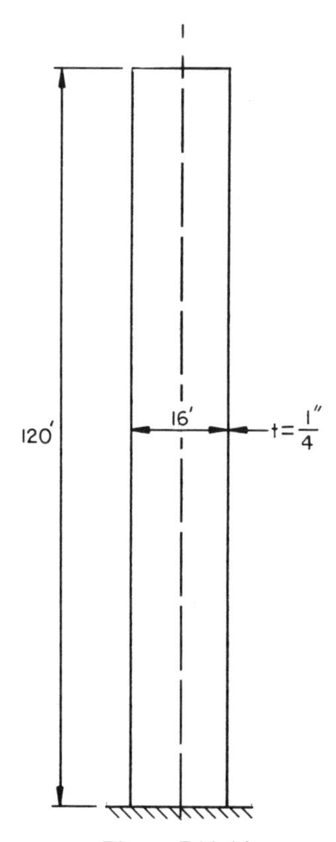

Figure P11-14.

12

Buckling of Shells of Revolution

12-1 Buckling of Spherical Shells

The buckling of spherical shells under external pressure has been investigated by numerous researchers. Von Karman (Von Karman and Tsien 1939) developed a solution that fits experimental data very closely. Taking a buckled section of a spherical shell (Fig. 12-1) he made the following assumptions:

1. The deflected shape is rotationally symmetric
2. The buckled length is small
3. The deflection of any element of the shell is parallel to axis of rotational symmetry
4. The effect of lateral contraction due to Poisson's ratio is neglected

Based on these assumptions and Fig. 12-1, the strain due to extension of the element is

$$\varepsilon = \frac{\dfrac{dr}{\cos\theta} - \dfrac{dr}{\cos\alpha}}{dr/\cos\alpha}$$

$$\varepsilon = \frac{\cos\alpha}{\cos\theta} - 1.$$

The strain energy of the extension of the element is

$$U_1 = \frac{ER^3}{2}(t/R)2\pi \int_0^{\beta} \left(\frac{\cos\alpha}{\cos\theta} - 1\right)^2 \sin\alpha\, d\alpha. \qquad (12\text{-}1)$$

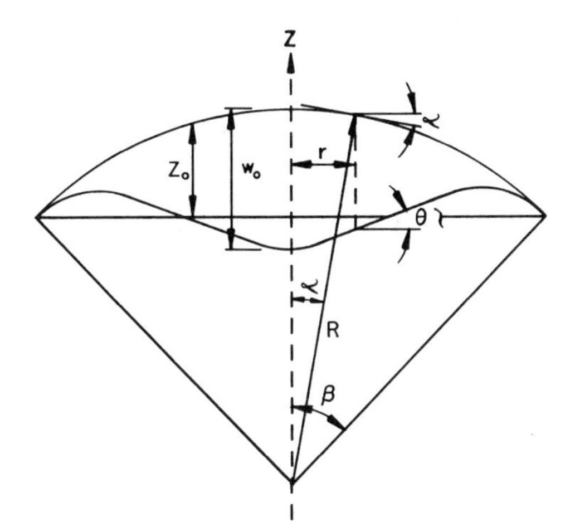

Figure 12-1.

The strain energy due to compression of the shell prior to buckling is

$$U_2 = \frac{-ER^3}{2}(t/R)2\pi \int_0^\beta \left(\frac{pR}{2Et}\right)^2 \alpha\, d\alpha. \tag{12-2}$$

Similarly, the strain energy due to bending is given by

$$U_3 = \frac{ER^3}{2}(t/R)^3 \frac{\pi}{6} \int_0^\beta \sin\alpha \left[\left(\frac{\cos\theta}{\cos\alpha}\frac{d\theta}{d\alpha} - 1\right)^2 + \left(\frac{\sin\theta}{\sin\alpha} - 1\right)^2\right] d\alpha. \tag{12-3}$$

The external work is equal to the applied pressure times the volume included between the deflected and original surfaces. This can be expressed as

$$W = R^3\pi \int_0^\beta \sin^2\alpha(\tan\theta - \tan\alpha)\cos\alpha\, d\alpha. \tag{12-4}$$

The total potential energy, Π, of the system is given by

$$\Pi = U_1 + U_2 + U_3 - W. \tag{12-5}$$

The terms in Eq. (12-5) can be simplified substantially by assuming β to be small. Then, expanding the sine and cosine expressions into a power series and neglecting terms higher than the third order, Eq. (12-5) becomes

$$\Pi = E(t/R) \int_0^\beta \left[\frac{1}{2}(\theta^2 - \alpha^2) - \frac{PR}{2Et} \right]^2 \alpha \, d\alpha$$
$$+ \frac{E(t/R)^3}{12} \int_0^\beta \left[\left(\frac{d\theta}{d\alpha} - 1 \right)^2 + \left(\frac{\theta}{\alpha} - 1 \right)^2 \right] \alpha \, d\alpha + p \int_0^\beta \alpha^2(\theta - \alpha) \, d\alpha. \tag{12-6}$$

Minimizing this equation with respect to θ results in an equation that is too cumbersome to solve. Accordingly, Von Karman assumed an expression for θ that satisfies the boundary condition $\theta = 0$ at $\alpha = 0$ and $\theta = \beta$ at $\alpha = \beta$, in the form

$$\theta = \{1 - K[1 - (\alpha/\beta)^2]\} \tag{12-7}$$

where K is a constant.

Substituting Eq. (12-7) into the expression

$$\frac{\partial \Pi}{\partial K} = 0$$

and utilizing the relationship $\sigma = pR/2t$ we get

$$\frac{\sigma}{E} = \frac{1}{70} \beta^2(28 - 21K + 4K^2) + \frac{4}{3}(t/R)^2 \frac{1}{\beta^2}. \tag{12-8}$$

The ordinate Z_o in Fig. (12-1) is calculated from

$$Z_o + \int_0^\beta (dz/dr) \, dr = 0$$

or

$$Z_o = R \int_0^\beta \tan\theta \cos\alpha \, d\alpha.$$

The ordinate of the shell at the center before deflection is given by $R(1 - \cos\beta)$. The deflection, w_o, at the center is then expressed by

$$w_o = R \int_0^\beta (\tan\alpha - \tan\theta)\cos\alpha \, d\alpha$$

or, assuming β to be small,

$$w_o = R \int_0^\beta (\alpha - \theta) \, d\alpha. \tag{12-9}$$

Substituting Eq. (12-9) into Eq. (12-7) gives

$$K = \frac{4w_o}{R\beta^2} \tag{12-10}$$

Substituting Eq. (12-10) into Eq. (12-8) results in

$$\frac{\sigma}{E} = \frac{2}{5}\beta^2 - \frac{6}{5}\frac{w_o}{R} + \left[\frac{32}{35}\frac{w_o^2}{R^2} + \frac{4}{3}\frac{t^2}{R^2}\right]\frac{1}{\beta^2}. \tag{12-11}$$

The minimum value of Eq. (12-11) is obtained by differentiating with respect to β^2 and equating to zero. This gives

$$\beta^2 = \sqrt{\frac{16}{7}(w_o/R)^2 + \frac{10}{3}(t/R)^2}$$

and Eq. (12-11) becomes

$$\frac{\sigma R}{Et} = \frac{4}{5}\left[\sqrt{\frac{16}{7}(w_o/t)^2 + \frac{10}{3}} - \frac{3}{2}(w_o/t)\right]. \tag{12-12}$$

A plot of this equation is shown in Fig. 12-2. The minimum buckling value is

$$\left.\begin{aligned} \sigma_{cr} &= 0.183Et/R \\ P_{cr} &= 0.366E\frac{t^2}{R^2} \end{aligned}\right]. \tag{12-13}$$

Example 12-1

A hemispherical shell with a radius of 72 inches is subjected to full vacuum. Determine the required thickness if a factor of safety (*F.S.*) of 4 is used for buckling. Let $E = 30{,}000$ ksi

Solution

From Eq. (12-13), with $F.S. = 4$, we get

$$\begin{aligned} t &= R\sqrt{\frac{4p}{0.366E}} \\ &= 72\sqrt{\frac{4 \times 14.7}{0.366 \times 30{,}000{,}000}} \\ &= 0.17 \text{ inch.} \end{aligned}$$

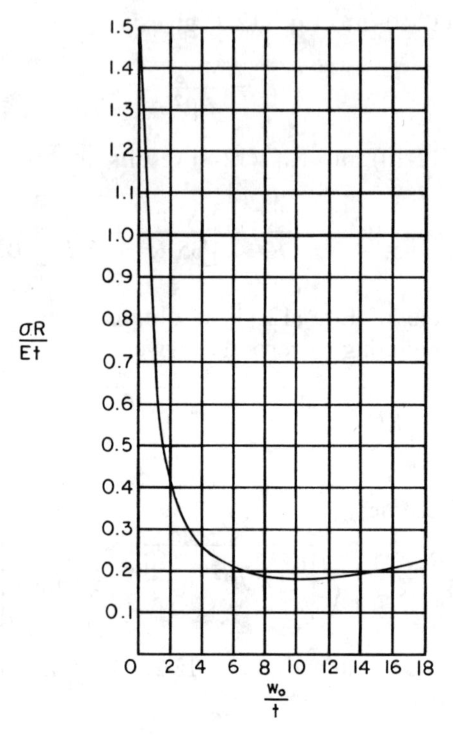

Figure 12-2.

Problems

12-1 Find the required thickness of the hemispherical shell in Problem 11-4. Use a factor of safety of 3.0.

12-2 A spherical aluminum diving chamber is under 1000 ft of water. Determine the required thickness. Let $R = 36$ inches, factor of safety $(F.S.) = 1.5$, and $E = 10{,}000$ ksi.

12-2 Buckling of Stiffened Spherical Shells

Equation (12-13) can be written in terms of buckling pressure as

$$p_{cr} = 0.366E(t/R)^2 \tag{12-14}$$

or

$$t = R\sqrt{\frac{p_{cr}}{0.366E}}. \tag{12-15}$$

Equation (12-15) shows that for a given external pressure and modulus of elasticity, the required thickness is proportional to the radius of the spherical section. As the radius gets larger, so does the thickness. One procedure for reducing the membrane thickness is by utilizing stiffened shells. Thus, for the shell shown in Fig. 12-13 with closely spaced stiffeners, the buckling pressure (Buchert 1964) is obtained by modifying Eq. (12-14) as

$$p_{cr} = 0.366E\left(\frac{t_m}{R}\right)^2\left(\frac{t_b}{t_m}\right)^{3/2} \tag{12-16}$$

in which

t_m = effective membrane thickness
$= t + (A/d)$
t_b = effective bending thickness
$= \left(\frac{12I}{d}\right)^{1/3}$

where

A = area of stiffening ring;
d = spacing between stiffeners;
I = moment of inertia of ring;
t = thickness of shell.

For large spherical structures such as large tank roofs and stadium domes, the stiffener spacing in Fig. 12-3 increases significantly. In this case, the composit buckling strength of shell and stiffeners (Buchert 1966) is expressed by

$$p_{cr} = 0.366E\left(\frac{t}{R}\right)^2\left(1 + \frac{12I}{dt^3}\right)^{1/2}\left(1 + \frac{A}{dt}\right)^{1/2} \tag{12-17}$$

Local buckling of the shell between the stiffeners must also be considered for large-diameter shells. One such equation is given by

$$p_{cr} = 7.42\,\frac{Et^3}{Rd^2}. \tag{12-18}$$

It should be noted that large edge rotations and deflections can reduce the value obtained by Eq. (12-17) significantly.

Equation (12-17) is also based on the assumption that the spacing of the stiffeners is the same in the circumferential and meridional directions. Other equations can be developed (Buchert 1966) for unequal spacing of stiffeners.

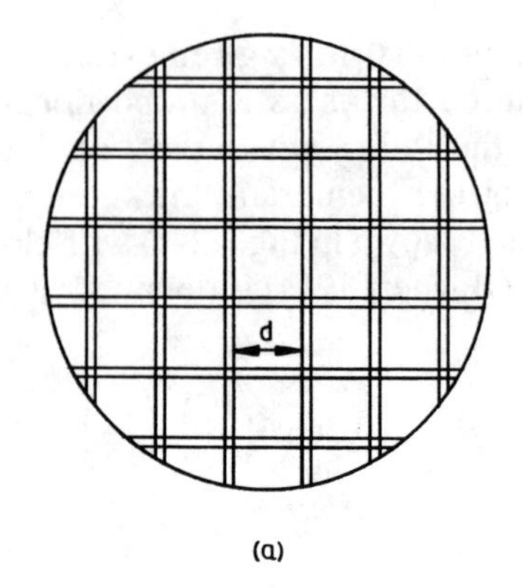

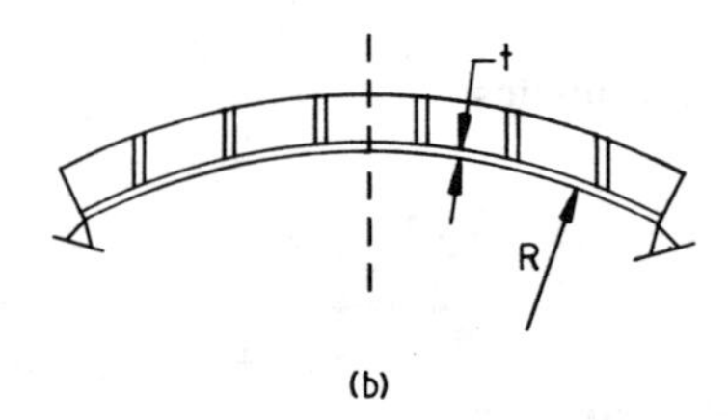

Figure 12-3.

Problems

12-3 Determine the allowable external pressure on the stiffened head in Fig. 12-3. Assume $R = 13$ ft, $t = 3/8$ inch, $d = 8$ inches, size of stiffeners is 4 inch $\times$ 3/8 inch, $E = 29{,}000$ ksi, and a factor of safety ($F.S.$) of 10.

12-4 Use Eqs. (12-17) and (12-18) to determine the required thickness of dome and the size and spacing of stiffeners. Let $R = 200$ ft, $E = 29{,}000$ ksi, $p = 85$ psf, and factor of safety $= 10$.

12-3 Buckling of Conical Shells

The derivation of the equations for the buckling of conical shells is fairly complicated and beyond the scope of this book. The derivation (Niordson 1947) for the buckling pressure of the cone shown in Fig. 12-4 consists of obtaining expressions for the work done by the applied pressure, membrane forces, stretching of the middle surface, and bending of the cone. The total work is then minimized to obtain a critical pressure expression in the form

$$p_{cr} = \frac{Et}{\rho_o a_o^2} f \frac{t^2 a_0^4}{(1 - \mu^2)\rho_0^2} \tag{12-19}$$

where

E = modulus of elasticity;
t = thickness of cone;
$a_o = \lambda\rho(1 - \beta/2)$;
$\lambda = \pi/\ell$;
$\rho_o = \rho(1 - \beta/2)$;
$\beta = \frac{\ell}{\rho} \tan \alpha$.

Equation (12-19) is very cumbersome to use due to the iterative process required. Seide showed (Seide 1962) that Eq. (12-19) for the buckling of a conical shell is similar to the equation for the buckling of a cylindrical shell having a length equal to the slant length of the cone and a radius equal to the average radius of the cone. He also showed that the buckling of a cone is affected by the function $f(1 - R_1/R_2)$ and is expressed as

$$p_{cr} = \bar{p}f(1 - R_1/R_2) \qquad (12\text{-}20)$$

where

$\bar{p}$ = pressure of equivalent cylinder as defined above;
f = cone function as defined in Fig. 12-5.

By various substitutions (Jawad 1980), it can be shown that Eq. (12-20) can be transferred to the form of Eq. (11-76) as

$$\begin{aligned} p_{cr} &= \frac{0.92E(t_e/R_2)^{2.5}}{L_e/R_2} \text{ for metallic cones} \\ p_{cr} &= \frac{0.87E(_e/R_2)^{2.5}}{L_e/R_2} \text{ for concrete cones} \end{aligned} \qquad (12\text{-}21)$$

where

t_e = effective thickness of cone

$$= t \cos \alpha \qquad (12\text{-}22)$$

t = thickness of cone

and

L_e = effective length of cone

$$= \frac{L}{2}(1 + R_1/R_2). \qquad (12\text{-}23)$$

Thus, conical shells subjected to external pressure may be analyzed as cylindrical shells with an effective thickness and length as defined by Eqs. (12-22) and (12-23).

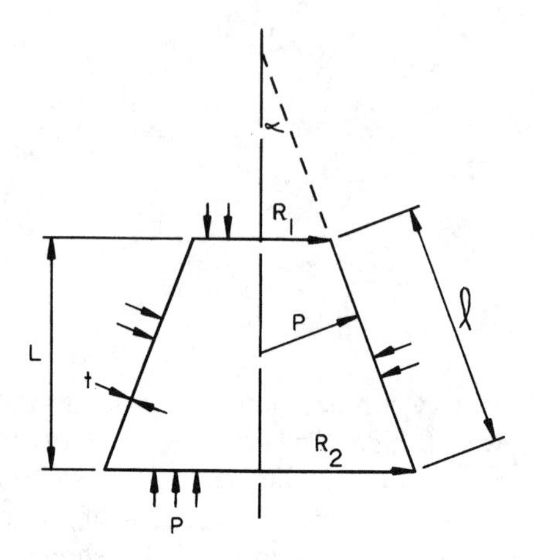

Figure 12-4.

Example 12-2

Determine the allowable external pressure, at room temperature, for the steel cone shown in Fig. 12-6 by using (a) Eq. (12-21) and (b) Eq. (11-73). Let $F.S. = 3.0$, and $E = 30{,}000$ ksi.

Solution

(a)

From Fig. 12-6, $\alpha = 17.35^o$

From Eq. (12-22), $t_e = (3/32)\,(\cos 17.35) = 0.0895$ inch.

From Eq. (12-23), $L_e = \dfrac{32}{2}\left(1 + \dfrac{20}{30}\right) = 26.67$ inch.

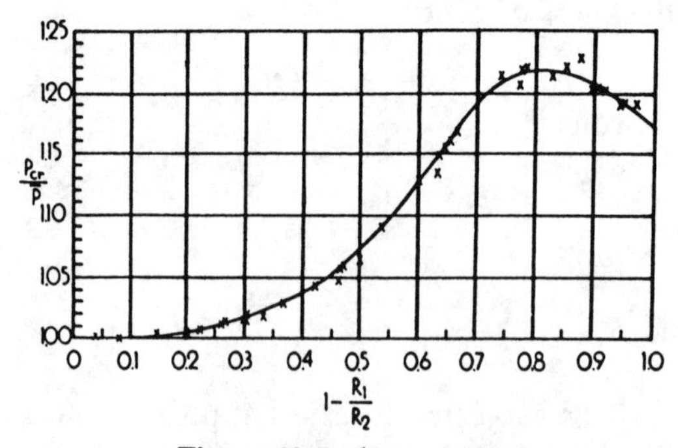

Figure 12-5. (Jawad 1980).

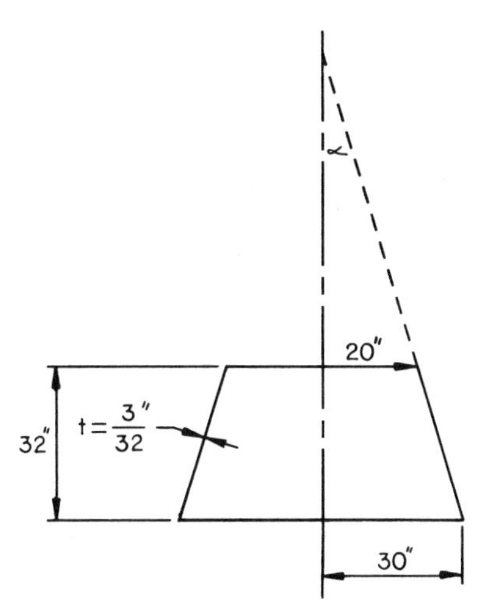

Figure 12-6.

Hence, from Eq. (12-21),

$$p = \frac{0.92 \times 30{,}000{,}000 \left(\dfrac{0.0895}{30}\right)^{2.5}}{(26.67/30)(3)}$$

$$= 5.03 \text{ psi.}$$

(b)

$$L_e/D_2 = 26.67/60 = 0.44$$
$$D_2/t_e = 60/0.0895 = 670$$

From Fig. 11-16, $\varepsilon = 0.00018$ inch/inch
From Fig. 11-17, $B = 2500$ psi

and from Eq. (11-73),

$$p = \frac{(2 \times 2500)(0.0895)}{(30)(3)}$$

$$= 4.97 \text{ psi.}$$

It should be remembered that there is an advantage in using method (b) in that the actual stress–strain diagram of the material is used. This takes into account the plastic region if σ falls in that region.

Problems

12-5 Determine the required thickness of the cone shown in Fig. P9-9 due to full vacuum. Use (a) Eq. (12-21) and (b) Eq. (11-73). Let E = 16,000 ksi and factor of safety ($F.S.$) = 2.50.
12-6 Determine the required thickness of the cone shown in Fig. 12-6 due to full vacuum. Let E = 30,000 ksi and use $F.S.$ = 3.0.

12-4 Design Considerations

Spherical Shells

Extensive tests on spherical shells (Kollar and Dulacska 1984) have shown that the buckling Eq. (12-13) must be reduced further to

$$\sigma_{cr} = 0.125Et/R \tag{12-24}$$

or, in terms of critical pressure,

$$p_{cr} = \frac{E}{4(R/t)^2}. \tag{12-25}$$

ASME uses Eq. (12-25) with a factor of safety of 4.0 to obtain the permissible external pressure on a spherical shell. Hence,

$$p = \frac{0.0625E}{(R/t)^2}. \tag{12-26}$$

Ellipsoidal Shells

An approximate equation for the buckling of ellipsoidal shells is similar to Eq. (12-13) for spherical shells and is given by

$$p_{cr} = 0.366E\frac{t^2}{r_1 r_2} \tag{12-27}$$

where r_1 and r_2 are defined in Fig. 6-8.

Shallow Heads

For shallow ellipsoidal shells under external pressure, the region near the knuckle area is in tension. The remaining surface can be approximated by a spherical shell and Eq. (12-27) is simplified by letting $r_1 = r_2$.

Conical Shells

Conical shells subjected to external pressure are designed as cylindrical shells with an effective thickness and length as given by Eqs. (12-22) and (12-23).

Hyperbolic Paraboloid Sheets

The general buckling of hyperbolic paraboloid sheets of revolution with stiffening rings at top and bottom can be expressed (Kollar and Dulacska 1984) by an equation of the form

$$p_{cr} = 0.070E\,\frac{t^2}{r_o^2} \tag{12-28}$$

where r_o is the smallest radius (Fig. 7-17) at the throat.

Other forms of buckling (Kollar and Dulacska 1984) such as free-edge, local, and axisymmetric buckling must also be investigated. Usually, these forms of buckling are less severe than that given by Eq. (12-28) except for specific conditions.

Various Shapes

Further theoretical and experimental research is still needed to establish buckling strength of various configurations and shapes. This includes eccentric cones, torispherical shells, toriconical shells, and stiffened shells. Additionally, the effect of out-of-roundness and edge conditions on the buckling strength of shells needs further investigation.

13

Roof Structures

13-1 Introduction

Many roof structures (Fig. 13-1) are designed in accordance with the membrane and bending theories of shells. Roofs such as hyperbolic paraboloids (Fig. 13-2a) and elliptic paraboloids (Fig. 13-2b) are analyzed in accordance with the membrane theory of shells of revolution developed in Chapter 6. Barrel roofs, (Fig. 13-3a) are treated as cylindrical segments in accordance with the bending theory discussed in Chapter 8 with discontinuity edge forces applied along the length of the cylindrical segments. Roofs in the shape of folded plates (Fig. 13-3b) are considered as beams connected at their ridges. All of these roofs are generally assumed to resist uniformly distributed dead and live loads.

In this chapter a brief discussion is presented for the analysis of each of the roof shapes mentioned above. A more complete treatment of the design and detail of these roof structures is discussed in the references cited in this chapter.

We begin the derivation of the equations needed in the analysis of hyperbolic and elliptic paraboloids by expressing the membrane forces in shells of double curvature (Fig. 6-4), in terms of rectangular coordinates (Fig. 13-4) rather than polar coordinates. Also, it will be assumed that the only significant forces acting on roof shells are those due to dead and live loads. These loads are assumed to act in the z-direction only for shallow shells. Projecting the forces of the infinitesimal element $ABCD$ in the plane $EFGH$ (Fig. 13-4), and summing forces in the x-, y-, and z-axes, respectively, gives

$$\frac{\partial N_x}{\partial x}\,dx\,\frac{dy}{\cos\alpha}\cos\phi + \frac{\partial N_{yx}}{\partial y}\,dy\,\frac{dx}{\cos\phi}\cos\phi = 0 \qquad (13\text{-}1)$$

Figure 13-1. Church of the Priority, St. Louis, MO. (Courtesy of Anthony Coombs.)

$$\frac{\partial N_y}{\partial y} dy \frac{dx}{\cos \phi} \cos \alpha + \frac{\partial N_{xy}}{\partial x} dx \frac{dy}{\cos \alpha} \cos \alpha = 0 \tag{13-2}$$

and

$$\frac{\partial N_x}{\partial x} dx \frac{dy}{\cos \alpha} \sin \phi + \frac{\partial N_{xy}}{\partial x} dx \frac{dy}{\cos \alpha} \sin \alpha + \frac{\partial N_y}{\partial y} dy \frac{dx}{\cos \phi} \sin \theta + \frac{\partial N_{xy}}{\partial y} dy \frac{dx}{\cos \phi} \sin \phi - p_z \frac{dx}{\cos \phi} \frac{dy}{\cos \alpha} \sin \beta = 0. \tag{13-3}$$

If we define

$$\left.\begin{aligned} \overline{N}_x &= N_x \frac{\cos \phi}{\cos \alpha} \\ \overline{N}_y &= N_y \frac{\cos \alpha}{\cos \phi} \\ \overline{p}_z &= p_z \frac{\sin \beta}{\cos \phi \cos \alpha} \end{aligned}\right] \tag{13-4}$$

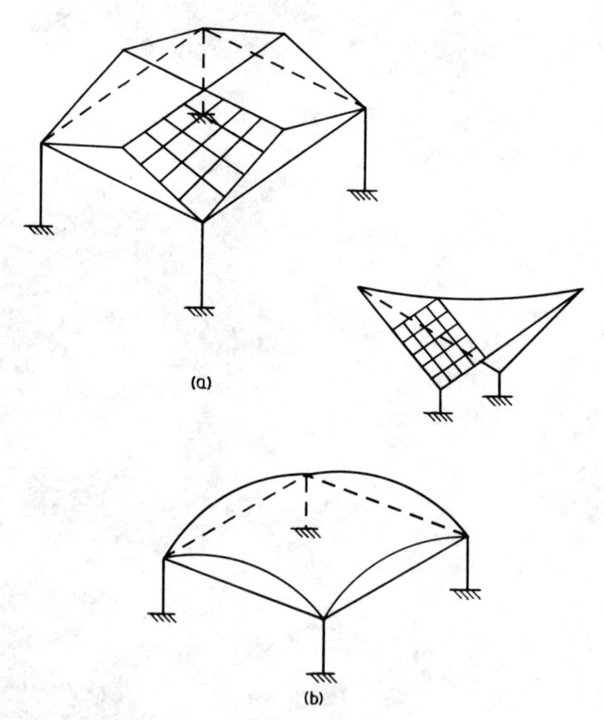

Figure 13-2.

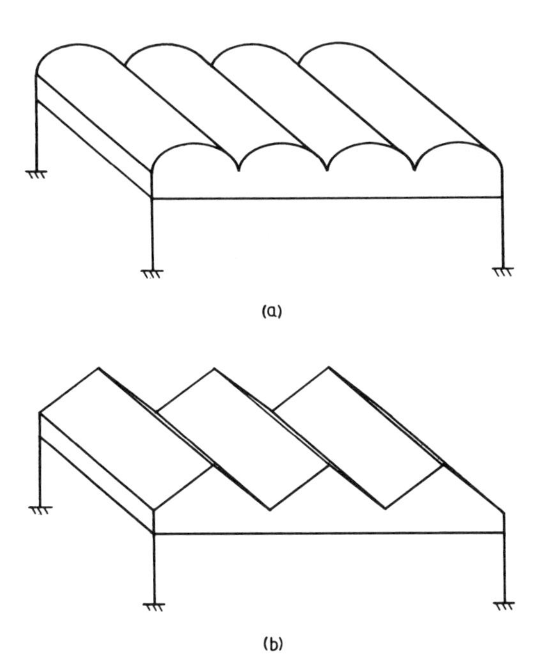

Figure 13-3.

and substitute these three expressions into Eqs. (13-1) through (13-3) we get

$$\frac{\partial \overline{N}_x}{\partial x} + \frac{\partial N_{xy}}{\partial y} = 0 \tag{13-5}$$

$$\frac{\partial \overline{N}_y}{\partial y} + \frac{\partial N_{xy}}{\partial x} = 0 \tag{13-6}$$

$$\frac{\partial \overline{N}_x}{\partial x} \tan \phi + \frac{\partial \overline{N}_y}{\partial y} \tan \alpha + \frac{\partial N_{xy}}{\partial x} \tan \alpha + \frac{\partial N_{xy}}{\partial y} \tan \phi = \overline{p}_z. \tag{13-7}$$

Equation (13-7) can further be simplified by substituting the quantities $\tan \phi = \partial z/\partial x$, $\tan \alpha = \partial z/\partial y$, and Eqs. (13-5) and (13-6) into it. This yields

$$\overline{N}_x \frac{\partial^2 z}{\partial x^2} + 2N_{xy} \frac{\partial^2 z}{\partial x\, \partial y} + \overline{N}_y \frac{\partial^2 z}{\partial y^2} = \overline{p}_z. \tag{13-8}$$

Equations (13-5), (13-6), and (13-8) are the three basic equations for analyzing structures such as hyperbolic paraboloids and elliptic paraboloids.

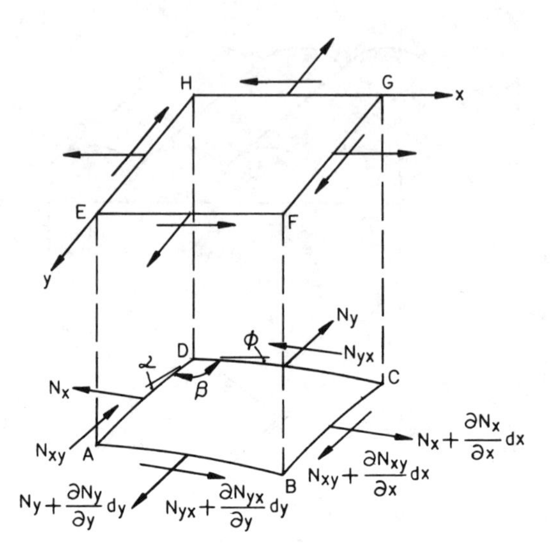

Figure 13-4.

Their solution is obtained by defining a stress function, F, of the form

$$N_{xy} = -\frac{\partial^2 F}{\partial x\, \partial y}. \tag{13-9}$$

Substituting this expression into Eqs. (13-5), (13-6), and (13-8) results in

$$\overline{N}_x = \frac{\partial^2 F}{\partial y^2} \tag{13-10}$$

$$\overline{N}_y = \frac{\partial^2 F}{\partial x^2} \tag{13-11}$$

$$\frac{\partial^2 F}{\partial y^2}\frac{\partial^2 z}{\partial x^2} - 2\frac{\partial^2 F}{\partial x\, \partial y}\frac{\partial^2 z}{\partial x\, \partial y} + \frac{\partial^2 F}{\partial x^2}\frac{\partial^2 z}{\partial y^2} = \overline{p}_z. \tag{13-12}$$

The solution of Eqs. (13-10) through (13-12) depends on the geometry of the specific shell being analyzed.

Problems

13-1 Derive Eqs. (13-1), (13-2), and (13-3).
13-2 Derive Eqs. (13-5), (13-6), and (13-7).
13-3 Derive Eq. (13-8).

13-2 Hyperbolic Paraboloid Shells

The equation for a hyperbolic paraboloid, (Fig. 13-5) is expressed as

$$z = \frac{\bar{y}^2}{2h_2} - \frac{\bar{x}^2}{2h_1} \tag{13-13}$$

where h_1 and h_2 are constants. At $\bar{x} = 0$, or alternatively at $\bar{y} = 0$, this equation reduces to a parabola. At z equal to a constant larger than zero, Eq. (13-13) reduces to a hyperbola. At $z = 0$, Eq. (13-13) reduces to an expression (Kelkar and Sewell 1987) that defines the relationship between two straight lines. These straight lines can be generated from a set of new axes, x and y, that are oriented (Fig. 13-6) with respect to the $\bar{x}-\bar{y}$ plane by angle

$$\tan \psi = \sqrt{\frac{h_2}{h_1}}. \tag{13-14}$$

For a rectangular shell, we let $h_1 = h_2 = h$, and the equation for the surface (Billington 1982) becomes

$$z = \frac{h}{ab} xy. \tag{13-15}$$

The load $\bar{p}_z$ in Eq. (13-12) is the projected load over the surface and is normally taken as a constant. Substituting Eq. (13-15) into Eq. (13-12) and letting $\bar{p}_z = p$ gives

$$\frac{\partial^2 F}{\partial x \, \partial y} = \frac{ab}{2h} p. \tag{13-16}$$

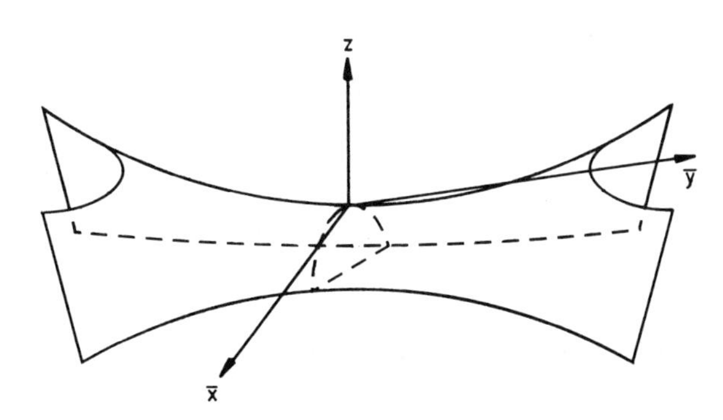

Figure 13-5.

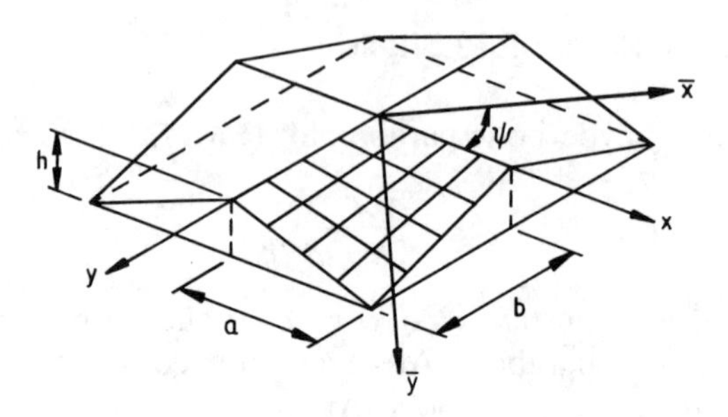

Figure 13-6.

From Eq. (13-9) we get the shear forces as

$$N_{xy} = -\frac{ab}{2h}p \tag{13-17}$$

Integrating Eq. (13-16) gives

$$F = \frac{ab}{2h}pxy + C_1(x) + C_2(y) \tag{13-18}$$

and from Eqs. (13-10) and (13-11) we get forces $\overline{N}_x$ and $\overline{N}_y$ as

$$\overline{N}_x = \frac{\partial^2 C(y)}{\partial y^2} \tag{13-19}$$

$$\overline{N}_y = \frac{\partial^2 C(x)}{\partial x^2}. \tag{13-20}$$

Most roof structures have either free edges or edges supported by beams that cannot resist any forces in their weak axis. For these structures, it is customary to assume $N_x = N_y = 0$ throughout the shell by letting C_1 and $C_2 = 0$. This assumes that the only stress in the structure is shear and is given by Eq. (13-17) as

$$\begin{aligned} N_{xy} &= \frac{-ab}{2h}p \\ &= \frac{-ab}{2h}p_z\frac{\sin\beta}{\cos\alpha\cos\phi}. \end{aligned} \tag{13-21}$$

The total vertical force on the structure due to dead and live loads is transferred to the supports through shear stress at the edges given by Eq. (13-21). The allowable shear stress is based on the critical shearing stress of a hyperbolic paraboloid which is approximately the same (Kollar and Dulacska 1984) as the critical shearing stress of a plate. The critical shearing stress of a plate is approximated (Timoshenko and Gere 1961) by the equation

$$\tau_{cr} = \frac{\pi^2 E}{12(1 - \mu^2)(b/t)^2}[5.35 + 4(b/a)^2] \qquad (13\text{-}22)$$

where b is the small length of the shell.

When the hyperbolic paraboloid is in the shape of a deep arch (Fig. 13-1), the loads are transferred in one direction only. In this case, the forces in the arches are determined by any Structural Analysis method.

Problems

13-4 Derive Eq. (13-15).

13-5 What is the critical buckling shearing stress in a reinforced concrete hyperbolic paraboloid roof (Fig. 13-6) if $a = 25$ ft, $b = 20$ ft, $h = 6$ ft, $t = 4$ inches, $E = 3{,}100$ ksi, and $\mu = 0.15$?

13-3 Elliptic Paraboloid Shells

The equation for an elliptic paraboloid (Fig. 13-7) is given by

$$z = \frac{x^2}{2h_1} + \frac{y^2}{2h_2}. \qquad (13\text{-}23)$$

When x = constant, or alternatively when y = constant, Eq. (13-23) becomes a parabola. When z is a constant, the equation becomes an ellipse. Substituting Eq. (13-23) into Eq. (13-12) and letting $\bar{p}_z = p$ gives

$$\frac{\partial^2 F}{\partial y^2}\frac{1}{h_1} + \frac{\partial^2 F}{\partial x^2}\frac{1}{h_2} = p. \qquad (13\text{-}24)$$

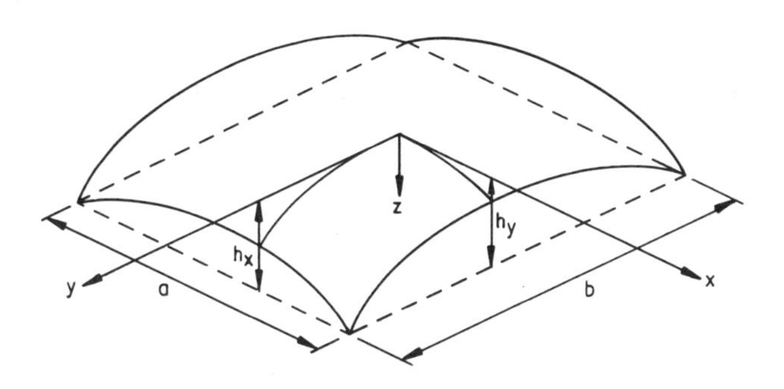

Figure 13-7.

For many elliptic paraboloid roofs, the edges are either free or supported by beams that cannot support any loads in their weak axes. Thus, N_x and N_y are zero at the boundary conditions. Solution of Eq. (13-24) is achieved by expressing F in a Fourier Series that vanishes at the boundary conditions and is of the form

$$F(x, y) = \sum_{n=1}^{\infty} f_n(x) \cos \frac{n\pi y}{b} \qquad n = 1, 3, 5, \ldots \tag{13-25}$$

The load p can also be expressed as

$$p = \sum_{n=1}^{\infty} p_n(x) \cos \frac{n\pi y}{b} \tag{13-26}$$

where

$$p_n(x) = \frac{4p}{b} \int_0^{b/2} \cos \frac{n\pi y}{b} \, dy$$

and Eq. (13-26) becomes

$$p = \sum_{n=1}^{\infty} -(-1)^{(n+1)/2} \frac{4p}{\pi n} \cos \frac{n\pi y}{b}. \tag{13-27}$$

Substituting Eqs. (13-25) and (13-27) into Eq. (13-24) yields

$$\frac{d^2 f_n}{dx^2} - k_n^2 f_n = (-1)^{(n+1)/2} \frac{4p}{n\pi} h_2 \tag{13-28}$$

where

$$k_n^2 = \frac{n^2\pi^2}{b^2} \frac{h_2}{h_1}.$$

Solving this equation for the homogeneous and particular solutions gives

$$f_n(x) = A \sinh kx + B \cosh kx + (-1)^{(n+1)/2} \frac{4pb^2}{n^3\pi^3} h_1 \tag{13-29}$$

Due to symmetry of loads, $A = 0$. Also, at $x = a/2$, $F = 0$. Thus, Eq. (13-29) can be solved for B as

$$B = (-1)^{(n+1)/2} \frac{4pb^2 h_1}{n^3\pi^3 \cosh k_n a/2}.$$

With B known, Eqs. (13-29) and (13-25) are solved and the values of N_x, N_y, and N_{xy} are obtained from Eqs. (13-4) and (13-9).

The allowable buckling load, p_{cr}, for elliptic paraboloid shells can be approximated (Kollar and Dulacska 1984) by

$$p_{cr} = 0.366 \frac{Et^2}{2} \left(\frac{1}{R_1^2} + \frac{1}{R_2^2} \right). \tag{13-30}$$

Problems

13-6 Derive Eq. (13-24).

13-7 Determine the values of N_x, N_y, and N_{xy} from Eqs. (13-4) and (13-9) using the expression given by Eqs. (13-25) and (13-29).

13-4 Folded Plates

Folded plate roofs (Fig. 13-8a and b) are commonly used in buildings where intermediate columns are undesirable. Also, bottom hoppers of rectangular storage tanks (Fig. 13-8c) can be analyzed using the folded plate theory. Methods of analyzing various folded structures vary greatly depending on the needed degree of accuracy. Usually a preliminary analysis is performed

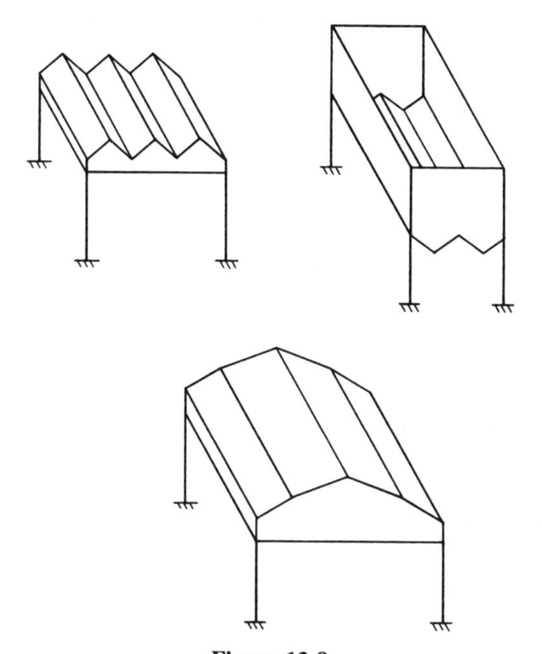

Figure 13-8.

first that is based on equilibrium considerations. Then, a more elaborate analysis is conducted that takes into consideration the rotation and deflection compatibility of the various components. The following assumptions are normally made when analyzing folded plates

1. The plate material is elastic, isotropic, and homogeneous.
2. The cross section of the plate is constant throughout its span.
3. The supporting end members have infinite stiffness in their own planes and are flexible normal to their own planes.
4. The plates carry loads transversely by bending normal to their planes.
5. The distribution of all loads in the longitudinal direction is the same on all plates.
6. The plates carry loads longitudinally by bending within their planes.
7. Longitudinal stresses vary linearly over the depth of each plate.
8. The torsional stiffness of the plates normal to their own planes is zero.
9. Displacements due to forces other than bending moments are neglected.

The preliminary analysis of folded plates proceeds as follows:

a. Calculate the transverse bending moment for each panel based on an assumed dead weight and applied loads. This is accomplished by assuming the edges of the panels (Fig. 13-9a) to be continuously supported in the transverse direction (Fig. 13-9b). The moments are determined by any Structural Analysis method. The reactions in the transverse supports are then determined.

b. Since the transverse supports are fictitious, the reactions must be eliminated by calculating their equivalent in-plane forces in the adjacent panels as illustrated in Fig. 13-9c.

Steps a and b can be combined into one step by using Matrix Analysis of a frame, taking into consideration axial forces.

c. Calculate the longitudinal bending stresses in each panel due to the in-plane forces from step b. The panels are assumed simply supported at the ends with maximum bending moment in the middle of the span.

d. The longitudinal bending stresses calculated in step c for adjacent panels will be different at the edge intersection because the loads in each panel are different. Thus, it is necessary to apply shearing forces (Fig. 13-9d) at the edges of connecting panels in order to have the longitudinal stresses at a given edge in equilibrium. The shearing forces are assumed parabolic in distribution and their magnitude is determined by solving a set of simultaneous equations.

e. The total stress in each panel is determined by combining the stresses determined in steps a, c, and d.

The method discussed in steps a through e satisfies the equilbrium equations across the edges. Compatibility of the deflections of adjacent panels at an edge due to in-plane loads was not considered. Neither was the

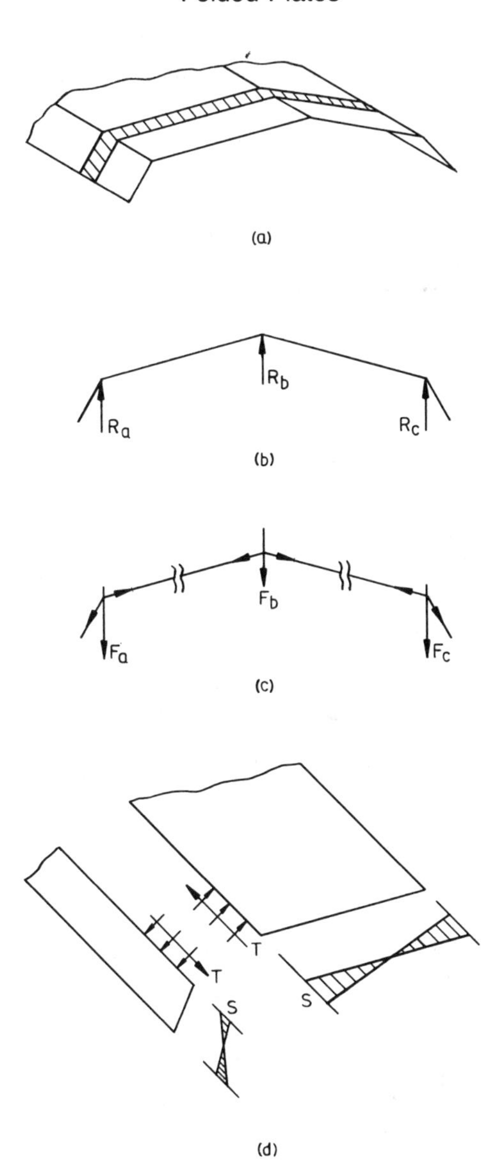

Figure 13-9.

transverse rotation of adjacent panels at the edges. These two conditions will have to be satisfied in order to complete the analysis of a folded plate.

The stresses obtained in steps a through e must be adjusted to take into consideration the deflection and rotation compatibility at each edge. We begin the derivation of the deflection expression by noticing that the shape of the moment diagram due to in-plane loads in the panels is parabolic as shown in Fig. 13-10a and is given by

$$M_x = PLx/2 - Px^2/2 \tag{13-31}$$

The distribution of the shear forces T_x must be assumed triangular in shape (Fig. 13-10) in order for the resulting parabolic moment to be compatible with the load bending moment. Thus, the shear force at any point x in Fig. 13-10b is expressed as

$$T_x = T_o(1 - 2x/L)$$

and the moment diagram is

$$\begin{aligned} M_x &= \int (d/2)(T_x)(L/2 - x)\,dx \\ &= (d/2)(T_o)(Lx/2 - x^2 + 2x^3/3L) \end{aligned}$$

or letting $T = (T_o)(L/2)(1/2)$

$$M_x = (2)(d)(T)(x/2 - x^2/L + 2x^3/3L^2). \tag{13-32}$$

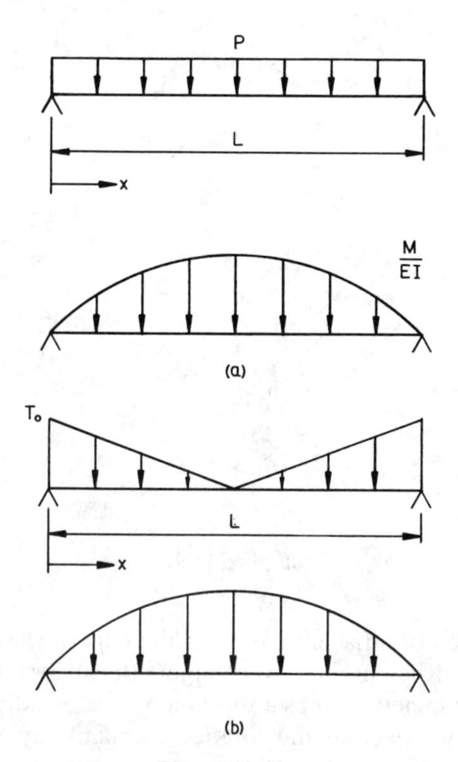

Figure 13-10.

This parabolic equation has a similar shape as Eq. (13-31). Hence, the expression for deflection of each panel can be calculated from the combination of Eqs. (13-31) and (13-32). In order to simplify the deflection calculations, the bending moments are expressed in terms of stresses. Referring to Fig. 13-11,

$$\frac{1}{R} = \frac{-\varepsilon_t + \varepsilon_b}{d}.$$

Also, from Eq. (1-6)

$$\chi = 1/R = \frac{d^2y}{dx^2}.$$

Hence, the equation

$$\frac{d^2y}{dx^2} = M/EI$$

becomes

$$M/EI = \frac{-\varepsilon_t + \varepsilon_b}{d}.$$

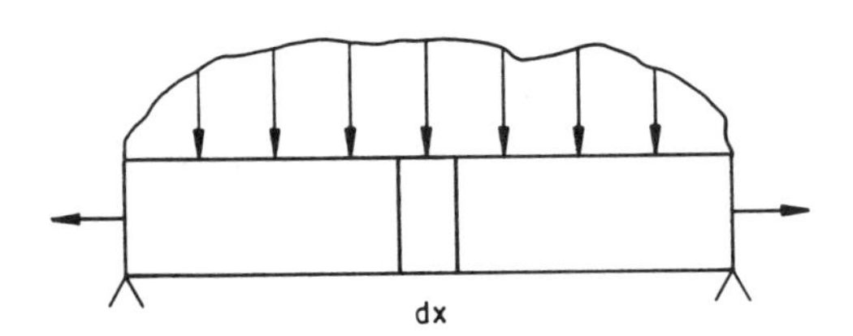

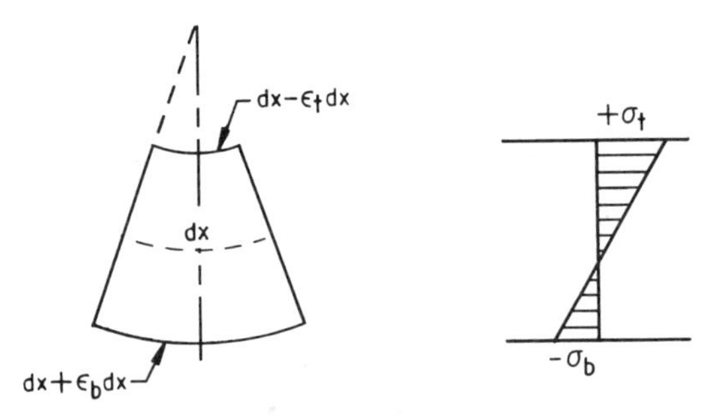

Figure 13-11.

Defining

$$\varepsilon = \sigma/E$$

gives

$$\frac{-\sigma_t + \sigma}{dE} = M/EI.$$

The deflection at the centerline of the panels can be determined from Structural Analysis methods by using Fig. 13-12. This gives

$$w = \frac{1}{9.6}(L^2)\left(\frac{-\sigma_t + \sigma_b}{dE}\right). \tag{13-33}$$

In this equation, the tensile stress is entered as a positive and the compressive stress as a negative quantity.

From the calculated deflections at each edge, the rotation of each panel is determined from geometry. The change in rotation difference of two adjacent panels at a given edge due to applied external forces is designated by θ_F.

The rotations θ_F at each of the internal edges are due to the fictitious hinges inserted in the structure in order to find the in-plane loads in the panels. These rotations must be eliminated as the edges in the actual structure are rigid. One method of eliminating the rotations θ_F is to apply correction moments at each of the inner edges as shown in Fig. 13-13. Accordingly, the procedure needed to satisfy the rotation compatibility of the panels is as follows

1. From the stresses obtained in step e above, calculate the deflection of each panel and then calculate the net rotation θ_F at each of the inner edges.

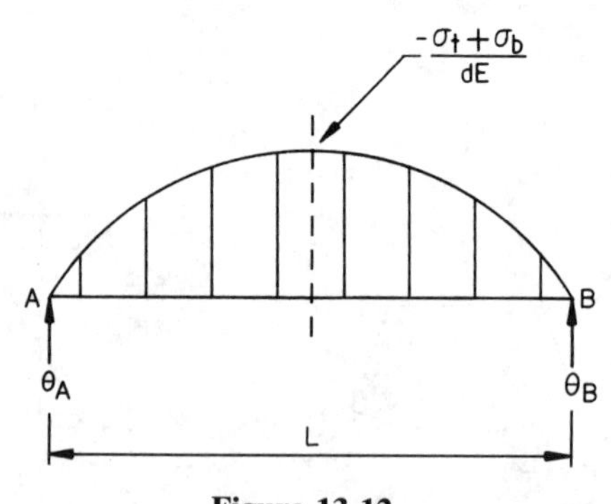

Figure 13-12.

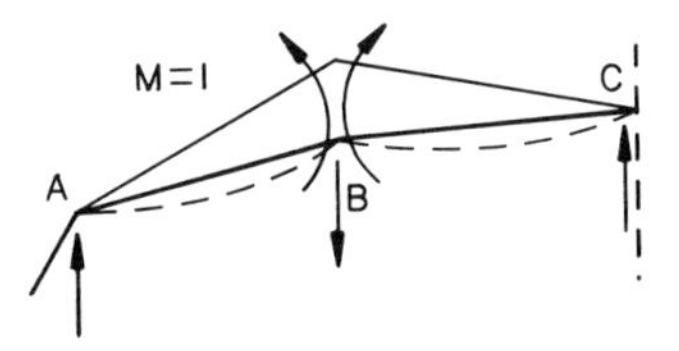

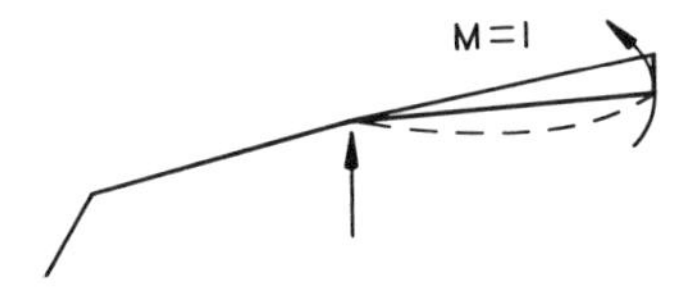

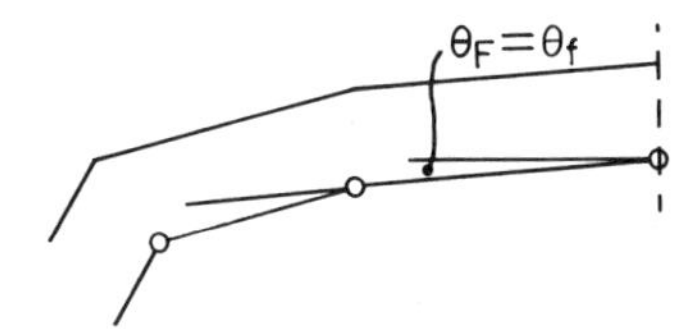

Figure 13-13.

2. Apply a correction moment, Mx, at each of the internal edges and calculate the rotation, θ'_{Mx}, at all the inner edges and the corresponding reactions. The shape of these reaction forces along the length of the panel must be the same as the shape of the rotation. Since the shape of the rotation due to Mx must be the same as the rotation due to applied loads, it follows that a parabolic shape is to be selected. However, an approximate shape that is easier to use is of the form

$$P_x = P_o \sin \frac{\pi x}{L}.$$

The expression for the moment is of the form

$$Mx = M_o \sin \frac{\pi x}{L}$$

and from the equations

$$\frac{d^2y}{dx^2} = \frac{-M}{EI}$$

and

$$\frac{d^4y}{dx^4} = \frac{-P}{EI}$$

we get

$$Mx = P_o \frac{L}{\pi^2} \sin \frac{\pi x}{L}. \tag{13-34}$$

3. Resolve the reactions to in-plane loads in the panels and use steps b through e to calculate stresses.

4. Then calculate deflections and rotations, θ''_{Mx}, at all inner edges. The expression for the deflection of a sinusoidal load is given by

$$w = \frac{L^2}{\pi^2}\left(\frac{-\sigma_t + \sigma_b}{dE}\right) \tag{13-35}$$

5. The net rotation at the selected inner edge is then given by

$$\theta_{Mx} = \theta'_{Mx} + \theta''_{Mx}$$

6. Steps 2 through 5 are repeated for each inner edge.

7. Since the net rotation of the panels at each edge is zero, a number of equations can be written as follows:

$$\begin{bmatrix} \theta_{F1} + k_1\theta_{Mx11} + k_2\theta_{Mx12} + k_3\theta_{Mx13} + \text{----} = 0 \\ \theta_{F2} + k_1\theta_{Mx21} + k_2\theta_{Mx22} + k_3\theta_{Mx23} + \text{----} = 0 \\ \vdots \\ \theta_{Fn} + k_1\theta_{Mxn1} + k_2\theta_{Mxn2} + k_3\theta_{Mxn3} + \text{----} = 0 \end{bmatrix} \tag{13-36}$$

where

θ_{F1} = rotation of edge 1 due to applied loads;
θ_{Mx11} = rotation of edge 1 due to moment M_x applied at edge 1;
θ_{Mx12} = rotation of edge 1 due to moment M_x applied at edge 2;
K_i = constants to be determined.

The final stresses are given by

$$\begin{bmatrix} S_1 = S_{1p} + k_1S_{11} + k_2S_{12} + \text{---} \\ S_2 = S_{2p} + k_1S_{21} + k_2S_{22} + \text{---} \\ \vdots \\ S_n = S_{np} + k_1S_{n1} + k_2S_{n2} + \text{---} \end{bmatrix} \tag{13-37}$$

13-5 Barrel Roofs

The governing differential equation for the bending of barrel roofs is derived in a similar manner as that of cylindrical shells in Chapter 8. The shell segment in Fig. 13-14 is subjected to uniform dead and live loads in the z-direction only. Equations for the equilibrium of the element (Gibson 1968) in the x-, y-, and z-directions are derived first. Then expressions are obtained for the stress–strain relationship. The relationship between the various deflections in the x-, y-, and z-axes are also determined from the geometry of the shell section. By a series of combinations and substitutions, and assuming a Poisson's ratio of zero, the equations are simplified into one governing equation of the form

$$r^2\left(\frac{\partial^2}{\partial x^2} + \frac{\partial^2}{r^2\,\partial\phi^2}\right)^4 w + \frac{12}{t^2}\frac{\partial^4 w}{\partial x^4} = \frac{12r^2}{Et^3}\left(\frac{\partial^2}{\partial x^2} + \frac{\partial^2}{r^2\,\partial\phi^2}\right)^2 p. \quad (13\text{-}38)$$

The homogeneous and particular solutions of this equation, taking advantage of loads and boundary conditions (Chattarjee 1971), yields an expression for the deflection with eight constants of integration. These constants are obtained from the boundary conditions of the roof.

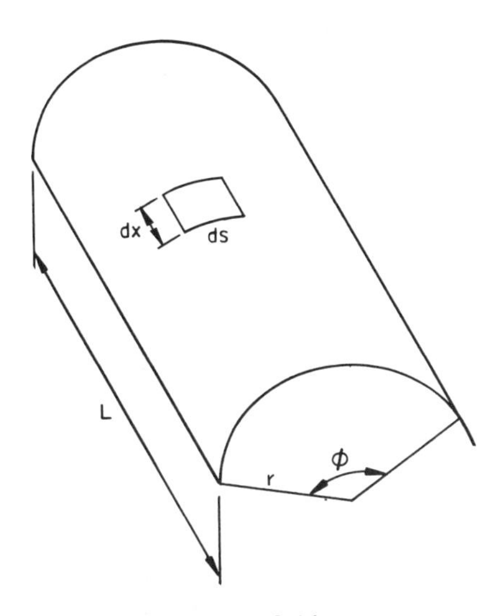

Figure 13-14.

The solutions of Eq. (13-38) as well as other corresponding moments and forces are found in many references. The ASCE Manual (ASCE 1960) gives comprehensive theoretical and tabular values for various boundary conditions. Gibson (Gibson 1968), Chatterjee (Chattarjee 1971), and Billington (Billington 1982) give many solved examples using theoretical and tabular values.

14

Basic Finite Element Equations

14-1 Definitions

The finite element method is a powerful tool for calculating stress in complicated shell and plate structures that are difficult to analyze by classical plate and shell theories. The method consists of subdividing a given domain into small elements connected at the nodal points as shown in Fig. 14-1. The mathematical formulation consists of combining the governing equations of each of the elements to form a solution for the domain that satisfies the boundary conditions. The approximations associated with finite element solutions depend on many variables such as the type of element selected, number of elements used to model the domain, and the boundary conditions.

The complete derivation of the various equations for one-, two-, and three-dimensional elements is beyond the scope of this book. However, a few equations are derived here to demonstrate the basic concept of Finite Element formulation and its applicability to the solution of plates and shells.

We begin the derivations by defining various elements (Fig. 14-2) and terms. Figure 14-2a shows a one-dimensional element in the x-direction with two nodal points, i and j. Figure 14-2b shows a two-dimensional triangular element in the x, y plane with nodal points i, j, and k. And Fig. 14-2c shows a three-dimensional rectangular brick element with eight nodal points.

Let the matrix $[\delta]$ define the displacements within an element. The size of the displacement matrix (Weaver and Johnston, 1984) depends on the complexity of the element being considered. The matrix $[q]$ defines nodal point displacements of an element and matrix $[F]$ defines the applied loads at the nodal points. The size of matrices $[q]$ and $[F]$ depends on the type and geometry of the element being considered. For the one-dimensional

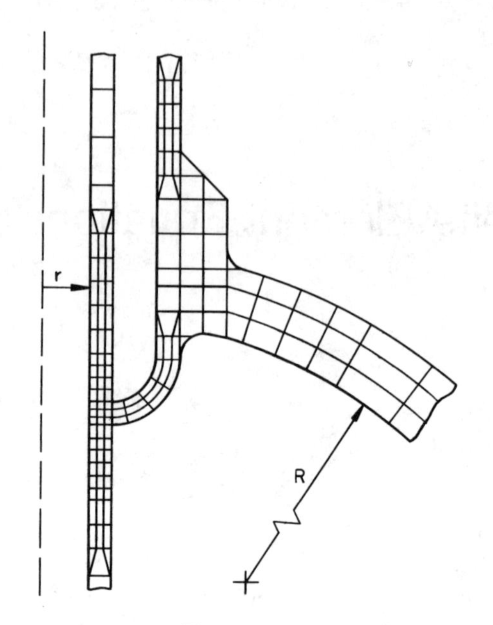

Figure 14-1.

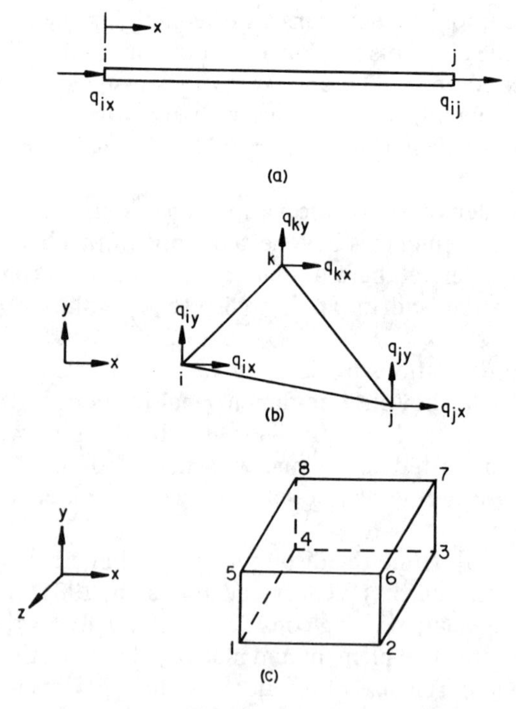

Figure 14-2.

element in Fig. 14-2a, the quantities $[q]$ and $[F]$ are defined as

$$\left.\begin{aligned} [q] &= \begin{bmatrix} q_{ix} \\ q_{jx} \end{bmatrix} \\ [F] &= \begin{bmatrix} F_{ix} \\ F_{jx} \end{bmatrix} \end{aligned}\right] . \tag{14-1}$$

Similarly, the $[q]$ and $[F]$ matrices for the two-dimensional triangular element in Fig. 14-2b are expressed by

$$[q] = \begin{bmatrix} q_{ix} \\ q_{iy} \\ q_{jx} \\ q_{jy} \\ q_{kx} \\ q_{ky} \end{bmatrix} \quad [F] = \begin{bmatrix} F_{ix} \\ F_{iy} \\ F_{jx} \\ F_{jy} \\ F_{kx} \\ F_{ky} \end{bmatrix} . \tag{14-2}$$

The shape function matrix $[N]$ defines the relationship between a function at the nodal points and the same function within the element. Thus, the relationship between the nodal deflection $[q]$ (Fig. 14-2a) and the general deflection δ at any point in the one-dimensional element is expressed as

$$\begin{aligned} \delta &= [N_i \quad N_j] \begin{bmatrix} q_i \\ q_j \end{bmatrix} \\ &= [N][q] \end{aligned}$$

while the relationship between the nodal displacements $[q]$ and the general displacements $[\delta]$ in Fig. 14-2b for a two-dimensional element is

$$\begin{bmatrix} u \\ v \end{bmatrix} = \begin{bmatrix} N_i & 0 & N_j & 0 & N_k & 0 \\ 0 & N_i & 0 & N_j & 0 & N_k \end{bmatrix} \begin{bmatrix} q_{ix} \\ q_{iy} \\ q_{jx} \\ q_{jy} \\ q_{kx} \\ q_{ky} \end{bmatrix}$$

or

$$[\delta] = [N][q]. \tag{14-3}$$

Let the strain–displacement matrix $[d]$ define the relationship between the strains in a continuum to the displacements in accordance with

$$[\varepsilon] = [d][\delta] \tag{14-4}$$

Strain in an element can also be expressed in terms of the deflection of the nodal points. Substituting Eq. (14-3) into Eq. (14-4) gives

$$[\varepsilon] = [d][N][q]. \tag{14-5}$$

Equation (14-5) can also be written as

$$[\varepsilon] = [B][q] \tag{14-6}$$

where

$$[B] = [d][N]. \tag{14-7}$$

The stress–strain relationship is obtained from Eq. (1-13) as

$$\left.\begin{aligned}[\sigma] &= [D][\varepsilon] - [D][\varepsilon_o] \\ &= [D][B][q] - [D][\varepsilon_o]\end{aligned}\right] \tag{14-8}$$

where $[\varepsilon_o]$ is the initial strain in a domain and $[\varepsilon]$ is the total strain.

With these definitions, the basic finite element equations can now be derived. Referring to Fig. 14-3, the strain energy for a differential element of volume dV is

$$dU = \frac{1}{2}[\varepsilon]^T[\sigma] - \frac{1}{2}[\varepsilon_o]^T[\sigma] \tag{14-9}$$

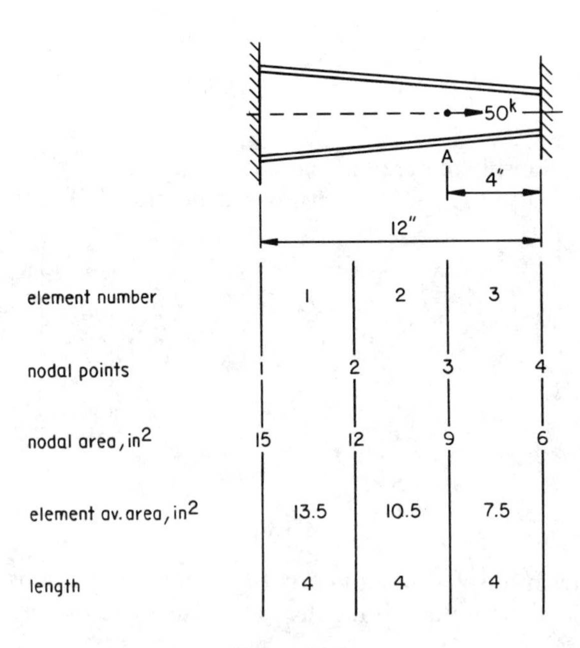

Figure 14-3.

The total strain energy is given by

$$U = \int_V \frac{1}{2}([\varepsilon]^T[\sigma] - [\varepsilon_o]^T[\sigma])\, dV \tag{14-10}$$

Substituting Eq. (14-8) into Eq. (14-10) yields

$$U = \frac{1}{2}\{[q]^T[B]^T[D][B][q] - 2[q]^T[B]^T[D][\varepsilon_o] + [\varepsilon_o]^T[D][\varepsilon_o]\}\, dV. \tag{14-11}$$

The external work due to the nodal loads $[F]$ is

$$W_F = [F]^T[q]. \tag{14-12}$$

The external work due to surface pressure, $[p]$, is

$$W_p = ([u][p])\, ds$$

or

$$W_p = \int_S ([q]^T[N]^T[p])\, ds. \tag{14-13}$$

The potential energy of one element is

$$\Pi = U - (W_F + W_p)$$

or for the whole system

$$\Pi = \sum_{e=1}^{E} [U^e - (W_F^e + W_p^e)] \tag{14-14}$$

where e refers to any given element.

The minimum potential energy is obtained from

$$\frac{\partial \Pi}{\partial q} = 0$$

or,

$$\frac{\partial \Pi}{\partial q} = \sum_{e=1}^{E} \left[\int_V [B_e]^T[D_e][B_e]\, dV\, [q] - \int_V [B_e]^T[D_e][\varepsilon_o]\, dV - \int_S [N_e]^T[p_e]\, ds \right] - F_e = 0. \tag{14-15}$$

The quantity

$$[B_e]^T[D_e][B_e]\, dV$$

is called the stiffness matrix of an element and is written as

$$[K_e] = \int_V [B_e]^T[D_e][B_e]\, dV. \tag{14-16}$$

Hence, the finite element equation becomes

$$\sum_{e=1}^{E} [K_e][q] = \sum_{e=1}^{E} \left[\int_V [B_e]^T[D_e][\varepsilon_o]\, dV + \int_S [N_e]^T[p_e]\, ds \right] + F_e \tag{14-17}$$

which can be abbreviated as

$$[K_e][q] = [F] \tag{14-18}$$

where

$[F]$ = applied forces.

Equation (14-17) is the basic finite element equation for a domain.

Problems

14-1 Write the matrices $[q]$ and $[F]$ for the three-dimensional element in Fig. 14-2c.
14-2 Derive Eq. (14-11).
14-3 Derive Eq. (14-15).

14-2 One-Dimensional Elements

In formulating the finite element equations, the shape of the element as well as other functions such as applied loads, deflections, strains, and stresses are approximated by a polynomial. The size of the polynomial depends on the degrees of freedom at the nodal points and the accuracy required. Hence, for the one-dimensional element shown in Fig. 14-2a, a polynomial for a function such as deflection (Grandin 1986) may be expressed as

$$\delta = C_1 + C_2 x \tag{14-19}$$

where

x = length along the x-axis;
C_1 and C_2 are constants.

The polynomial given by Eq. (14-19) can be written as a function of two matrices $[g]$ and $[C]$ as

$$\delta = [g][C] \tag{14-20}$$

$$\delta = [1 \quad x]\begin{bmatrix} C_1 \\ C_2 \end{bmatrix}.$$

At the nodal points x_i and x_j, Eq. (14-20) becomes

$$\begin{bmatrix} q_i \\ q_j \end{bmatrix} = \begin{bmatrix} 1 & x_i \\ 1 & x_j \end{bmatrix}\begin{bmatrix} C_1 \\ C_2 \end{bmatrix} \tag{14-21}$$

Define $[h]$ as the relationship between $[C]$ and $[q]$ at the nodal points. Thus, Eq. (14-21) becomes

$$\begin{bmatrix} q_i \\ q_j \end{bmatrix} = [h][C]$$

where

$$[h] = \begin{bmatrix} 1 & x_i \\ 1 & x_j \end{bmatrix}.$$

Solving for the matrix $[C]$ gives

$$\begin{bmatrix} C_1 \\ C_2 \end{bmatrix} = [h]^{-1}\begin{bmatrix} q_i \\ q_j \end{bmatrix}$$

$$\begin{bmatrix} C_1 \\ C_2 \end{bmatrix} = \frac{1}{x_i - x_j}\begin{bmatrix} x_j & -x_i \\ -1 & 1 \end{bmatrix}\begin{bmatrix} q_i \\ q_j \end{bmatrix} \tag{14-22}$$

or, from Eq. (14-20) the function at any point is

$$\delta = [g][h]^{-1}\begin{bmatrix} q_i \\ q_j \end{bmatrix}$$

or

$$\delta = \frac{1}{x_j - x_i}[(x_j - x) \quad (-x_i + x)]\begin{bmatrix} q_i \\ q_j \end{bmatrix}. \tag{14-23}$$

The quantity $[g][h]^{-1}$ relates the deflection at the nodal points to that within the element. It is called the shape function and is designated as $[N]$. Thus,

$$[N] = [g][h]^{-1}. \tag{14-24}$$

Then Eq. (14-23) can be written as

$$\delta = [N]\begin{bmatrix} q_i \\ q_j \end{bmatrix} \qquad (14\text{-}25)$$

where

$$[N] = \frac{1}{x_j - x_i}[(x_j - x) \quad (-x_i + x)]. \qquad (14\text{-}26)$$

Once the shape function for a linear element is established, the governing stiffness expression, Eq. (14-18), can also be determined. Thus, for the one-dimensional element shown in Fig. 14-2a, the general deflection function is expressed by Eq. (14-19) and the shape function $[N]$ is given by Eq. (14-26). From Hook's law, the strain in an axial member is expressed as

$$\varepsilon = \frac{d}{dx}u$$

and from Eqs. (14-5) and (14-6)

$$[\varepsilon] = \frac{d}{dx}[N][q]$$

or

$$[\varepsilon] = [B]\begin{bmatrix} q_i \\ q_j \end{bmatrix} \qquad (14\text{-}27)$$

where

$$[B] = \frac{[-1 \quad 1]}{x_j - x_i} = \frac{1}{L}[-1 \quad 1].$$

For a uniaxial body without initial strain,

$$[\sigma] = [E][\varepsilon]$$

Hence,

$$[D] = [E]$$

From Eq. (14-16) the value of the stiffness matrix $[K]$ becomes

$$[K] = \frac{AE}{L}\begin{bmatrix} 1 & -1 \\ -1 & 1 \end{bmatrix}. \qquad (14\text{-}28)$$

From Eq. (14-17), the first term on the righthand side is due to the thermal effect and reduces to

$$\alpha EA(\Delta T)\begin{bmatrix}-1\\1\end{bmatrix}. \tag{14-29}$$

The second term on the righthand side of Eq. (14-17) is for the surface loads. In this case, the surface loads can be applied only at the nodal points i and j. Hence, when p_x is applied at node i,

$$\int_S [N]^T[p_x]\,ds = p_x\begin{bmatrix}1\\0\end{bmatrix}\int ds = p_xA_i\begin{bmatrix}1\\0\end{bmatrix}. \tag{14-30}$$

When p_x is applied at node j,

$$\int_S [N]^T[p_x]\,ds = p_xA_j\begin{bmatrix}1\\0\end{bmatrix}. \tag{14-31}$$

The complete finite element equation for one-dimensional elements is obtained by combining Eqs. (14-28) through (14-31)

$$\frac{AE}{L}\begin{bmatrix}1 & -1\\-1 & 1\end{bmatrix}\begin{bmatrix}q_i\\q_j\end{bmatrix} = \alpha EA(\Delta T)\begin{bmatrix}-1\\1\end{bmatrix} + A_ip_x\begin{bmatrix}1\\0\end{bmatrix} + A_jp_x\begin{bmatrix}1\\0\end{bmatrix} + F$$

or

$$[K][q] = [F]. \tag{14-32}$$

Stress in the member is obtained from Eq. (14-8) as

$$\sigma = [E][B][q] = \frac{E}{L}[-1 \quad 1]\begin{bmatrix}q_i\\q_j\end{bmatrix} - E\alpha(\Delta T). \tag{14-33}$$

Example 14-1

Find the stress in the tapered conical shell (Fig. 14-3). The shell is subjected to a force of 50 kips at point A. The cross sectional area of the cone increases from 6 square inches at the right end to 15 square inches at the left end. The shell is also subjected to a uniform decrease in temperature of 50°F. Assume the shell to be subdivided into three equal lengths and let the coefficient of thermal expansion be equal to 7×10^{-6} inch/inch/°F. Also, let the modulus of elasticity equal 30×10^6 psi and Poisson's ratio equal 0.3.

Solution

Element 1

The stiffness matrix from Eq. (14-28) is

$$[K_1] = \frac{\overline{A}E}{L}\begin{bmatrix} 1 & -1 \\ -1 & 1 \end{bmatrix} = 10^7\begin{bmatrix} 10.125 & -10.125 \\ -10.125 & 10.125 \end{bmatrix}.$$

The thermal force from Eq. (14-29) is

$$\alpha\overline{A}E(-\Delta T)\begin{bmatrix} -1 \\ 1 \end{bmatrix} = 10^3\begin{bmatrix} 141.75 \\ -141.75 \end{bmatrix}$$

and Eq. (14-18) gives

$$10^7\begin{bmatrix} 10.125 & -10.125 \\ -10.125 & 10.125 \end{bmatrix}\begin{bmatrix} q_1 \\ 2_2 \end{bmatrix} = 10^3\begin{bmatrix} 141.75 \\ -141.75 \end{bmatrix}.$$

Element 2

The governing Eq. (14-18) for element 2 is

$$10^7\begin{bmatrix} 7.875 & -7.875 \\ -7.875 & 7.875 \end{bmatrix}\begin{bmatrix} q_2 \\ q_3 \end{bmatrix} = 10^3\begin{bmatrix} 110.25 \\ -110.25 \end{bmatrix}.$$

Element 3

$$[K_3] = \frac{\overline{A}E}{L}\begin{bmatrix} 1 & -1 \\ -1 & 1 \end{bmatrix} = 10^7\begin{bmatrix} 5.625 & -5.625 \\ -5.625 & 5.625 \end{bmatrix}$$

The thermal force is

$$\alpha\overline{A}\mathrm{E}(-\Delta T)\begin{bmatrix} -1 \\ 1 \end{bmatrix} = 10^3\begin{bmatrix} 78.75 \\ -78.75 \end{bmatrix}.$$

The nodal forces are

$$= 10^3\begin{bmatrix} 50 \\ 0 \end{bmatrix}$$

and Eq. (14-18) becomes

$$10^7\begin{bmatrix} 5.625 & -5.625 \\ -5.625 & 5.625 \end{bmatrix}\begin{bmatrix} q_3 \\ q_4 \end{bmatrix} = 10^3\begin{bmatrix} 128.75 \\ -78.75 \end{bmatrix}.$$

Combining the matrices for elements 1, 2, and 3 (Table 14-1) gives

$$10^7\begin{bmatrix} 10.125 & -10.125 & 0 & 0 \\ -10.125 & 18.000 & -7.875 & 0 \\ 0 & -7.875 & 13.500 & -5.625 \\ 0 & 0 & -5.625 & 5.625 \end{bmatrix}\begin{bmatrix} q_1 \\ q_2 \\ q_3 \\ q_4 \end{bmatrix} = 10^3\begin{bmatrix} 141.75 \\ -31.50 \\ 18.50 \\ -78.75 \end{bmatrix}.$$

Because the deflections q_1 and q_4 are zero at the supports, we can delete the first and last rows and columns from the stiffness matrix and the above matrix reduces to

$$10^7\begin{bmatrix} 18.000 & -7.875 \\ -7.875 & 13.500 \end{bmatrix}\begin{bmatrix} q_2 \\ q_3 \end{bmatrix} = 10^3\begin{bmatrix} -31.50 \\ 18.50 \end{bmatrix}.$$

Solving for the values of q_2 and q_3 results in

$$\begin{bmatrix} q_2 \\ q_3 \end{bmatrix} = \frac{1}{10^4}\begin{bmatrix} -1.545 \\ 0.469 \end{bmatrix}.$$

Table 14-1. Total stiffness and force matrices

Stiffness Matrix [*K*]

F/q	1	2	3	4
1	10.125	−10.125		
2	−10.125	10.125 7.875	 −7.875	
3		−7.875	7.875 5.625	 −5.625
4			−5.625	5.625

Load Matrix [*F*]

Node	Force
1	141.75
2	−141.75 110.25
3	−110.25 128.75
4	−78.75

The strain expression in each element is given by Eq. (14-27) as

$$\varepsilon = \frac{1}{L}(-q_i + q_j)$$

$$\varepsilon_1 = -0.386E\text{–}4$$

$$\varepsilon_2 = 0.504E\text{–}4$$

$$\varepsilon_3 = -0.117E\text{–}4.$$

The stress is obtained from Eq. (14-33) as

$$\sigma = E\varepsilon - \alpha E(-\Delta T)$$
$$= 30 \times 10^6\varepsilon + 10{,}500$$

$$\begin{bmatrix} \sigma_1 \\ \sigma_2 \\ \sigma_3 \end{bmatrix} = \begin{bmatrix} 9340 \\ 12{,}010 \\ 10{,}150 \end{bmatrix}$$

A classical theoretical solution of this simple problem can be obtained for comparison purposes. We can let the right end grow freely due to temperature and applied load. We then apply a load at the end to let the deflection at the right end be equal to zero. Using for the deflection due to loads the equation

$$w = \int \frac{F\,dx}{EA_x}$$

$$= \frac{F}{E}\int \frac{dx}{15 - \dfrac{9x}{L}}$$

we get $q_2 = -1.574$ $\quad q_3 = 0.449$

$$\begin{bmatrix} \sigma_1 \\ \sigma_2 \\ \sigma_3 \end{bmatrix} = \begin{bmatrix} 9290 \\ 11{,}930 \\ 10{,}040 \end{bmatrix}$$

The calculated stress in Example 14-1 is different in each of the elements. This causes a discontinuity in stress at internal nodal points joining two elements. To overcome this, a procedure (Segerlind 1976), called the conjugate stress method is used to average the stresses at the nodal points. It calculates an approximate average stress value at the nodal points from the following equation

$$[Q][\bar{\sigma}] = [R] \tag{14-34}$$

where

$[Q]$ = a function of the $[N]$ matrix;
$[\bar{\sigma}]$ = stress at the nodal points, called conformal stress;
$[R]$ = a function of the element stress, called conjugate stress.

In accordance with the theory of conjugate stress approximations, the matrices $[Q]$ and $[R]$ for an element are calculated from the quantities

$$[Q] = \int_V [N]^T [N] \, dV \tag{14-35}$$

and

$$[R] = \int_V [\sigma][N]^T \, dV \tag{14-36}$$

where $[\sigma]$ is the stress in the element.

In many applications the member axis, which is used to determine the stiffness and load matrices, does not coincide with the global axis of the structure as illustrated in Fig. 14-4. In order to accommodate this condition, the member orientation with respect to the global axes needs to be taken into consideration. The resulting stiffness matrix (Wang 1986) is of the form

$$[K_G] = \frac{AE}{L} \begin{bmatrix} k_1 & k_2 & -k_1 & -k_2 \\ & k_3 & -k_2 & -k_3 \\ & & k_1 & k_2 \\ & \text{symmetric} & & k_3 \end{bmatrix} \tag{14-37}$$

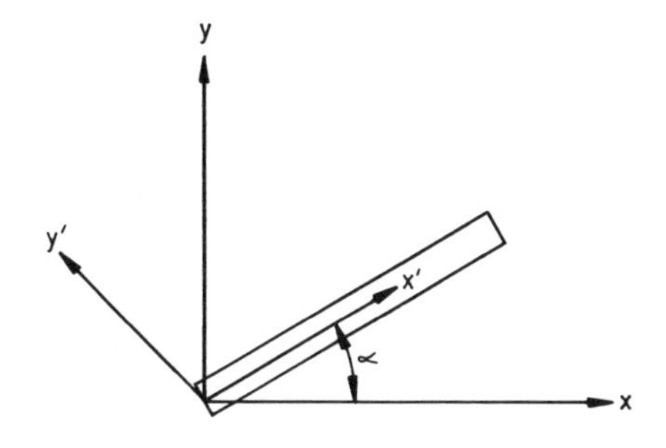

Figure 14-4.

where

$k_1 = \cos^2 \alpha \qquad k_2 = \cos \alpha \sin \alpha \qquad k_3 = \sin^2 \alpha$
α = angle shown in Fig. 14-4 and measured counterclockwise from the positive x-axis,

and load in each member is expressed as

$$F = \frac{EA}{L}(-q_{ix} \cos \alpha - q_{iy} \sin \alpha + q_{jx} \cos \alpha + q_{jy} \sin \alpha) \qquad (14\text{-}38)$$

Problems

14-4 What is the thermal stress in the conical shell shown in Example 14-1 if the applied axial load is equal to zero?
14-5 What is the stress in the conical shell shown in Example 14-1 if the change in temperature is equal to zero?

14-3 Linear Triangular Elements

From Fig. 14-2b it is seen that each element has three nodal points and each nodal point has two degrees of freedom. Hence the displacement within the element is expressed by the following polynomial

$$u = C_1 + C_2x + C_3y$$
$$v = C_4 + C_5x + C_6y \qquad (14\text{-}39)$$

where u and v are the deflection in the x- and y-axes, respectively. In matrix form these equations are written as

$$\begin{bmatrix} u \\ v \end{bmatrix} = [g][C]$$

where

$$[g] = \begin{bmatrix} 1 & x & y & 0 & 0 & 0 \\ 0 & 0 & 0 & 1 & x & y \end{bmatrix} \qquad (14\text{-}40)$$

and

$$[C] = \begin{bmatrix} C_1 \\ \cdot \\ \cdot \\ C_6 \end{bmatrix}.$$

The shape matrix $[N]$ is obtained from Eq. (14-24) and is expressed as

$$[N] = \begin{bmatrix} N_i & 0 & N_j & 0 & N_k & 0 \\ 0 & N_i & 0 & N_j & 0 & N_k \end{bmatrix} \tag{14-41}$$

where N_i, N_j, and N_k are defined as

$$N_i = \frac{1}{2\Delta}(a_i + b_i x + c_i y)$$

$$N_j = \frac{1}{2\Delta}(a_j + b_j x + c_j y)$$

$$N_k = \frac{1}{2\Delta}(a_k + b_k x + c_k y)$$

$$a_i = x_j y_k - x_k y_j, \quad b_i = y_j - y_k, \quad c_i = x_k - x_j$$

$$a_j = x_k y_i - x_i y_k, \quad b_j = y_k - y_i, \quad c_j = x_i - x_k$$

$$a_k = x_i y_j - x_j y_i, \quad b_k = y_i - y_j, \quad c_k = x_j - x_i$$

$$\Delta = \frac{1}{2}\begin{vmatrix} 1 & x_i & y_i \\ 1 & x_j & y_j \\ 1 & x_k & y_k \end{vmatrix}$$

2Δ = area of triangle with coordinates $x_i y_i$, $x_j y_j$, $x_k y_k$.
The u and v expressions within the element are

$$\begin{bmatrix} u \\ v \end{bmatrix} = [N] \begin{bmatrix} q_{ix} \\ q_{iy} \\ q_{jx} \\ q_{jy} \\ q_{kx} \\ q_{ky} \end{bmatrix}.$$

The strain–deflection relationship is obtained from Eq. (2-16) as

$$\begin{bmatrix} \varepsilon_x \\ \varepsilon_y \\ \gamma_{xy} \end{bmatrix} = \begin{bmatrix} \frac{\partial}{\partial x} & 0 \\ 0 & \frac{\partial}{\partial y} \\ \frac{\partial}{\partial y} & \frac{\partial}{\partial x} \end{bmatrix} \begin{bmatrix} u \\ v \end{bmatrix}. \tag{14-42}$$

This strain expression can be designated as

$$[\varepsilon] = [d]\begin{bmatrix} u \\ v \end{bmatrix} = [d][N][q]$$

or

$$[\varepsilon] = [B][q]$$

where

$$\begin{aligned}[B] &= [d][N] \\ &= \frac{1}{2\Delta}\begin{bmatrix} b_i & 0 & b_j & 0 & b_k & 0 \\ 0 & c_i & 0 & c_j & 0 & c_k \\ c_i & b_i & c_j & b_j & c_k & b_k \end{bmatrix}. \end{aligned} \tag{14-43}$$

For a plane-stress formulation the stiffness matrix $[D]$ was derived in Chapter 1 and is given by Eq. (1-14) as

$$[D] = \frac{E}{1 - \mu^2}\begin{bmatrix} 1 & \mu & 0 \\ \mu & 1 & 0 \\ 0 & 0 & \dfrac{1 - \mu}{2} \end{bmatrix}.$$

For a plane-strain formulation, Eq. (1-13) is used. Letting

$$\varepsilon_z = \gamma_{yz} = \gamma_{xz} = 0$$

$$[D] = \frac{E(1 - \mu)}{(1 + \mu)(1 - 2\mu)}\begin{bmatrix} 1 & \dfrac{\mu}{1 - \mu} & 0 \\ \dfrac{\mu}{1 - \mu} & 1 & 0 \\ 0 & 0 & \dfrac{1 - 2\mu}{2(1 - \mu)} \end{bmatrix}. \tag{14-44}$$

The stiffness matrix is calculated from Eq. (14-16). The result can be expressed as

$$K = \frac{Et}{4\Delta}\begin{bmatrix} k_{11} & k_{12} & k_{13} & k_{14} & k_{15} & k_{16} \\ & k_{22} & k_{23} & k_{24} & k_{25} & k_{26} \\ & & k_{33} & k_{34} & k_{35} & k_{36} \\ & & & k_{44} & k_{45} & k_{46} \\ & \text{symmetric} & & & k_{55} & k_{56} \\ & & & & & k_{66} \end{bmatrix} \tag{14-45}$$

where

$$k_{11} = k_1(y_2 - y_3)^2 + k_3(x_3 - x_2)^2$$
$$k_{12} = k_2(x_3 - x_1)(y_2 - y_3) + k_3(x_3 - x_2)(y_2 - y_3)$$
$$k_{13} = k_1(y_2 - y_3)(y_3 - y_1) + k_3(x_3 - x_2)(x_1 - x_3)$$
$$k_{14} = k_2(x_1 - x_3)(y_2 - y_3) + k_3(x_3 - x_2)(y_3 - y_1)$$
$$k_{15} = k_1(y_1 - y_2)(y_2 - y_3) + k_3(x_2 - x_1)(x_3 - x_2)$$
$$k_{16} = k_2(x_2 - x_1)(y_2 - y_3) + k_3(x_3 - x_2)(y_1 - y_2)$$
$$k_{22} = k_1(x_3 - x_2)^2 + k_3(y_2 - y_3)^2$$
$$k_{23} = k_2(x_3 - x_2)(y_3 - y_1) + k_3(x_1 - x_3)(y_2 - y_3)$$
$$k_{24} = k_1(x_3 - x_2)(x_1 - x_3) + k_3(y_2 - y_3)(y_3 - y_1)$$
$$k_{25} = k_2(x_3 - x_2)(y_1 - y_2) + k_3(x_2 - x_1)(y_2 - y_3)$$
$$k_{26} = k_1(x_2 - x_1)(x_3 - x_2) + k_3(y_1 - y_2)(y_2 - y_3)$$
$$k_{33} = k_1(y_3 - y_1)^2 + k_3(x_1 - x_3)^2$$
$$k_{34} = k_2(x_1 - x_3)(y_3 - y_1) + k_3(x_1 - x_3)(y_3 - y_1)$$
$$k_{35} = k_1(y_1 - y_2)(y_3 - y_1) + k_3(x_1 - x_3)(x_2 - x_1)$$
$$k_{36} = k_2(x_2 - x_1)(y_3 - y_1) + k_3(x_1 - x_3)(y_1 - y_2)$$
$$k_{44} = k_1(x_1 - x_3)^2 + k_3(y_3 - y_1)^2$$
$$k_{45} = k_2(x_1 - x_3)(y_1 - y_2) + k_3(x_2 - x_1)(y_3 - y_1)$$
$$k_{46} = k_1(x_1 - x_3)(x_2 - x_1) + k_3(y_1 - y_2)(y_3 - y_1)$$
$$k_{55} = k_1(y_1 - y_2)^2 + k_3(x_2 - x_1)^2$$
$$k_{56} = k_2(x_2 - x_1)(y_1 - y_2) + k_3(x_2 - x_1)(y_1 - y_2)$$
$$k_{66} = k_1(x_2 - x_1)^2 + k_3(y_1 - y_2)^2$$

and

For Plane-Stress

$$k_1 = \frac{1}{1 - \mu^2} \qquad k_2 = \frac{\mu}{1 - \mu^2} \qquad k_3 = \frac{1}{2(1 + \mu)}$$

For Plane-Strain

$$k_1 = \frac{1 - \mu}{(1 + \mu)(1 - 2\mu)} \qquad k_2 = \frac{\mu}{(1 + \mu)(1 - 2\mu)}$$
$$k_3 = \frac{1}{2(1 + \mu)}$$

The forces are calculated from Eq. (14-17).

Example 14-2

The triangular plate (Fig. 14-5a) is stiffened at the edges as shown. Find the stress in the various components. Let $E = 30{,}000$ ksi and $\mu = 0.3$.

Solution

The various nodal points are numbered as shown in Fig. 14-5b. The stiffness matrices, K, for members A, B, and C are obtained from Eq. (14-37) as

Member *A* with $\alpha = 0°$

With nodal points q_1, q_2, q_3, q_4

$$K_A = \frac{0.5 \times 3000 \times 10^4}{30} \begin{bmatrix} 1.00 & 0.00 & -1.00 & 0.00 \\ 0.00 & 0.00 & 0.00 & 0.00 \\ -1.00 & 0.00 & 1.00 & 0.00 \\ 0.00 & 0.00 & 0.00 & 0.00 \end{bmatrix}$$

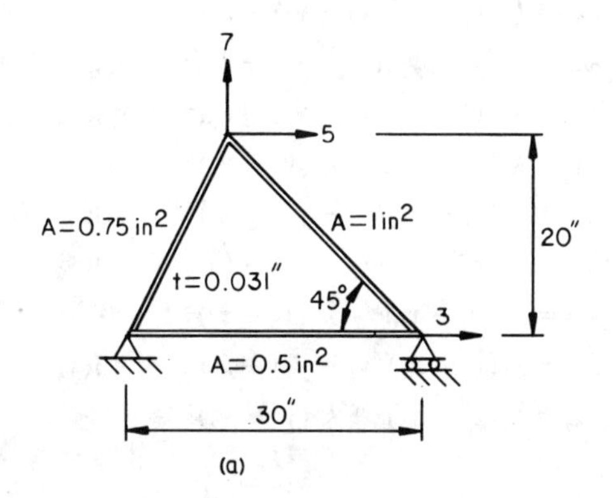

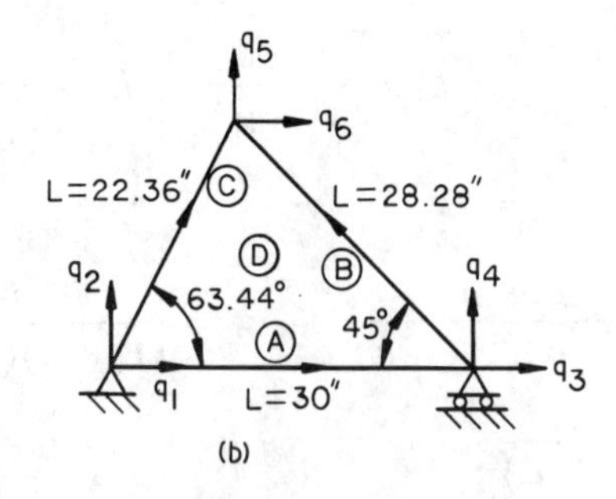

Figure 14-5.

or, since q_1, q_2, and q_4 are zero, we eliminate rows and columns 1, 2, and 4 and we get

$$K_A = 10^4[50].$$

Member *B* with α = 135°

With nodal points q_1, q_2, q_5, q_6

$$K_B = \frac{1 \times 3000 \times 10^4}{28.28} \begin{bmatrix} 0.50 & -0.50 & -0.50 & 0.50 \\ -0.50 & 0.50 & 0.50 & -0.50 \\ -0.50 & 0.50 & 0.50 & -0.50 \\ 0.50 & -0.50 & -0.50 & 0.50 \end{bmatrix}$$

or, since q_4 is zero,

$$K_B = 10^4 \begin{bmatrix} 53.04 & -53.04 & 53.04 \\ -53.04 & 53.04 & -53.04 \\ 53.04 & -53.04 & 53.04 \end{bmatrix}$$

Member C with α = 63.44°

With nodal points q_1, q_2, q_5, q_6

$$K_C = \frac{0.75 \times 3000 \times 10^4}{22.36} \begin{bmatrix} 0.20 & 0.40 & -0.20 & -0.40 \\ 0.40 & 0.80 & -0.40 & -0.80 \\ -0.20 & -0.40 & 0.20 & 0.40 \\ -0.40 & -0.80 & 0.40 & 0.80 \end{bmatrix}$$

or, since q_1 and q_2 are zero

$$K_C = 10^4 \begin{bmatrix} 20.125 & 40.25 \\ 40.125 & 80.50 \end{bmatrix}$$

Member *D*

With q_1, q_2, and q_4 equal to zero

From Eq. (14-45)

$$K_D = \frac{Et}{4\Delta} \begin{bmatrix} K_{33} & K_{35} & K_{36} \\ & K_{55} & K_{56} \\ & & K_{66} \end{bmatrix}$$

$$K_D = 10^4 \begin{bmatrix} 74.11 & -17.90 & 30.69 \\ -17.90 & 53.71 & 0.00 \\ 30.69 & 0.00 & 153.31 \end{bmatrix}$$

From Table 14-2, the total matrix is

$$K = 10^4 \begin{bmatrix} 177.15 & -70.94 & 83.73 \\ -70.94 & 126.88 & -12.79 \\ 83.73 & -12.79 & 286.85 \end{bmatrix}$$

and the force matrix is

$$F = \begin{bmatrix} 3.0 \\ 5.0 \\ 7.0 \end{bmatrix} \text{ kips.}$$

From Eq. (14-32),

$$Kq = F$$

or

$$\begin{bmatrix} q_3 \\ q_5 \\ q_6 \end{bmatrix} = \begin{bmatrix} 3.235 \\ 5.927 \\ 1.760 \end{bmatrix} \times 10^{-6} \text{ inch.}$$

The stresses in members A, B, and C are obtained from Eq. (14-38) as

$$\sigma_A = 3.24 \text{ ksi}, \qquad \sigma_B = -0.7 \text{ ksi}, \qquad \sigma_C = 5.67 \text{ ksi}.$$

The stress in plate D is obtained from Eq. (14-8) as

$$\begin{bmatrix} \sigma_x \\ \sigma_y \\ \tau_{xy} \end{bmatrix} = \begin{bmatrix} 8.85 \\ 7.94 \\ 5.59 \end{bmatrix} \text{ ksi.}$$

Table 14-2. Total stiffness matrix (multiplied by 10^4)

F/q	3	5	6
3	50.00		
	53.04	−53.04	53.04
	74.11	−17.90	30.69
5	−53.04	53.04	−53.04
		20.13	40.25
	−17.90	53.71	0.00
6	53.04	−53.04	53.04
		40.25	80.50
	30.69	0.0	153.31

14-4 Axisymmetric Triangular Linear Elements

Many plate and shell configurations (Fig. 14-1) are modeled as axisymmetric triangular elements. Axisymmetric triangular elements (Fig. 14-6) have the same size N matrix as that defined by Eq. (14-41) for plane elements. The strain–stress relationship given by Eq. (14-42) for plane elements must be modified for axisymmetric elements to include the hoop strain ε_θ. Thus, Eq. (14-42) becomes

$$\begin{bmatrix} \varepsilon_r \\ \varepsilon_z \\ \varepsilon_\theta \\ \gamma_{rz} \end{bmatrix} = \begin{bmatrix} \dfrac{\partial}{\partial r} & 0 \\ 0 & \dfrac{\partial}{\partial z} \\ \dfrac{1}{r} & 0 \\ \dfrac{\partial}{\partial z} & \dfrac{\partial}{\partial r} \end{bmatrix} \begin{bmatrix} u \\ v \end{bmatrix} \tag{14-46}$$

and the $[B]$ matrix becomes

$$[B] = \frac{1}{2\Delta} \begin{bmatrix} b_i & 0 & b_j & 0 & b_k & 0 \\ 0 & c_i & 0 & c_j & 0 & c_k \\ \dfrac{2\Delta N_i}{r} & 0 & \dfrac{2\Delta N_j}{r} & 0 & \dfrac{2\Delta N_k}{r} & 0 \\ c_i & b_i & c_j & b_j & c_k & b_k \end{bmatrix} \tag{14-47}$$

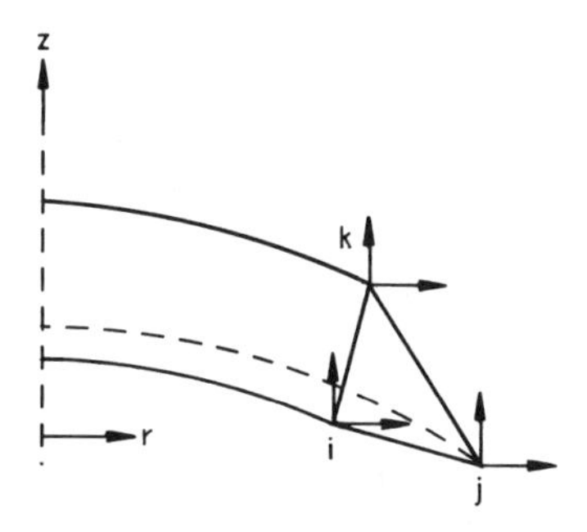

Figure 14-6.

where

$$\begin{bmatrix} \varepsilon_r \\ \varepsilon_z \\ \varepsilon_\theta \\ \gamma_{rz} \end{bmatrix} = [B] \begin{bmatrix} q_{ir} \\ q_{iz} \\ q_{jr} \\ q_{jz} \\ q_{kr} \\ q_{kz} \end{bmatrix} \tag{14-48}$$

The stress–strain relationship is obtained from Eq. (1-13) with $\gamma_{rz} = \gamma_{r\theta}$. This gives

$$[D] = \frac{E(1-\mu)}{(1+\mu)(1-2\mu)} \begin{bmatrix} 1 & \frac{\mu}{1-\mu} & \frac{\mu}{1-\mu} & 0 \\ \frac{\mu}{1-\mu} & 1 & \frac{\mu}{1-\mu} & 0 \\ \frac{\mu}{1-\mu} & \frac{\mu}{1-\mu} & 1 & 0 \\ 0 & 0 & 0 & \frac{1-2}{2(1-\mu)} \end{bmatrix}. \tag{14-49}$$

The stiffness matrix is determined from Eq. (14-16) as

$$[K_e] = \int_V [B_e]^T[D_e][B_e]\, dV.$$

The evaluation of the integral $(B^T DB)\, dV$ in axisymmetric problems is complicated by the fact that the matrix $[B]$ contains the variable $1/r$. A common procedure for integrating this quantity (Zienkiewicz 1977) is to use the radius $\bar{r}$ at the centroid of the element. Also, we can substitute for the quantity dV the value $(2\pi\bar{r}A)$ where A is the area of the element. Hence, the stiffness matrix $[K]$ becomes

$$[K] = [B]^T[D][B]2\bar{r}A. \tag{14-50}$$

14-5 Higher Order Elements

Equations derived for the linear triangular elements in Sections 14-3 and 14-4 can also be established for the linear rectangular elements shown in Fig. 14-7. The equations for the rectangular elements are slightly more complicated than those for triangular elements (Rockey et al. 1975) due to the additional fourth nodal point. In both cases the strain is constant throughout the element. This is a disadvantage in areas where a large strain

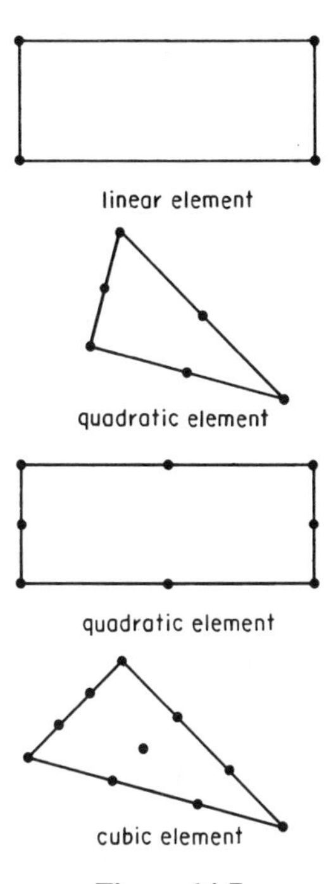

Figure 14-7.

gradient exists because a large number of elements is needed. Accordingly, higher order elements are normally utilized. The higher order elements have additional nodal points in the sides, and sometimes in the interior. With more nodal points in an element, the strain becomes more complex within an element and fewer elements are needed to define a complex geometry or an area with large strain gradients.

The shape function $[N]$ needed to define higher order elements is more complicated than that of linear elements and its derivation requires more sophisticated methods using natural coordinate systems (Weaver and Johnston 1984). Also, the stiffness matrix, which is a function of $[N]$, requires numerical integration which is cumbersome to evaluate without a computer. The accuracy of the results depends, in part, on the method used for the numerical integration.

Finite element formulation of a plate element, as well as finite element formulation of a shell element, have also been derived (Gallagher 1975) and are based on various polynomial approximations. The accuracy of these formulations depends on the particular plate or shell theory being used.

Finite element formulation of three-dimensional brick elements is also available in the literature. The equations become cumbersome for elements higher than quadratic.

APPENDIX A

Fourier Series

A-1 General Equations

A periodic function (Wylie 1960) can be represented by a series that is expressed as

$$f(x) = 0.5A_o + A_1 \cos x + A_2 \cos 2x + \cdots + A_m \cos mx$$
$$+ B_1 \sin x + B_2 \sin 2x + \cdots + B_m \sin mx$$

or

$$f(x) = 0.5A_o + \sum_{m=1}^{\infty} A_m \cos mx + \sum_{m=1}^{\infty} B_m \sin mx. \qquad \text{(A-1)}$$

The series given by Eq. (A-1) is known as a Fourier Series and is used to express periodic functions such as those shown in Fig. A-1. The coefficients A and B in Eq. (A-1) are evaluated over a 2π period starting at a given point d. The value of A_o can be obtained by integrating Eq. (A-1) from $x = d$ to $x = d + 2\pi$.

Thus,

$$\int_d^{d+2\pi} f(x)\,dx = 0.5A_o \int_d^{d+2\pi} dx + A_1 \int_d^{d+2\pi} \cos x\,dx + \cdots$$
$$+ A_m \int_d^{d+2\pi} \cos mx\,dx + B_1 \int_d^{d+2\pi} \sin x\,dx + \cdots$$
$$+ B_m \int_d^{d+2\pi} \sin mx\,dx.$$

The first term in the righthand side of the equation gives πA_o. All other

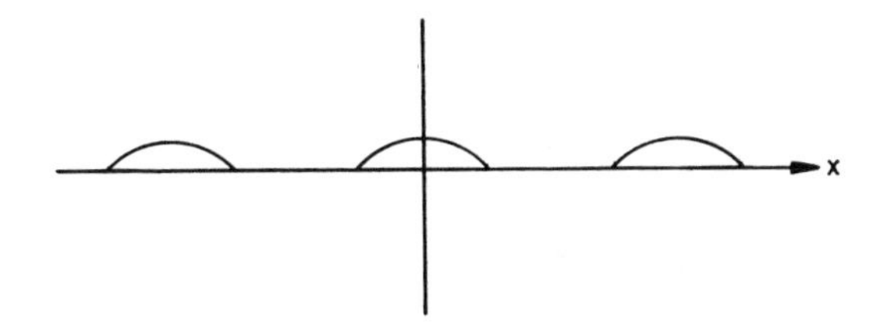

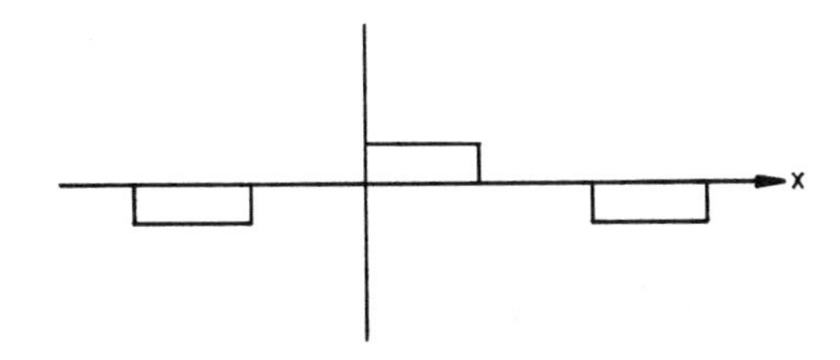

Figure A-1.

terms on the righthand side are zero because of the relationships

$$\int_d^{d+2\pi} \cos mx \, dx = 0 \qquad m \neq 0$$

$$\int_d^{d+2\pi} \sin mx \, dx = 0.$$

Hence,

$$A_o = \frac{1}{\pi} \int_d^{d+2\pi} f(x) \, dx. \tag{A-2}$$

The A_m term in Eq. (A-1) can be obtained by multiplying both sides of the equation by cos mx.

$$\int_d^{d+2} f(x) \cos mx \, dx = \frac{1}{2} A_o \int_d^{d+2\pi} \cos mx \, dx$$

$$+ A_1 \int_d^{d+2\pi} \cos x \cos mx \, dx + \cdots$$

$$+ A_m \int_d^{d+2\pi} \cos mx \cos mx \, dx + B_1 \int_d^{d+2\pi} \sin x \cos mx \, dx$$

$$+ \cdots + B_m \int_d^{d+2\pi} \sin mx \cos mx \, dx. \tag{A-3}$$

Since

$$\int_{d}^{d+2\pi} \cos mx \cos nx \, dx = 0 \qquad m \neq n$$

$$\int_{d}^{d+2\pi} \cos^2 mx \, dx = \pi \qquad m \neq 0$$

$$\int_{d}^{d+2\pi} \cos mx \sin nx \, dx = 0.$$

Equation (A-3) becomes

$$\int_{d}^{d+2\pi} f(x) \cos mx \, dx = A_m \pi$$

or

$$A_m = \frac{1}{\pi} \int_{d}^{d+2\pi} f(x) \cos mx \, dx. \tag{A-4}$$

Similarly the values of B_m can be found by multiplying both sides of Eq. (A-1) by sin mx. Using the expressions

$$\int_{d}^{d+2\pi} \sin mx \sin nx \, dx = 0 \qquad m \neq n$$

and

$$\int_{d}^{d+2\pi} \sin^2 mx \, dx = \pi$$

the equation becomes

$$B_m = \frac{1}{\pi} \int_{d}^{d+2\pi} f(x) \sin mx \, dx. \tag{A-5}$$

Accordingly, we can state that for a given periodic function $f(x)$, a Fourier expansion can be written as shown in Eq. (A-1) with the various constants obtained from Eqs. (A-2), (A-4), and (A-5).

Example A-1

Express the function shown in Fig. A-2 in a Fourier Series

Solution

$$f(x) = 0 \qquad -\pi < x < 0$$
$$f(x) = p_o \qquad 0 < x < \pi$$
$$d = -\pi$$

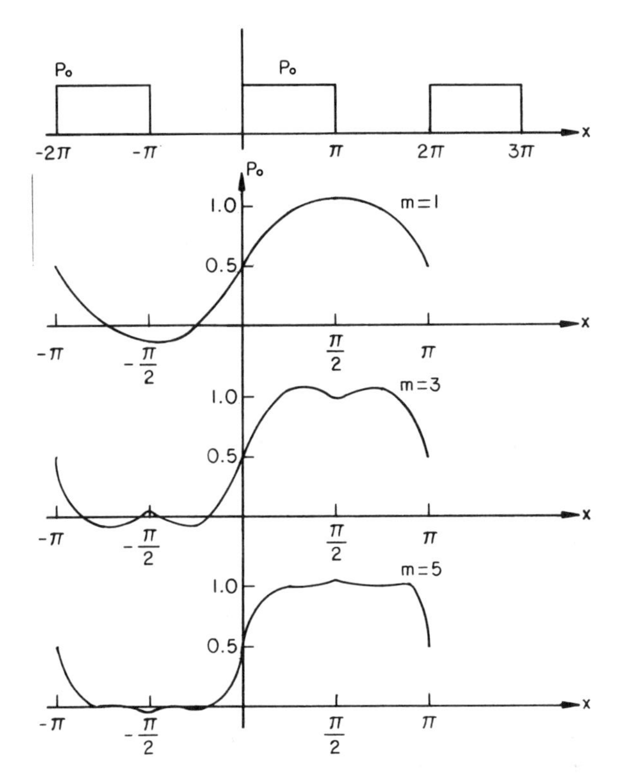

Figure A-2.

From Eq. (A-2),

$$A_o = \frac{1}{\pi}\int_{-\pi}^{0} (0)\,dx + \frac{1}{\pi}\int_{0}^{\pi} p_o\,dx$$

or

$$A_o = p_o.$$

From Eq. (A-4)

$$A_m = \frac{1}{\pi}\int_{0}^{\pi} p_o \cos mx\,dx$$

$$A_m = 0.$$

From Eq. (A-5),

$$
\begin{aligned}
B_m &= \frac{1}{\pi}\int_0^{\pi} p_o \sin mx \, dx \\
&= \frac{p_o}{m\pi}(-\cos mx)\Big|_0^{\pi} \\
&= \frac{-p_o}{m\pi}(\cos m\pi - 1) \\
B_m &= \frac{2p_o}{m\pi} \qquad \text{when } m \text{ is odd} \\
&= 0 \qquad \text{when } m \text{ is even.}
\end{aligned}
$$

Therefore, the expansion of the function shown in Fig. A-2 is expressed as

$$
f(x) = 0.5p_o + \frac{2p_o}{\pi}\sum_{m=1,3,\ldots}^{\infty}\frac{1}{m}\sin mx.
$$

A plot of this equation with $m = 1, 3, 5$ is shown in Fig. A-2.

Problems

A-1 What is the Fourier expansion of the function shown in Fig. PA-1?
A-2 What is the Fourier expansion of the function shown in Fig. PA-2?

A-2 Interval Change

In applying the Fourier Series to plate and shell problems, it is more convenient to specify intervals other than 2π. Defining the new interval

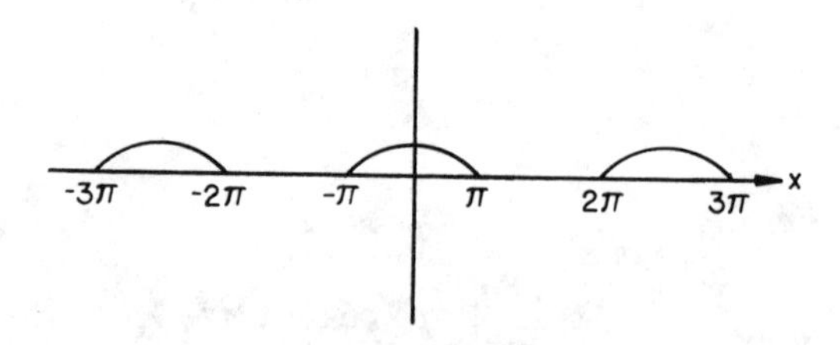

Figure PA-1.

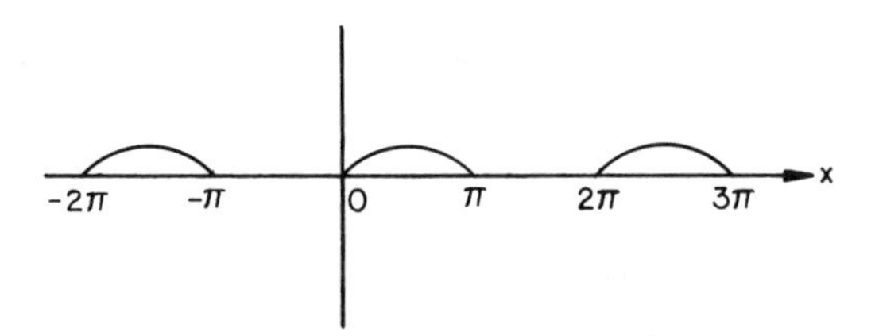

Figure PA-2.

as $2p$, Eqs. (A-2), (A-4), and (A-5) can be written as

$$A_o = \frac{1}{p}\int_{d}^{d+2p} f(x)\, dx \tag{A-6}$$

$$A_m = \frac{1}{p}\int_{d}^{d+2p} f(x) \cos \frac{m\pi x}{p}\, dx \tag{A-7}$$

$$B_m = \frac{1}{p}\int_{d}^{d+2p} f(x) \sin \frac{m\pi x}{p}\, dx \tag{A-8}$$

where

$2p$ = period of function,

and the series can be written as

$$f(x) = \frac{1}{2}A_o + \sum_{m=1}^{\infty} A_m \cos \frac{m\pi x}{p} + \sum_{m=1}^{\infty} B_m \sin \frac{m\pi x}{p}. \tag{A-9}$$

Parts of the equation under the summation signs can be solved by available computer programs. A short computer program called "SNGLSUM" is listed in Table A-1.

Example A-2

Find the Fourier expansion of the Function $f(x) = \cos x$ as shown in Fig. A-3.

Solution

The period $2p$ is equal to π. Thus, $p = \pi/2$ and $d = -\pi/2$.

$$A_o = \frac{1}{\pi/2}\int_{-\pi/2}^{\pi/2} \cos x\, dx = 4/\pi$$

$$A_m = \frac{2}{\pi}\int_{-\pi/2}^{\pi/2} \cos x \cos \frac{m\pi x}{\pi/2}\, dx = \frac{4}{\pi}\sum_{m=1}^{\infty} \frac{(-1)^{m+1}}{(4m^2 - 1)}$$

$$B_m = 0.$$

Table A-1. Program "SNGLSUM"

```
10 REM THE NAME OF THIS PROGRAM IS SINGLE SUMMATION,
   "SNGLSUM"
20 REM THIS PROGRAM CALCULATES THE SUM OF A SINGLE
   SERIES
30 REM INPUT EQUATION TO BE EVALUATED IN LINE 210
40 REM STATEMENTS 210 TO 219 ARE RESRV'D FOR EQUAT'N TO BE
   SOLV'D
50 REM INPUT THE MAXIMUM SUMMATION VALUE IN LINE 200
60 PRINT "IS THE M SUMMATION ODD (O), EVEN (E), OR
   CONTINUOUS (C)"
70 INPUT A$
80 IF A$ = "O" THEN 120
90 IF A$ = "E" THEN 150
100 IF A$ = "C" THEN 180
110 GOTO 60
120 M1 = 1
130 M2 = 2
140 GOTO 200
150 M1 = 2
160 M2 = 2
170 GOTO 200
180 M1 = 1
190 M2 = 1
200 FOR M = M1 TO 30 STEP M2
210 Y = 1 / (M)^2
220 LET T = T + Y
230 NEXT M
240 PRINT "SUMMATION OF SERIES IS = "; T
250 END
```

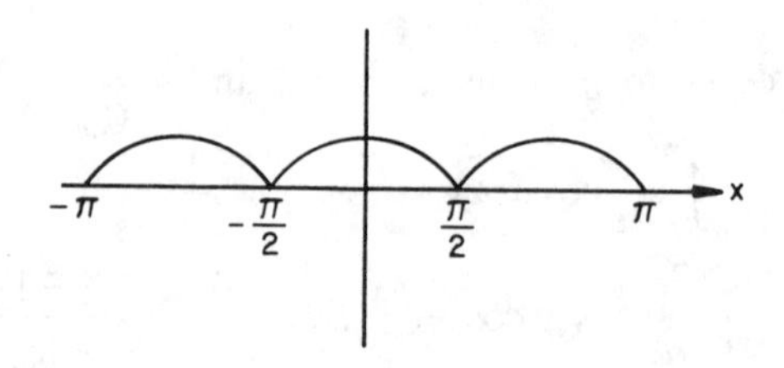

Figure A-3.

and from Eq. (A-9),

$$f(x) = \frac{2}{\pi} + \frac{4}{\pi} \sum_{m=1}^{\infty} \frac{(-1)^{m+1}}{4m^2 - 1} \cos 2mx.$$

A-3 Half-Range Expansions

If a function is symmetric with respect to the axis of reference as shown in Fig. A-4, then the coefficient integral can be simplified by integrating over one-half the period. This integration can be performed as an even or an odd function. Hence,

For an even periodic function

$$\left.\begin{aligned} A_o &= \frac{2}{p} \int_0^p f(x)\, dx \\ A_m &= \frac{2}{p} \int_0^p f(x) \cos \frac{m\pi x}{p}\, dx \\ B_m &= 0. \end{aligned}\right] \qquad \text{(A-10)}$$

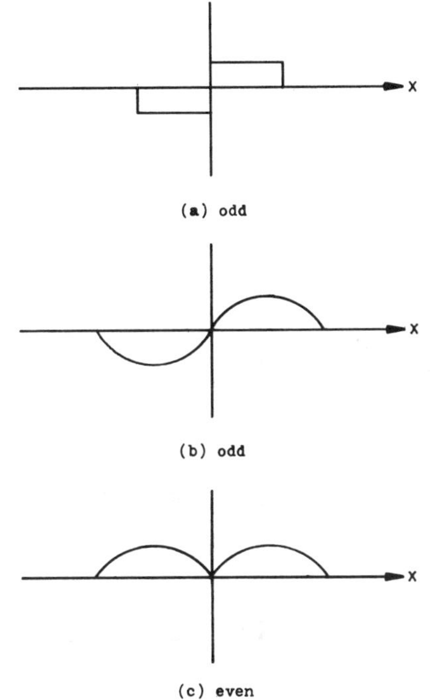

Figure A-4.

For an odd periodic function

$$\left.\begin{aligned} A_o &= A_m = 0 \\ B_m &= \frac{2}{p}\int_0^p f(x)\sin\frac{m\pi x}{p}\,dx \end{aligned}\right] \tag{A-11}$$

It should be noted that the even and odd functions defined by Eqs. (A-10) and (A-11) and Fig. A-4 do not refer necessarily to the shape of the function but rather to the reference line from which they are defined. This can best be illustrated by the following example.

Example A-3

Figure A-5 shows a plot of the function $y = x - x^2$. Obtain and plot the Fourier Series expansion of this function (a) from $y = -1$ to $y = 1$; (b) as an even series from $y = 0$ to $y = 1$; (c) as an odd series from $y = 0$ to $y = 1$.

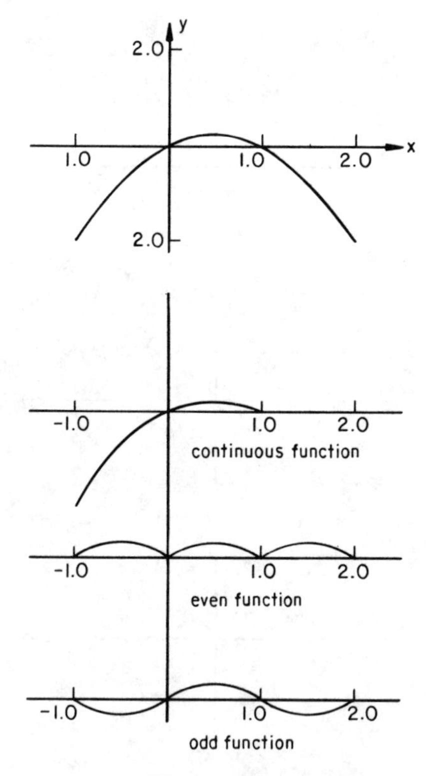

Figure A-5.

Solution

(a)

$$d = -1, \quad 2p = 2 \quad \text{or} \quad p = 1$$

$$A_o = \int_{-1}^{1} (x - x^2)\, dx = -2/3$$

$$A_m = \int_{-1}^{1} (x - x^2) \cos \frac{m\pi x}{1}\, dx = -\frac{4 \cos m\pi}{m^2\pi^2}$$

$$B_m = \int_{-1}^{1} (x - x^2) \sin \frac{m\pi x}{1}\, dx = -\frac{2 \cos m\pi}{m\pi}$$

$$f(x) = -1/3 - \sum_{m=1}^{\infty} \frac{4 \cos m\pi}{m^2\pi^2} \cos m\pi x - \sum_{m=1}^{\infty} \frac{2 \cos m\pi}{m\pi} \sin m\pi x$$

$$= -1/3 - \frac{4}{\pi^2} \sum_{m=1}^{\infty} \frac{(-1)^m}{m^2} \cos m\pi x - \frac{2}{\pi} \sum_{m=1}^{\infty} \frac{(-1)^m}{m} \sin m\pi x$$

(b)

$$A_o = 2 \int_{0}^{1} (x - x^2)\, dx = 1/3$$

$$A_m = 2 \int_{0}^{1} (x - x^2) \cos \frac{m\pi x}{1}\, dx = -\frac{2(1 + \cos m\pi)}{m^2\pi^2}$$

$$B_m = 0$$

$$f(x) = 1/6 - \sum_{m=1}^{\infty} \frac{2(1 + \cos m\pi)}{m^2\pi^2} \cos m\pi x$$

(c)

$$A_o = A_m = 0$$

$$B_m = 2 \int_{0}^{1} (x - x^2) \sin \frac{m\pi x}{1}\, dx = \frac{4(1 - \cos m\pi)}{m^3\pi^3}$$

$$f(x) = \sum_{m=1}^{\infty} \frac{4(1 - \cos m\pi)}{m^3\pi^3} \sin m\pi x$$

A-4 Double Fourier Series

In solving rectangular plate problems of length a and width b, it is customary to express the applied loads in terms of a single or double series. The double Fourier series is normally expressed as an odd periodic function with a half range period given between 0 and a for one side of the plate and 0 to b for the other side. Thus,

$$f(x, y) = \sum_{m=1}^{\infty} \sum_{n=1}^{\infty} B_{mn} \sin \frac{m\pi x}{a} \sin \frac{n\pi y}{b} \tag{A-12}$$

where

$$B_{mn} = \frac{4}{ab} \int_0^b \int_0^a f(x, y) \sin \frac{m\pi x}{a} \sin \frac{n\pi y}{b} \, dx \, dy. \tag{A-13}$$

Equation (A-12) can be solved by many available commercial programs. One simplified such program, called "DBLSUM," is listed in Table A-2.

Table A-2. Program "DBLSUM"

```
10 REM THE NAME OF THIS PROGRAM IS DOUBLE SUMMATION,
   "DBLSUM"
20 REM THIS PROGRAM CALCULATES THE SUM OF A DOUBLE
   SERIES
30 REM INPUT EQUATION TO BE EVALUATED IN LINE 370
40 REM STATEMENTS 370 TO 379 ARE RESRV'D FOR EQUAT'N TO BE
   SOLV'D
45 REM LINES 350 AND 360 GIVE THE MAX. SUMMATION VALUE
50 PRINT "IS THE M SUMMATION ODD (O), EVEN (E), OR
   CONTINUOUS (C)"
60 INTPUT A$
70 IF A$ = "O" THEN 110
80 IF A$ = "E" THEN 140
90 IF A$ = "C" THEN 170
100 GOTO 50
110 M1 = 1
120 M2 = 2
130 GOTO 200
140 M1 = 2
150 M2 = 2
160 GOTO 200
170 M1 = 1
180 M2 = 1
190 GOTO 200
200 PRINT "IS THE N SUMMATION ODD (O), EVEN (E), OR
   CONTINUOUS (C)"
210 INPUT A$
220 IF A$ = "O" THEN 260
230 IF A$ = "E" THEN 290
240 IF A$ = "C" THEN 320
250 GOTO 200
260 N1 = 1
```

Table A-2. Continued

```
270 N2 = 2
280 GOTO 350
290 N1 = 2
300 N2 = 2
310 GOTO 350
320 N1 = 1
330 N2 = 1
340 GOTO 350
350 FOR M = M1 TO 30 STEP M2
360 FOR N = N1 TO 30 STEP N2
370 U (-1)^((M + N)/2 - 1)
371 V = M*N*(M^2 + N^2)^2
372 Y = U/V
380 LET T = T + Y
390 NEXT N
400 NEXT M
410 PRINT "SUMMATION OF SERIES IS = "; T
420 END
```

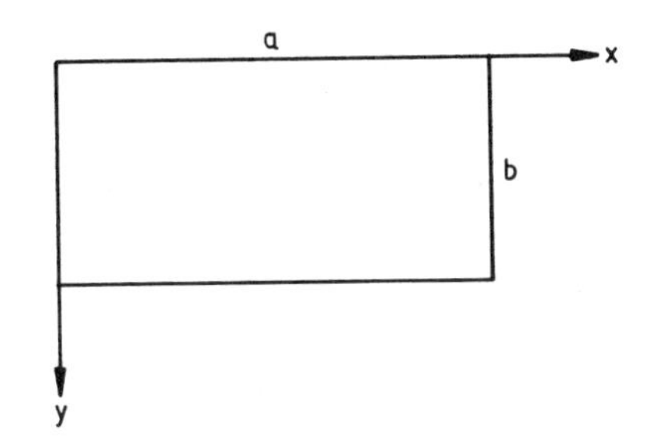

Figure A-6.

Example A-4

The rectangular plate shown in Fig. A-6 is subjected to a uniform pressure p_o. Determine the Fourier expansion for the pressure.

Solution

From Eq. (A-13),

$$B_{mn} = \frac{4p_o}{ab} \int_0^b \int_0^a \sin \frac{m\pi x}{a} \sin \frac{n\pi y}{b} \, dx \, dy$$

$$= \frac{16p_o}{\pi^2 mn} \qquad m, n \text{ are odd functions}$$

$$f(x, y) = \frac{16p_o}{\pi^2} \sum_{m=1,3,\ldots}^{\infty} \sum_{n=1,3,\ldots}^{\infty} \frac{1}{mn} \sin \frac{m\pi x}{a} \sin \frac{n\pi y}{b}$$

APPENDIX B

Bessel Functions

B-1 General Equations

In many plate and shell applications involving circular symmetry, the resulting differential equations are solved by means of a power series known as Bessel functions. Some of these functions are discussed in this appendix.

The differential equation

$$\frac{d^2y}{dx^2} + \frac{1}{x}\frac{dy}{dx} + y = 0 \tag{B-1}$$

is referred to as Bessel's equation of zero order. Its solution (Bowman 1958) is given by the following power series

$$y = C_1 J_o(x) + C_2 Y_o(x) \tag{B-2}$$

where

C_1 and C_2 = constants obtained from boundary conditions; $J_o(x)$ = Bessel function of the first kind of zero order.

$$J_o(x) = 1 - \frac{x^2}{2^2} + \frac{x^4}{2^2 \,.\, 4^2} - \frac{x^6}{2^2 \,.\, 4^2 \,.\, 6^2} + \cdots$$

$$= \sum_{m=0}^{\infty} \frac{(-1)^m}{(m!)^2}\left(\frac{x}{2}\right)^{2m}$$

$Y_o(x)$ = Bessel function of the second kind of zero order.

$$Y_o(x) = J_o(x) \int \frac{dx}{xJ_o^2(x)}$$

$$= J_o(x) \ln x + \frac{x^2}{2^2} - \frac{x^4}{2^2 \,.\, 4^2}(1 + 1/2)$$

$$+ \frac{x^6}{2^2 \,.\, 4^2 \,.\, 6^2}(1 + 1/2 + 1/3) - \cdots$$

Equation (B-1) is usually encountered in a more general form as

$$x^2 \frac{d^2y}{dx^2} + x \frac{dy}{dx} + (x^2 - k^2)y = 0. \qquad \text{(B-3)}$$

The solution of this equation (Hildebrand 1964) is

$$y = C_1 J_k(x) + C_2 J_{-k}(x)$$

when k is not zero or a positive integer, or

$$y = C_1 J_k(x) + C_2 Y_k(x)$$

when k is zero or a positive integer. And where,

$$\begin{aligned}
J_k(x) &= \text{Bessel function of the first kind of order } k \\
&= \sum_{m=0}^{\infty} \frac{(-1)^m}{(m!)(m+k)!} (x/2)^{2m+k} \\
J_{-k}(x) &= \text{Bessel function of the first kind of order } k \\
&= \sum_{m=0}^{\infty} \frac{(-1)^m}{(m!)(m-k)!} (x/2)^{2m-k} \\
Y_k(x) &= \text{Bessel function of the second kind of order } k \\
&= \frac{2}{\pi}\left[(\ln(x/2) + \gamma)J_k(x) - \frac{1}{2}\sum_{m=0}^{k-1} \frac{(k-m-1)!}{m!} (x/2)^{2m-k}\right. \\
&\qquad \left. + \frac{1}{2}\sum_{m=0}^{\infty} (-1)^{m+1}[h(m) + h(m+k)] \frac{(x/2)^{2m+k}}{m!(m+k)!}\right] \\
h(m) &= \sum_{r=1}^{m} 1/r \qquad m > 1. \\
\gamma &= 0.5772
\end{aligned}$$

A plot of $J_o(x)$, $J_1(x)$, $Y_o(x)$, and $Y_1(x)$ is shown in Fig. B-1. Also, Table B-1 gives some values of $J(x)$ and $Y(x)$.

A different form of Eq. (B-3) that is encountered often in plate and shell theory is

$$x^2 \frac{d^2y}{dx^2} + x \frac{dy}{dx} - (x^2 + k^2)y = 0. \qquad \text{(B-4)}$$

The solution of this equation (Dwight 1972) is

$$y = C_1 I_k(x) + C_2 I_{-k}(x)$$

when k is not zero or a positive integer, or

$$y = C_1 I_k(x) + C_2 K_k(x)$$

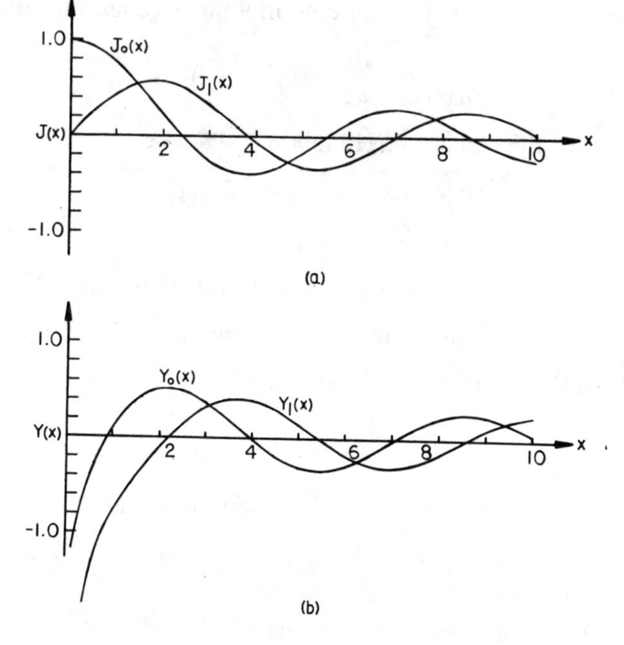

Figure B-1.

Table B-1. Values of J_0, J_1, Y_0, and Y_1

x	$J_0(x)$	$J_1(x)$	$Y_0(x)$	$Y_1(x)$
0.0	1.0000	0.0000	$-\infty$	$-\infty$
0.5	0.9385	0.2423	−0.4445	−1.4715
1.0	0.7652	0.4401	0.0883	−0.7812
1.5	0.5118	0.5579	0.3825	−0.4123
2.0	0.2239	0.5767	0.5104	−0.1070
2.5	−0.0484	0.4971	0.4981	0.1459
3.0	−0.2601	0.3391	0.3769	0.3247
3.5	−0.3801	0.1374	0.1890	0.4102
4.0	−0.3972	−0.0660	−0.0169	0.3979
4.5	−0.3205	−0.2311	−0.1947	0.3010
5.0	−0.1776	−0.3276	−0.3085	0.1479
5.5	−0.0068	−0.3414	−0.3395	−0.0238
6.0	0.1507	−0.2767	−0.2882	−0.1750
6.5	0.2601	−0.1538	−0.1732	−0.2741
7.0	0.3001	−0.0047	−0.0260	−0.3027
7.5	0.2663	0.1353	0.1173	−0.2591
8.0	0.1717	0.2346	0.2235	−0.1581
8.5	0.0419	0.2731	0.2702	−0.0262
9.0	−0.0903	0.2453	0.2499	0.1043
9.5	−0.1939	0.1613	0.1712	0.2032
10.0	−0.2459	0.0435	0.0557	0.2490

when k is zero or a positive integer and where

$$
\begin{aligned}
I_k(x) &= \text{modified Bessel function of the first kind of order } k \\
&= \sum_{m=1}^{\infty} \frac{(x/2)^{2m+k}}{m!(m+k)!}. \\
K_k(x) &= \text{modified Bessel function of the second kind of order } k \\
&= (-1)^{k+1}[\ln(x/2) + \gamma]I_k(x) \\
&\quad + \frac{1}{2}\sum_{m=0}^{k-1} \frac{(-1)^m(k-m-1)!}{m!}(x/2)^{2m-k} \\
&\quad + \frac{1}{2}\sum_{m=0}^{\infty} \frac{(-1)^k(x/2)^{2m+k}}{m!(m+k)!}[(1 + 1/2 + \cdots + 1/m) \\
&\quad + (1 + 1/2 + \cdots + 1/(m+k))].
\end{aligned}
$$

A plot of $I_o(x)$, $I_1(x)$, $K_o(x)$, and $K_1(x)$ is shown in Fig. B-2.

Another equation that is often encountered in plate and shell theory is given by

$$x^2 \frac{d^2y}{dx^2} + x\frac{dy}{dx} - (ix^2 + k^2)y = 0. \tag{B-5}$$

The solution of this equation (Hetenyi 1964) for the important case of $k = 0$ is given by

$$y = C_1Z_1(x) + C_2Z_2(x) + C_3Z_3(x) + C_4Z_4(x) \tag{B-6}$$

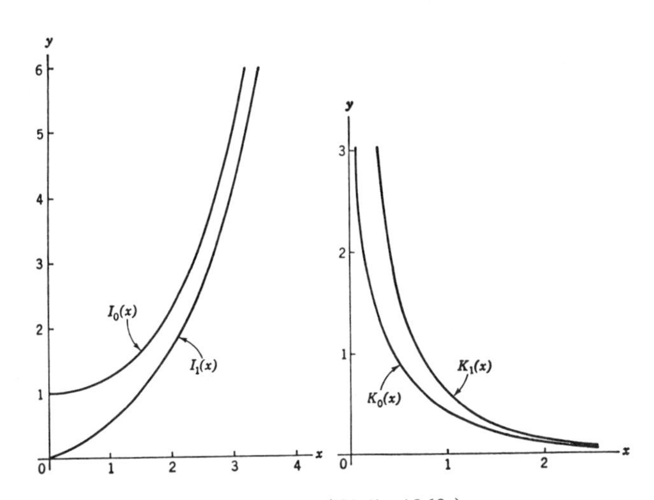

Figure B-2. (Wylie 1960.)

where

$$Z_1(x) = \operatorname{ber}(x) = \sum_{m=0}^{\infty} (-1)^m \frac{(x/2)^{4m}}{[(2m)!]^2}$$

$$Z_2(x) = -\operatorname{bei}(x) = -\sum_{m=0}^{\infty} (-1)^m \frac{(x/2)^{4m+2}}{[(2m+1)!]^2}$$

$$Z_3(x) = -\frac{2}{\pi} \operatorname{kei}(x) = \frac{Z_1(x)}{2} - \frac{2}{\pi} [R_1 + (\ln(\gamma x/2))(Z_2(x))]$$

$$Z_4(x) = -\frac{2}{\pi} \ker(x) = \frac{Z_2(x)}{2} + \frac{2}{\pi} [R_2 + (\ln(\gamma x/2))(Z_1(x))]$$

$$R_1 = (x/2)^2 - \frac{h(3)}{(3!)^2}(x/2)^6 + \frac{h(5)}{(5!)^2}(x/2)^{10} - \cdots$$

$$R_2 = \frac{h(2)}{(2!)^2}(x/2)^4 - \frac{h(4)}{(4!)^2}(x/2)^8 + \frac{h(6)}{(6!)^2}(x/2)^{12} - \cdots$$

$$h(n) = 1 + 1/2 + 1/3 + \cdots + 1/n$$

$$\gamma = 0.5772.$$

A plot of $Z_1(x)$, $Z_2(x)$, $Z_3(x)$, $Z_4(x)$, and their derivatives is shown in Fig. B-3.

B-2 Some Bessel Identities

The derivatives and integrals of Bessel functions follow a certain pattern. The identities given here are needed to solve some of the problems given in this text.

$$\frac{d}{dx}[xJ_1(x)] = xJ_o(x)$$

$$\frac{d}{dx}J_o(x) = -J_1(x)$$

$$\frac{d}{dx}[x^n J_n(x)] = x^n J_{n-1}(x)$$

$$\frac{d}{dx}\left[\frac{J_n(x)}{x^n}\right] = \frac{-J_{n+1}(x)}{x^n}$$

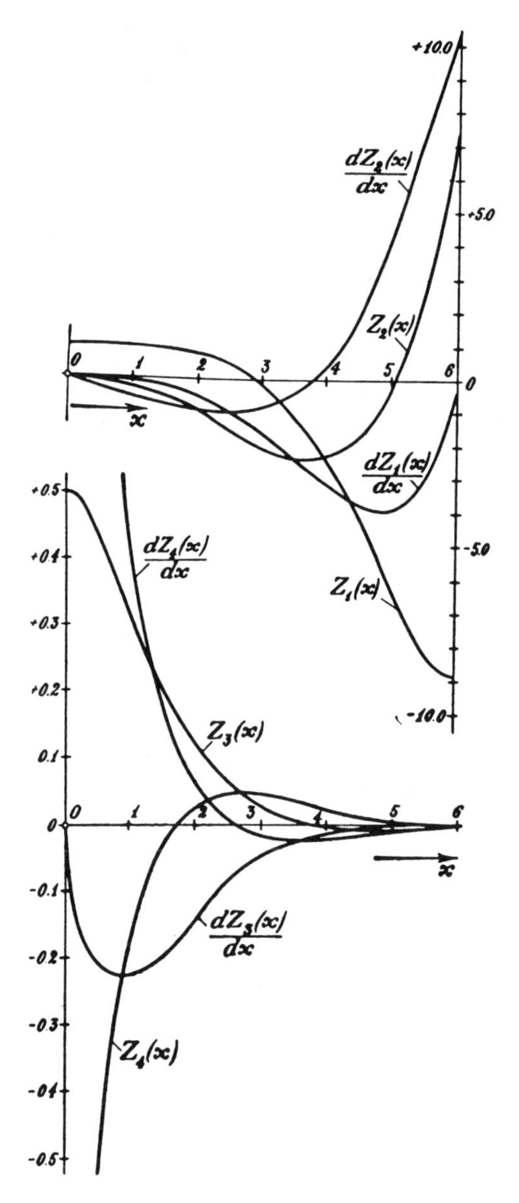

Figure B-3. (Hetenyi 1964.)

$$\frac{d^2Z_1(x)}{dx^2} = Z_2(x) - \frac{1}{x}\frac{dZ_1(x)}{dx}$$

$$\frac{d^2Z_2(x)}{dx^2} = -Z_1(x) - \frac{1}{x}\frac{dZ_2(x)}{dx}$$

$$\frac{d^2Z_3(x)}{dx^2} = Z_4(x) - \frac{1}{x}\frac{dZ_3(x)}{dx}$$

$$\frac{d^2Z_4(x)}{dx^2} = -Z_3(x) - \frac{1}{x}\frac{dZ_4(x)}{dx}$$

The last four equations are needed in the solution of circular plates on elastic foundation. In these equations the value of (kx) is needed rather than (x) in the Z functions. In this case, these equations take on the form

$$k^2Z_1''(kx) = k^2Z_2(x) - \frac{k}{x}Z_1'(kx)$$

$$k^2Z_2''(kx) = -k^2Z_1(x) - \frac{k}{x}Z_2'(kx)$$

$$k^2Z_3''(kx) = k^2Z_4(x) - \frac{k}{x}Z_3'(kx)$$

$$k^2Z_4''(kx) = -k^2Z_3(x) - \frac{k}{x}Z_4'(kx)$$

B-3 Simplified Bessel Functions

As x approaches zero, the various Bessel functions can be expressed as

$$J_k(x) = \frac{x^k}{(2^k)(k!)}$$

$$Y_k(x) = \frac{-2^k(k-1)!}{\pi}x^{-k} \qquad k \neq 0$$

$$Y_o(x) = \frac{2}{\pi}\ln x$$

$$I_k(x) = \frac{x^k}{2^k k!}$$

$$K_k(x) = (2^{k-1})[(k-1)!]x^{-k} \qquad k \neq 0$$

$$K_o(x) = -\ln x$$

$$Z_1(x) = 1.0 \qquad Z_2(x) = -x^2/4$$

$$Z_3(x) = 1/2 \qquad Z_4(x) = \frac{2}{\pi} \ln \frac{\gamma x}{2}$$

$$\frac{dZ_1(x)}{dx} = -x^3/16 \qquad \frac{dZ_3(x)}{dx} = \frac{x}{\pi} \ln \frac{\gamma x}{2}$$

$$\frac{dZ_2(x)}{dx} = -x/2 \qquad \frac{dZ_4(x)}{dx} = \frac{2}{\pi x}$$

As x approaches infinity, the various Bessel functions can be expressed as

$$J_k(x) = \sqrt{\frac{2}{\pi x}} \cos(x - \xi_k) \qquad \xi_k = (2k + 1)\frac{\pi}{4}$$

$$Y_k(x) = \sqrt{\frac{2}{\pi x}} \sin(x - \xi_k)$$

$$I_k(x) = \frac{e^x}{\sqrt{2\pi x}}$$

$$K_k(x) = \frac{e^{-x}}{\sqrt{2x/\pi}}$$

$$Z_1(x) = \eta \cos \sigma \qquad Z_2(x) = -\eta \sin \sigma$$

$$Z_3(x) = \beta \sin \tau \qquad Z_4(x) = -\beta \cos \tau$$

$$\eta = \frac{1}{\sqrt{2\pi x}} e^{x/\sqrt{2}} \qquad \beta = \sqrt{\frac{2}{\pi x}} e^{-x/\sqrt{2}}$$

$$\sigma = \frac{x}{\sqrt{2}} - \pi/8 \qquad \tau = \frac{x}{\sqrt{2}} + \pi/8$$

$$\frac{dZ_1(x)}{dx} = \frac{\eta}{\sqrt{2}} (\cos \sigma - \sin \sigma)$$

$$\frac{dZ_2(x)}{dx} = \frac{-\eta}{\sqrt{2}} (\cos \sigma + \sin \sigma)$$

$$\frac{dZ_3(x)}{dx} = \frac{\beta}{\sqrt{2}} (\cos \tau - \sin \tau)$$

$$\frac{dZ_4(x)}{dx} = \frac{-\beta}{\sqrt{2}} (\cos \tau + \sin \tau)$$

APPENDIX C

Conversion Factors

Pressure Units

	1 psi	1 N/mm^2	1 bar	1 KPa	1 kgf/cm^2
psi	1.0000	145.0	14.50	0.1450	14.22
N/mm^2	0.006895	1.000	0.1000	0.0010	0.09807
bars	0.06895	10.000	1.000	0.0100	0.9807
KPa	6.895	1000.0	100.00	1.000	98.07
kgf/cm^2	0.0703	10.20	1.020	0.0102	1.000

1 N/mm^2 = 1 MPa.

Modulus of Elasticity Units

	1 ksi	1 KN/mm^2	1 MPa	1 kgf/mm^2
ksi	1.000	145.0	0.1450	1.422
KN/mm^2	0.006895	1.000	0.001	0.009807
MPa	6.895	1000.00	1.000	9.807
kgf/mm^2	0.7033	102.0	0.1020	1.000

Force Units

	1 lb	1 kgf	1 N
lb	1.000	2.205	0.2248
kgf	0.454	1.000	0.1020
N	4.448	9.807	1.0000

Answers to Selected Problems

1-2 $\sigma_1 = 11{,}945$ psi, $\sigma_2 = -1360$ psi, $\tau_{12} = 6650$ psi
1-4 $\sigma_x = 92.2$ MPa, $\sigma_y = 124.9$ MPa
1-15 $t = 2.74$ inch
2-5

$$M_x = p \sum_{m=1,3}^{\infty} \left[\frac{m^2\pi^2}{a^2} H_m(1 - \mu) \cosh \frac{m\pi y}{a} + \frac{m^2\pi^2}{a^2} G_m H_m y(1 - \mu) \sinh \frac{m\pi y}{a} - \frac{2\mu m\pi}{a} G_m H_m \cosh \frac{m\pi y}{a} + \frac{4a^2}{m^3\pi^3} \right] \sin \frac{m\pi x}{a}$$

$$M_y = p \sum_{m=1,3}^{\infty} \left[\frac{m^2\pi^2}{a^2} H_m(\mu - 1) \cosh \frac{m\pi y}{a} + \frac{m^2\pi^2}{a} G_m H_m y(\mu - 1) \sinh \frac{m\pi y}{a} - \frac{2m\pi}{a} G_m H_m \cosh \frac{m\pi y}{a} + \frac{4\mu a^2}{m^3\pi^3} \right] sin \frac{m\pi x}{a}$$

$$H_m = \frac{-4a^4}{m^5\pi^5\left(\cosh \frac{m\pi b}{2a} + G_m \frac{b}{2} \sinh \frac{m\pi b}{2a}\right)}$$

$$G_m = -\frac{\frac{m\pi}{a} \sinh \frac{m\pi b}{2a}}{\sinh \frac{m\pi b}{2a} + \frac{m\pi}{a} \cosh \frac{m\pi b}{2a}}$$

2-6

$$w = \sum_{m=1}^{\infty} \left(B_m \cosh \frac{m\pi y}{a} + C_m y \sinh \frac{m\pi y}{a} \right) \sin \frac{m\pi x}{a}$$

where

$$B_m = E_m/C_1, \qquad C_m = -\frac{2}{b} B_m \coth \frac{m\pi b}{2a}$$

$$C_1 = \left(\frac{m^2\pi^2}{a^2} - \frac{2m\pi}{a} C_2 \right) \cosh \frac{m\pi b}{2a} - \frac{m^2\pi^2}{a^2} C_2 \frac{b}{2} \sinh \frac{m\pi b}{2a}$$

$$C_2 = \frac{2}{b} \coth \frac{m\pi b}{2a}$$

2-12 $M = 7{,}620$ inch-lb/inch
3-3 $= 8670$ psi, Max $w = 0.23$ inch.
3-4 $= 12{,}330$ psi
3-6

$$M_r = \frac{p}{16}(3 + \mu)(a^2 - r^2) - \frac{pb^2}{4}(K_1 + K_2 - K_3 - K_4 - 1)$$

where

$$K_1 = (1 + \mu)\left(\frac{3 + \mu}{2(1 + \mu)} - \frac{b^2}{a^2 - b^2} \ln b/a - 1/2 \right)$$

$$K_2 = \frac{1 - \mu}{r} \left(\frac{1 + \mu}{1 - \mu} \frac{a^2 b^2}{a^2 - b^2} \ln b/a \right)$$

$$K_3 = (1 + \mu) \ln r/a$$

$$K_4 = \frac{(3 + \mu)}{4} \left(\frac{a^2 - r^2}{r^2} \right)$$

$$M_t = \frac{p}{16}[a^2(3 + \mu) - r^2(1 + 3\mu)] - \frac{pb^2}{4}(K_1 + K_2 - K_3 + K_5 - \mu)$$

$$K_5 = \frac{(3 + \mu)}{4} \left(\frac{a^2 + r^2}{r^2} \right)$$

4-4 $M_p = pL^2/8$
4-6 $M_p = pL^2/144$
4-11 $M_p = 157.3\ p$
6-4 $t = 0.90$ inch

6-5 $N_\phi = -p_o R/2$

$N_\theta = -p_o R(\cos^2 \phi - 1/2)$

6-6

$$N_\phi = \frac{-\gamma R^2}{6}\left(\frac{3H}{R} + 1 - \frac{2\cos^2\phi}{1 + \cos\phi}\right)$$

$$N_\theta = \frac{-\gamma R^2}{6}\left(\frac{3H}{R} - 1 - \frac{6 - 4\cos^2\phi}{1 + \cos\phi}\right)$$

6-9

$$\max N_s = \frac{-847\gamma L^2 \sin\alpha}{432} \qquad \text{at } s = L/12$$

$$\max N_\theta = \frac{\gamma L^2}{4}\sin\alpha \qquad \text{at } s = L/2$$

7-4 $A = 5.76\ \text{in}^2$
7-5

$t_1 = 0.21$ inch, $t_2 = 0.82$ inch

$t_3 = 0.98$ inch, $t_4 = 0.69$ inch

$A = 1.22\ \text{in}^2$

8-1 Max $M_x = 0.322\ Q_o/\beta$ at $x = 0.61\sqrt{rt}$
8-2 At Section *a-a* $M_o = 0$ and $H_o = 0.0195D\beta^3$
8-3 $M_a = 14.95/\beta$ and $M_b = 44.97/\beta$
8-6

$p = 3.49$ psi

$M_o = 0.1074$P inch-lb/inch, $Q_o = 1.855$P lb/inch.

8-7 Discontinuity moments and forces at junction are

moment in top cylinder = 595.4 inch-lb/inch (comp. on outside)
moment in bottom cylinder = 163.1 inch-lb/inch (comp. on outside)
moment in plate = 758.5 inch-lb/inch (tension on top)
horizontal force in top cylinder = 495.9 lbs compressive
horizontal force in bottom cylinder = 109.2 lbs tensile
horizontal force in plate = 386.7 lbs tensile

9-2 $M_o = 1886$ inch-lb/inch, $H_o = 1200$ lb/inch
11-3 $t = 5/16$ inch.
12-2 $t = 0.73$ inch.
12-3 $p = 148.9$ psi.
A-2

$$f(x) = \frac{1}{\pi} - \frac{2}{\pi} \sum_{m=2,4}^{\infty} \frac{\cos mx}{(m-1)(m+1)} + \frac{1}{2} \sin x$$

References

American Association of State Highway and Transportation Officials. 1992. Standard Specifications for Highway Bridges. Washington, DC: AASHTO.

American Concrete Institute. 1989. Building Code Requirements for Reinforced Concrete. ACI-318. Chicago: ACI.

American Institute of Steel Construction. 1991. Manual of Steel Construction-Allowable Stress Design. Chicago: AISC.

American Iron and Steel Institute. 1981. Steel Penstocks and Tunnel Liners. Washington, DC: AISI.

American Petroleum Institute. 1991. Recommended Rules for Design and Construction of Large, Welded, Low-Pressure Storage Tanks—API 620. Washington, DC: API.

American Society of Civil Engineers. 1960. Design of Cylindrical Concrete Shell Roofs. Manual No. 31. New York: ASCE.

American Society of Mechanical Engineers. 1992a. Pressure Vessel Code, Section VIII, Division 1. New York: ASME.

American Society of Mechanical Engineers. 1992b. Pressure Vessel Code-Alternate Rules, Section VIII, Division 2. New York: ASME.

American Society of Mechanical Engineers. 1992c. Fiber-Reinforced Plastic Pressure Vessels, Section X. New York: ASME.

Baker, E. H., Cappelli, A. P., Kovalevsky, L., Rish, F. L., Verette, R. M. 1968. Shell Analysis Manual—NASA 912. Washington, DC: National Aeronautics and Space Administration.

Becker, H. July 1957. Handbook of Structural Stability—Part II—Buckling of Composite Elements NACA PB 128 305. Washington, DC: National Advisory Committee for Aeronautics.

Beer, F. P., and Johnson Jr., E. R. 1981. Mechanics of Materials. New York: McGraw-Hill.

Billington, D. P. 1982. Thin Shell Concrete Structures. New York: McGraw-Hill.

Bowman, F. 1958. Introduction to Bessel Functions. New York: Dover Publications.

Buchert, K. P. 1964. Stiffened Thin Shell Domes. AISC Engineering Journal. Chicago: AISC.

Buchert, K. P. 1966. Buckling Considerations in the Design and Construction of Doubly Curved Space Structures. International Conference on Space Structures, 1966—F8. England: University of Surrey.

Chattarjee, B. K. 1971. Theory and Design of Concrete Shells. New York: Gordon and Breach.

Dwight, H. B. 1972. Tables of Integrals and Other Mathematical Data. New York: Macmillan.

Flugge, W. 1967. Stresses in Shells. New York: Springer-Verlag.

Gallagher, R. H. 1975. Finite Element Analysis. Englewood Cliffs, NJ: Prentice-Hall.

Gerard, G., August 1957. Handbook of Structural Stability—Part IV—Failure of Plates and Composite Elements. NACA N62-55784. Washington, DC: National Advisory Committee for Aeronautics.

Gerard, G. August 1957. Handbook of Structural Stability—Part V—Compressive Strength of Flat Stiffened Panels NACA PB 185 629. Washington, DC: National Advisory Committee for Aeronautics.

Gerard, G. 1962. Introduction to Structural Stability Theory. New York: McGraw-Hill.

Gerard, G. and Becker, H. July 1957. Handbook of Structural Stability—Part I—Buckling of Flat Plates. NACA PB 185 628. Washington, DC: National Advisory Committee for Aeronautics.

Gerard, G. and Becker, H. 1957. Handbook of Structural Stability—Part III—Buckling of Curved Plates and Shells. NACA TN 3783. Washington, DC: National Advisory Committee for Aeronautics.

Gibson, J. E. 1965. Linear Elastic Theory of Thin Shells. New York: Pergamon Press.

Gibson, J. E. 1968. The Design of Shell Roofs. London: E & F. N. Spon Ltd.

Gould, P. L. 1988. Analysis of Shells and Plates. New York: Springer-Verlag.

Grandin, H. Jr. 1986. Fundamentals of The Finite Element Method. New York: Macmillan.

Hetenyi, M. 1964. Beams on Elastic Foundation. Ann Arbor, Michigan: University of Michigan Press.

Hildebrand, F. 1964. Advanced Calculus for Applications. Englewood Cliffs, NJ: Prentice-Hall.

Iyengar, N. G. R. 1988. Structural Stability of Columns and Plates. New York: John Wiley & Sons.

Jawad, M. H. 1980. Design of Conical Shells Under External Pressure. Journal of Pressure Vessel Technology, Volume 102. New York: ASME.

Jawad, M. H. and Farr, J. R. 1989. Structural Analysis and Design of Process Equipment. New York: John Wiley & Sons.

Jones, L. L. 1966. Ultimate Load Analysis of Reinforced and Prestressed Concrete Structures. New York: F. Unger Publishing.

Jones, R. M. 1975. Mechanics of Composite Materials. New York: Harper & Row.

Kelkar, V. S. and Sewell, R. T. 1987. Fundamentals of the Analysis and Design of Shell Structures. Englewood Cliffs, NJ: Prentice-Hall.

Kellogg, The M. W. Company. 1961. Design of Piping Systems. New York: John Wiley & Sons.

Kollar, L. and Dulacska, E. 1984. Buckling of Shells for Engineers. New York: John Wiley & Sons.

Love, A. E. H. 1944. A Treatise on the Mathematical Theory of Elasticity. New York: Dover Publications.

McFarland, D. E., Smith, B. L., and Bernhart, W. D. 1972. Analysis of Plates. New York: Macmillan.

Moody, W. T. 1970. Moments and Reactions for Rectangular Plates—Engineering Monograph No. 27. Washington, DC: U.S. Department of the Interior.

Moy, S. S. J. 1981. Plastic Methods for Steel and Concrete Structures. New York: John Wiley & Sons.

Niordson, F. I. N. 1947. Buckling of Conical Shells Subjected to Uniform External Lateral Pressure. Transactions of the Royal Institute of Technology, No. 10. Stockholm: Royal Institute of Technology.

O'Donnell, W. J. and Langer, B. F. 1962. Design of Perforated Plates. Journal of Engineering for Industry. New York: ASME.

Perry, C. L. 1950. The Bending of Thin Elliptic Plates. Proceedings of Symposia in Applied Mathematics. Volume III, p. 131. New York: McGraw-Hill.

Pilkey, W. and Pin Yu Chang. 1978. Modern Formulas for Statics and Dynamics. New York: McGraw-Hill.

Potter, M. C. 1978. Mathematical Methods in the Physical Sciences. Englewood Cliffs, NJ: Prentice-Hall.

Raetz, R. V. 1957. An Experimental Investigation of the Strength of Small-Scale Conical Reducer Sections Between Cylindrical Shells Under External Hydrostatic Pressure. U.S. Department of the Navy, David Taylor Model Basin. Report 1187. Washington, DC: U.S. Navy.

Roark, R. J. and Young, W. C. 1975. Formulas for Stress and Strain. New York: McGraw-Hill.

Rockey, K. C., Evans, H. R., Griffiths, D. W., Nethercot, D. A. 1975. The Finite Element Method. New York: John Wiley & Sons.

Segerlind, L. J. 1976. Applied Finite Element Analysis. New York: John Wiley & Sons.

Seide, P. 1962. A Survey of Buckling Theory and Experiment for Circular Conical Shells of Constant Thickness. NASA Publication IND-1510. Ohio: NASA.

Seide, P. 1981. Stability of Cylindrical Reinforced Concrete Shells. In Concrete Shell Buckling. ACI SP-67. Chicago: American Concrete Institute.

Shenk, A. 1979. Calculus and Analytic Geometry. California: Goodyear Publishing.

Sokolnikoff, I. S. 1956. Mathematical Theory of Elasticity. New York: McGraw-Hill.

Sturm, R. G. 1941. A Study of the Collapsing Pressure of Thin-Walled Cylinders. Engineering Experiment Station Bulletin 329. Urbana, IL: The University of Illinois.

Swanson, H. S., Chapton, H. J., Wilkinson, W. J., King, C. L., and Nelson, E. D. June 1955. Design of Wye Branches for Steel Pipe. New York: Journal of AWWA.

Szilard, R. 1974. Theory and Analysis of Plates-Classical and Numerical Methods. Englewood Cliffs, NJ: Prentice-Hall.

Timoshenko, S. P. 1983. History of Strength of Materials. New York: Dover Publications.

Timoshenko, S. P. and Gere, J. M. 1961. Theory of Elastic Stability. New York: McGraw-Hill.

Timoshenko, S. P. and Goodier, J. N. 1951. Theory of Elasticity. New York: McGraw-Hill.

Timoshenko, S. P. and Woinowsky-Krieger, S. 1959. Theory of Plates and Shells. New York: McGraw-Hill.

Troitsky, M. S. 1987. Orthotropic Bridges-Theory and Design. Cleveland: The James Lincoln Arc Welding Foundation.

Tubular Exchanger Manufacturers Association. 1988. Standards of the Tubular Exchanger Manufacturers Association. Tarrytown, NY: TEMA.

Ugural, A. C. 1981. Stresses in Plates and Shells. New York: McGraw-Hill.

Von Karman, Th. and Tsien, Hsue-Shen. 1939. The Buckling of Spherical Shells by External Pressure. Journal of Aeronautical Sciences.

Wang, C. K. 1986. Structural Analysis on Microcomputers. New York: Macmillan.

Weaver, W. Jr. and Johnston, P. R. 1984. Finite Elements for Structural Analysis. Englewood Cliffs, NJ: Prentice-Hall.

Winter, G., Urquhart, L. C., O'Rourke, C. E., Nilson, A. H. 1964. Design of Concrete Structures. New York: McGraw-Hill.

Wood, R. H. 1961. Plastic and Elastic Design of Slabs and Plates. New York: The Ronald Press Company.

Wylie, C. R. Jr. 1960. Advanced Engineering Mathematics. New York: McGraw-Hill.

Zick, L. P. and St. Germain, A. R. May 1963. Circumferential Stresses in Pressure Vessel Shells of Revolution. Journal of Engineering for Industry. New York: ASME.

Zienkiewicz, O. C. 1977. The Finite Element Method. New York: McGraw-Hill.

Index